Computer Methods and Experimental Measurements for Surface Effects and Contact Mechanics VIII

WIT*PRESS*

WIT Press publishes leading books in Science and Technology.
Visit our website for new and current list of titles.
www.witpress.com

WIT*eLibrary*

Home of the Transactions of the Wessex Institute.
Papers presented at Computer Methods and Experimental Measurements for Surface Effects and Contact Mechanics VII are archived in the WIT eLibrary in volume 55 of WIT Transactions on Engineering Sciences (ISSN 1743-3533).

The WIT eLibrary provides the international scientific community with immediate and permanent access to individual papers presented at WIT conferences.
http://library.witpress.com

EIGHTH INTERNATIONAL CONFERENCE ON
COMPUTER METHODS AND EXPERIMENTAL MEASUREMENTS FOR
SURFACE EFFECTS AND CONTACT MECHANICS

CONTACT AND SURFACE 2007

CONFERENCE CHAIRMEN

J.T.M. De Hosson
University of Groningen, The Netherlands

C.A. Brebbia
Wessex Institute of Technology, UK

S-I Nishida
Saga University, Japan

INTERNATIONAL SCIENTIFIC ADVISORY COMMITTEE

Organised by
Wessex Institute of Technology, UK

Sponsored by
WIT Transactions on Engineering Sciences

WIT Transactions on Engineering Sciences

Transactions Editor

Carlos Brebbia
Wessex Institute of Technology
Ashurst Lodge, Ashurst
Southampton SO40 7AA, UK
Email: carlos@wessex.ac.uk

Editorial Board

Computer Methods and Experimental Measurements for Surface Effects and Contact Mechanics VIII

Editors:

J.T.M. De Hosson
University of Groningen, The Netherlands

C.A. Brebbia
Wessex Institute of Technology, UK

S-I Nishida
Saga University, Japan

WIT*PRESS* Southampton, Boston

Editors:

J.T.M. De Hosson
University of Groningen, The Netherlands

C.A. Brebbia
Wessex Institute of Technology, UK

S-I Nishida
Saga University, Japan

Published by

WIT Press
Ashurst Lodge, Ashurst, Southampton, SO40 7AA, UK
Tel: 44 (0) 238 029 3223; Fax: 44 (0) 238 029 2853
E-Mail: witpress@witpress.com
http://www.witpress.com

For USA, Canada and Mexico

Computational Mechanics Inc
25 Bridge Street, Billerica, MA 01821, USA
Tel: 978 667 5841; Fax: 978 667 7582
E-Mail: infousa@witpress.com
http://www.witpress.com

British Library Cataloguing-in-Publication Data

A Catalogue record for this book is available
from the British Library

ISBN: 978-1-84564-073-6
ISSN: 1746-4471 (print)
ISSN: 1743-3533 (on-line)

The texts of the papers in this volume were set individually by the authors or under their supervision. Only minor corrections to the text may have been carried out by the publisher.

Printed in Great Britain by Cambridge Printing

Preface

This book contains most of the papers presented at the Eighth International Conference on Surface Treatment and Contact Mechanics held in 2007 at the Wessex Institute of Technology in Ashurst, UK. Previous conferences in the same series were organized in Southampton(1993), Milano(1995), Oxford(1997), Assisi (1999), Sevilla (2001), Crete (2003) and Bologna (2005). This series of conferences is aimed at encouraging international collaboration among the participants and the exchange of new ideas. In particular the book deals with the interplay between applied physics, materials science, computational mechanics and mechanical engineering.

The combination surface treatment and contact mechanics is an important one. The surface of a component is usually the most important engineering factor. While it is in use it is often the surface of a work-piece that is subjected to wear and corrosion. The complexity of the tribological properties of materials and the economic aspects of friction and wear justify an increasing research effort. To an increasing degree, therefore, the search is on for surface modification techniques, which can increase the wear resistance of materials. It is worth noting here that wear resistance is a property, not of materials but of systems, since the material of the work-piece always wears against some other medium. It is its relation to its environment – e.g. lubrication, speed of sliding or rotation - that determines the wear resistance of the material in a given construction. In this book various new developments are highlighted, both from an experimental and computational viewpoint. Special emphasis is given to the application of advanced theoretical and experimental approaches.

Thanks are due to the authors for their contributions. The editors are also grateful to the members of the International Scientific Advisory Committee, who helped in the reviewing process to ensure the quality of the conference and this book.

The Editors,
Ashurst, UK
2007.

Contents

Section 1: Surface treatments

Section 2: Thin coatings

Section 3: Surface problems in contact mechanics

Section 4: Contact mechanics

Section 7: New applications

Section 1
Surface treatments

Fatigue strength improvement of Ti alloy with DLC coating

S.-I. Nishida[1], N. Hattori[1], Y. Nakabaru[1] & A. Tsuchiyama[2]
[1]*Faculty of Science & Engineering, Saga University, Saga City, Japan*
[2]*Fukuoka Industrial Technology Center, Yahatanishi-ku, Kitakyushu, Japan*

Abstract

As is generally known, Titanium alloy has excellent properties for its lightweight, high strength ratio, high anti-corrosive resistance etc. On the other hand, as Ti alloy is not so good at wear resistance, there occasionally appear some problems at the contact area with the other metals. Therefore, surface treatment of Ti alloy would be one good solution for the above problems. Diamond-like carbon (DLC) would be one of the most practical methods for compensating the defect of Ti alloy's properties due to its high hardness, electrical insulation, resistance to chemical attacks, superior smooth surface, excellent wear resistance etc. Rotating bending fatigue tests had been performed in order to investigate the effect of DLC coating on the fatigue properties of Ti-6Al-4V alloy. The DLC films have been deposited on Ti alloy specimens using an ionization deposition method. Four kinds of specimens coated for different coating times were prepared for the fatigue test. The result obtained in this test shows that the fatigue limit and internal hardness improved as the coating time becomes longer and the fatigue limit was improved by 100MPa as compared to the specimen without surface treatment. It is considered that this result would be due to compressive residual stress generated in the specimen's surface during the process of the DLC coating.
Keywords: Ti alloy, fatigue strength, DLC film, hardness, fatigue crack initiation and propagation.

WIT Transactions on Engineering Sciences, Vol 55, © 2007 WIT Press
www.witpress.com, ISSN 1743-3533 (on-line)
doi:10.2495/SECM070011

1 Introduction

Titanium alloys have been widely used in various kinds of industries because of their light and high specific tensile strength, excellent corrosion resistance etc. On the other hand, there exists a big limitation for their applications due to the high processing cost of Ti and its alloys [1]. One of this cost effective use will be to utilize its unique properties by combining with another techniques e.g. surface treatments. Many tests have been carried out in trying to improve the mechanical properties of Ti alloys by various kinds of surface treatments such as case hardening, plating, selective hardening, ion plating etc. [2-6]. In addition, some of the authors have also tried to improve the fatigue strength of Ti-6Al-4V alloy by ion-nitride, plasma flame, electron plating and plasma immersion DLC [7-9] and clarified that the fatigue limit of DLC specimens were improved by about 20% than that of the conventional Ti-6Al-4V. Therefore, the authors have focused to DLC coating on Ti-6Al-4V changing the coating time and analyzed the mechanism of fatigue strength improvement.

2 Experimental procedure

2.1 Specimen

The specimen used in this test is the most representative titanium alloy Ti-6Al-4V alloy, of which chemical compositions are listed in Table1. Fig.1 shows the shape and dimensions of the fatigue specimen. All of the specimens were polished with emery paper (#400-3000) and annealed at 600 °C for half an hour.

2.2 Surface treatment

The specimens were DLC coated with C_6H_6 gas changing the coating time. Table 2 lists the surface treatment conditions of DLC film. In addition, the specimen B1 was contaminated with Al substrate during Ar gas bombardment and this contaminated layer is considered to be the intermediate one whose thickness is 0.2 μm.

Table 1: Chemical composition (mass%).

Al	V	Fe	O	C	N	H
6.1	4.2	0.15	0.14	0.011	0.010	0.0043

2.3 Testing machines and surface observation

Ono-type rotating bending fatigue testing machine (14.7N-m) was used in this test under the repetition of 3000 rpm. The fracture surfaces were observed with an optical microscope and a scanning electron microscope (SEM). In addition, the vertical section of each modified surface was observed using SEM. The micro-Vickers hardness was measured under a load of 0.245N.

Table 2: Surface treatment conditions of DLC film.

	NST	B1	D1	D2	D3
Ar bombardment		20 minutes per side without bias voltage			
Source gas			Benzene		
Benzene flow rate, F / sccm			10		
Anode voltage, V_a / V			50		
Reflector voltage, V_c / V			20		
Filament current, I_f / A			30		
Working pressure, P / Pa			2.7×10^{-1}		
Coating time, t / min			18	35	53

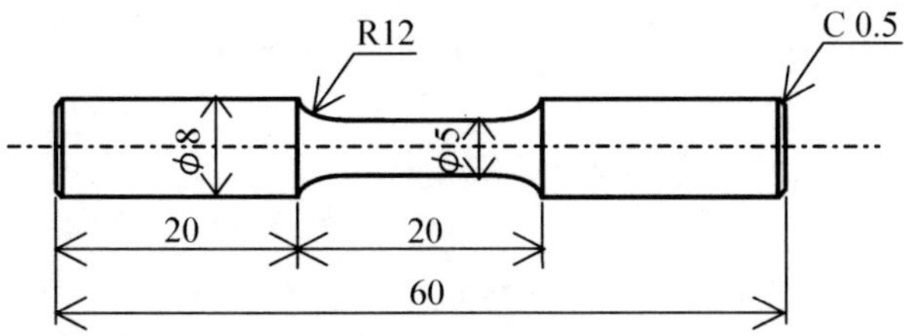

Figure 1: Shape and dimensions of fatigue specimen.

3 Results and discussions

3.1 Hardness distribution and S-N curves

Fig.2 shows the hardness testing results from a specimen's surface to its core under the load of 0.245N. The hardness number of NST (Non-surface treatment) specimen is HV320 and that of DLC specimen increased at the specimen's surface by about 1.5 times. From the hardness distribution pattern, it increased according to its coating time.

Fig.3 shows the S-N curves. As the fatigue limit of NST specimen is 390MPa, all of the specimens with DLC coating improve their fatigue limit, which also tends to increases with the coating time. The fatigue limit of D3 specimen shows the highest value being 500MPa, which is higher than that of NST one by 30%.

Fig.4 shows the relation between fatigue limit or hardness and coating time. The hardness means the value at 10μm from the specimen's surface. According to this figure, the fatigue limit and hardness tend to be increased with coating time.

It is considered that the increase of hardness could be due to compressive residual stress which appeared in the surface layer with DLC coating. This compressive residual stress retards the crack propagation up to the inside of the specimen and improves the fatigue strength of coated specimens. This subject will be also discussed in section 3.4 later.

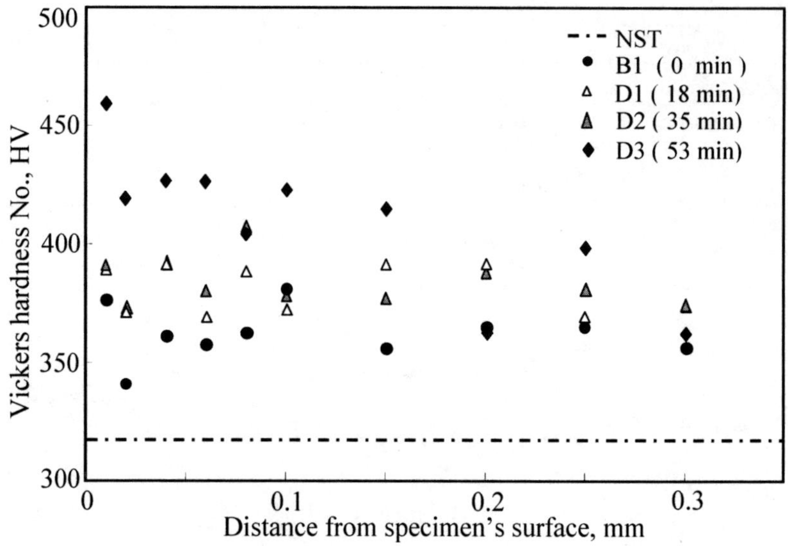

Figure 2: Hardness distribution from the specimen's surface.

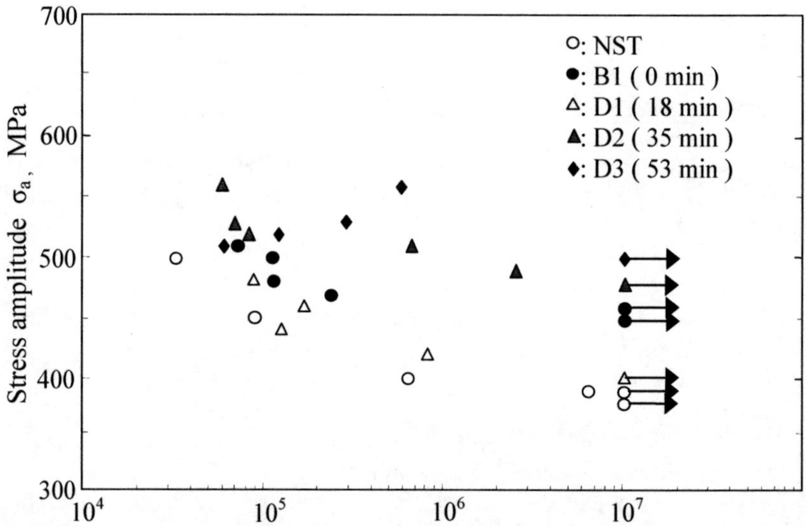

Figure 3: S-N curves.

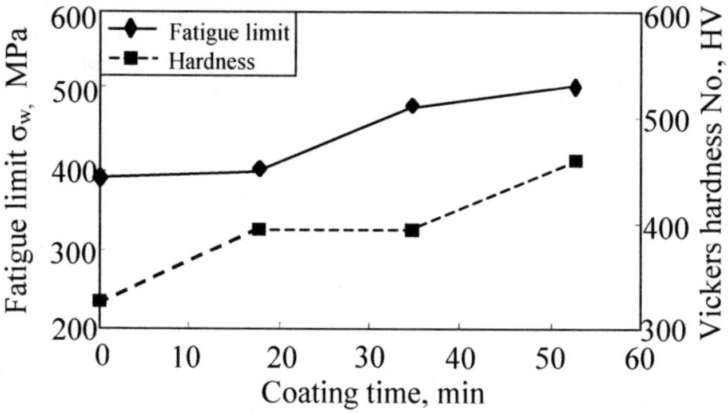

Figure 4: Relation between fatigue limit or hardness and coating time.

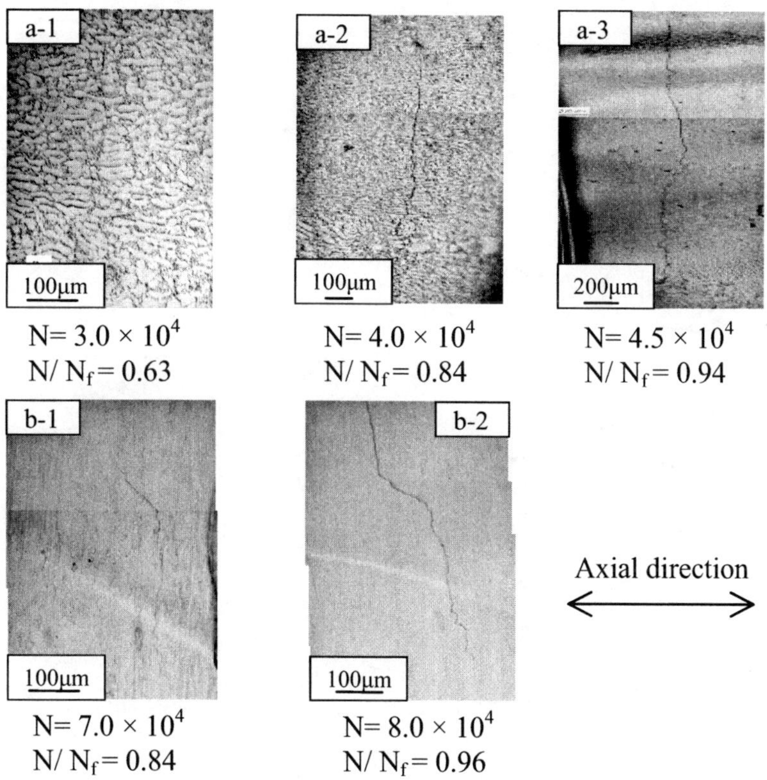

Figure 5: Fatigue crack of (a) NST specimen (σ_a = 600MPa, N_f = 4.8×10^4 cycles) and (b) D2 specimen (σ_a= 520MPa, N_f= 8.34×10^4 cycles).

3.2 Fatigue crack propagation

Fig.5 (a) shows the fatigue crack propagation of NST specimen. The fatigue micro-cracks are initiated at the cycle ratio about 53% and propagates to be a final fracture.

Fig.5 (b) shows the fatigue crack propagation of D2 specimen by the successive observation method at the specimen's surface. The fatigue micro-cracks are initiated at the cycle ratio of about 84% and propagates to be a final fracture. It is considered that this cycle ratio is very larger than that of NST specimen due to retarding effect by the compressive residual stress. It can be confirmed that the fatigue micro-cracks of D2 specimen are initiated by about 45 degree to the axial direction and propagate by about 90 degree to the axial direction. This feature is also observed in the NST specimen' surface. This result indicates that the fatigue micro-cracks of DLC coated specimen are initiated from the surface of the specimen at first, and then the cracks propagated into the core.

3.3 Observation results of fracture surface

Fig.6 shows the fracture surface of B1 specimen. The fatigue cracks initiate in the specimen's surface and propagate into the core direction.

Fig.7 shows the fracture surface of DLC coated specimen. Fig.7(b) is the magnified one of (a) indicated by arrow mark and this is coincided with the position as shown in Fig.5. The fatigue cracks initiate from the specimen's surface being the same as B1 specimen.

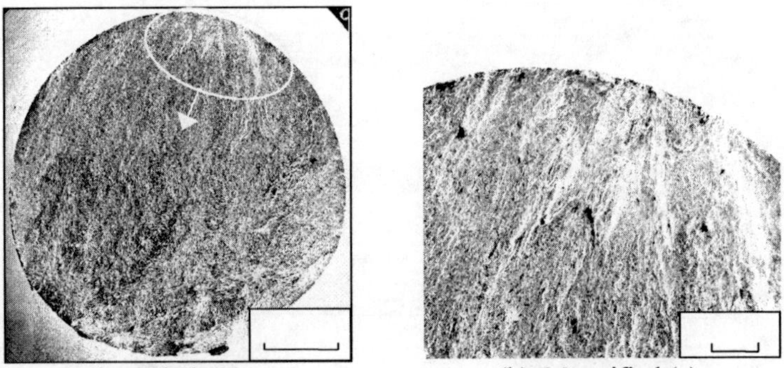

(a) Whole view (b) Magnified (a)

Figure 6: Fracture surface of B1 specimen (σ_a= 470MPa, N_f = 2.4×10^5 cycles).

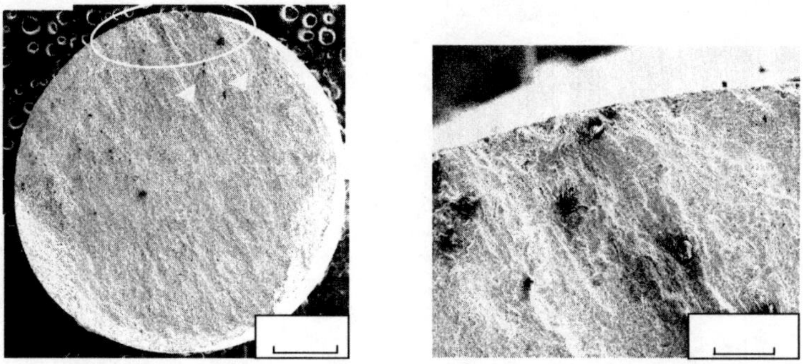

(a) Whole view (b) Magnified (a)

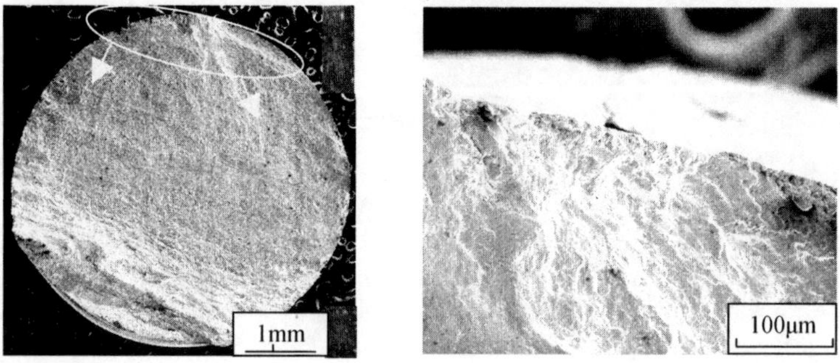

(c) Whole view (d) Magnified (c)

Figure 7: Fracture surface of DLC coated specimen. (a) and (b) are D2 specimen (σ_a= 520MPa, N_f = 8.34×10^4 cycles), (c) and (d) are D3 specimen (σ_a= 520MPa, N_f= 8.34×10^4 cycles).

3.4 Compressive residual stress

The authors have tried to measure the internal stress in the DLC film. But it was impossible to measure accurately due to the problem of specimen shape. Therefore, the case of DLC coated on the Si substrate was consulted in this paper. DLC film on the Si wafer had been deposited with the coating equipment being same as our testing condition. Fig.8 shows the influence of the anode voltage on the internal stress in DLC films as a function of the film thickness [10]. The compressive residual stress of films deposited under the anode voltage of 50V increases from about 2.5MPa to the maximum value of about 4.1MPa. In contract with the case of 75V and 100V, the internal stress becomes smaller according to the film thickness. These results indicate that DLC films prepared at anode voltage of 50V could increase the compressive residual stress.

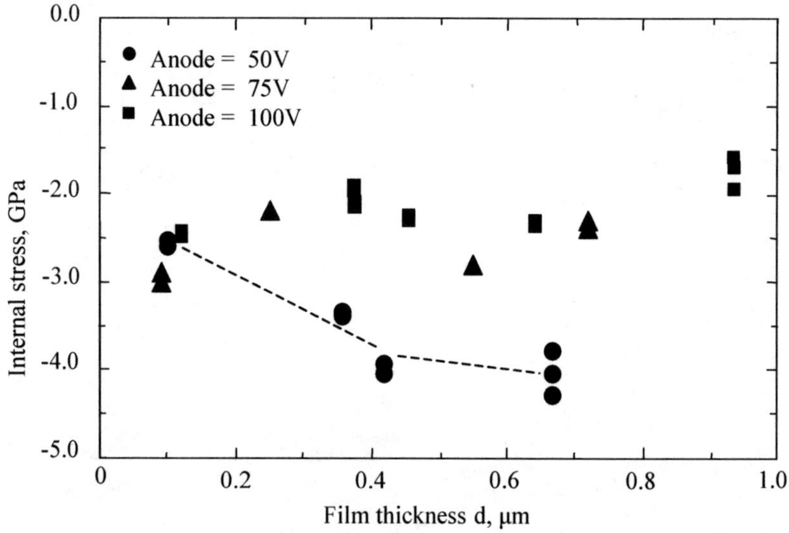

Figure 8: Relation between and film thickness and internal hardness [10].

As DLC films in our test were deposited under the voltage of 50V, it is considered that compressive residual stress was generated on the specimen's surface. In addition, the film thickness was increased with the coating time and the largest compressive residual stress was generated in D3 specimen's surface among all of the specimens.

4 Conclusions

The effect of DLC coating time on the fatigue properties of Ti-6Al-4V has been investigated and tried to analyze the mechanism of the improvement for the fatigue strength. The main results obtained in this test are as follows:

(1) From the Vickers hardness results, the DLC coated specimen show higher value than that of the conventional Ti-6Al-4V alloy due to the compressive residual stress. Especially, D3 specimen shows the highest value.

(2) DLC-2 and 3 specimens show the higher fatigue strength than that of the conventional Ti-6Al-4V by about 20% and 30%, respectively. The fatigue limit and hardness tend to be increased with the coating time.

(3) The fatigue micro-cracks of DLC specimen are initiated from the specimen's surface and then propagate into the core. The crack initiation ratio is very larger than that of NST specimen due to the retardation effect by compressive residual stress.

References

[1] M.j.J.Donachie Jr. A Titanium technical guide, ASM International, (1988), p.9-19

[2] J.M.Wiliuma & R.A.Buchanan, Materials Science and Engineering, Vol.69, (1985), p.237-246
[3] E.J.Lee, R.G.Bayer, Metal finishing, (1985), p.39-42
[4] T.I.Wu and J.K.Wu, Metallurgical Transactions, Vol.24A, p.1181-1185
[5] J.A.Mock, Materials Engineering, Vol.80, (1974), p.101-108
[6] H.J.Gossman, D.J.Eaglesham et al, Applied Physics Letters, Vol.74, (1999), p.2435-2437
[7] S.Nishida, N.Hattori, Proc.of International Conference of Asian Pacific for Fracture & Strength, 96, (1996), p.447-452
[8] S.Nishida, N.Hattori, Proc. Of International Conference on surface Treatment'97, (1997), p.199-209
[9] S.Nishida, S.Young, N.Hattori, & A.Tsuchiyama, Proc.of Internal Conference on Surface Treatment'03, (2003), p1-12.
[10] A.Tsuchiyama, Y.Shima, H.Hasuyama, Proc.of Internal Conference on Surface Treatment'03, (2003), p41-49.

Thick tool steel coatings with laser cladding

V. Ocelík, U. de Oliveira & J. Th. M. De Hosson
Department of Applied Physics, Netherlands Institute for Metals Research, University of Groningen, Groningen, The Netherlands

Abstract

This paper concentrates on thick and crack-free laser clad coatings (up to 3 mm). The coating material is a chromium-molybdenum-tungsten-vanadium alloyed high-speed steel that shows high wear resistance, high compressive strength, good toughness, very good dimensional stability on heat treatment and very good temper resistance. It will be demonstrated that laser cladding of MicroMelt 23 powder offers a relatively wide processing window resulting in the formation of thick, microstructurally uniform, hard, crack- and defect- free coating on ordinary steel substrates. Microstructural observations using light and field emission gun scanning electron microscopy with EDS and EBSD attachments together with internal strain measurements using diffraction of X-rays revealed the reason for low susceptibility to crack formation. An intensive martensitic transformation inside small austenitic cells surrounded by hard carbides following the rapid solidification process compensates the tensile strain and finally results in compressive stresses at the coating surface. Laser cladding on different steels substrate geometries will be demonstrated together with hardness profiles and their dependence on cladding conditions.
Keywords: tool steel, laser cladding, microstructure, residual stress.

1 Introduction

Laser surface treatment includes several different techniques utilizing the heat of a laser beam acting at the surface to modify the composition and microstructure and produce a wide range of metallurgical effects [1]. Laser cladding using the powder blowing technique [2, 3] comprises fusion of an alloy powder layer to a substrate with minimum melting of the substrate. Melting starts at the surface and the particles being heated and melted when passing the laser beam are trapped in the melt pool. This technique is used for the deposition of alloys on

WIT Transactions on Engineering Sciences, Vol 55, © 2007 WIT Press
www.witpress.com, ISSN 1743-3533 (on-line)
doi:10.2495/SECM070021

turbine blades, engine valves, valve seats and drilling components. Although a wide variability of clad materials and substrates are reported in literature only a few shows results of laser cladding of tool steel powders on ordinary steel or cast iron substrates. Yellup [2] reported about laser deposited tool steel among other coatings with low dilution, high integrity and fine structures. Also Mazumder and co-workers [4] have shown the possibility to build 3D part from H13 tool steel using the so-called direct metal deposition technique, which is an advanced laser cladding technique with coaxial blowing powder. Reparation of moulds which have experienced severe damage during their service lifetime by local laser cladding of high-speed steel with high molybdenum content was recently reported by Navas at al. [5]. Abrasive wear behaviour of several laser clad tool steel coatings studied in work by Wang at al. [6] showed that the laser track overlapping results in different properties inside a re-heated zone in comparison with an area when material in the as clad condition exists.

In this work we present a laser clad coating made from tool steel powder originally used to form parts via powder metallurgy processes. This material is commercially known as Vanadis 23 (Udeholm) or MicroMelt 23 (Carpenter Powder Products). It is a chromium-molybdenum-tungsten-vanadium alloyed high-speed steel that shows a high wear resistance, high compressive strength, good toughness, very good dimensional stability on heat treatment and very good tempering resistance [7]. It will be demonstrated, that the laser cladding of both these powders offers a wide processing window resulting in the formation of thick, microstructurally uniform, hard, crack- and defect- free coatings on ordinary steel substrates.

2 Experimental

Laser cladding with a side blowing powder [3] was used in our experimental set-up. Nd:YAG Rofin Sinar laser with the output power of 1.8 kW working in continuous regime was used as a heat source scanning the surface of the substrate materials with scanning speeds between 5 and 10 mm/s. Laser beam was defocused to form a circular beam spot with a diameter of 4 mm. Vanadis 23 and MicroMelt 23 powders with chemical composition shown in Table 1 and particle size between 45 and 125 μm were delivered to the processing zone using Sulzer Metco Twin 10C powder feeding system with a powder feeding rate between 150 and 220 mg/s.

Table 1: Chemical composition of MicroMelt 23 powder from Carpenter Powder Product inspection certificate.

Element	Fe	W	Mo	Cr	V	C	Si	Co	Ni	Mn	Cu
wt%	bal	6.45	5.0	4.2	3.1	1.26	0.6	0.6	0.28	0.37	0.16

During laser cladding the Argon was used as a carrier as well as a shielding gas in amounts of 3 and 15 l/min, respectively. 30% overlapping of individual laser tracks was used to build continuous laser coatings. Samples for microstructural observation were prepared by longitudinal and transversal cross-sectioning, mechanical grinding and polishing. Light microscopy, as well as enhanced scanning electron microscopy based on field emission gun Philips XL model microscopes equipped with EDS, and Oriented Image Microscopy (TSL) together with XRD were used to study the coating microstructure. CSM Revetest scratch tester equipped with Vicker's indenter was used to measure the hardness at load of 4 N.

The (311) fcc plane reflection in austenite is chosen as a suitable reflection for the characterization of macroscopic strains in our experiments because it corresponds to a high 2θ angle when CuKα radiation is applied. $Sin^2\Psi$ experiments were carried out in reflection mode with a Phillips X'pert X-ray system equipped with a Cu radiation source and side angle Ψ range from -60° to +60° scanned in three different axis: φ = 0° (longitudinal), φ = 45° (diagonal) and φ = 90° (perpendicular) to laser tracks cladding direction. All other experimental details as well as method for calculation of strain and stress from such measurements can be found in [8].

3 Results and discussion

3.1 Laser processing and coatings production

Both powders used in laser cladding experiments show a very stable behaviour during feeding and laser cladding and offer a relatively broad laser processing window in which they provide a regular single laser track shape and homogeneous pore-less coating after 30% overlapping of consecutive laser tracks. Steel substrates with different compositions (C45, 100Cr6 and bearing steel) and geometries were tested.

Figure 1: Laser cladding of Vanadis23 powder on C45 steel substrate. a) Single laser track profile on transversal cut; b) 3 mm thick coating prepared by 30% overlapping of individual laser tracks in two cladding layers; c) and d) 20 mm wide rings on 60 mm diameter bar coated by single and double layer laser cladding; e) Surface of coated ring from bar on c) after cutting, machining and grinding.

Figure 1 demonstrates this behaviour showing the profile of an individual laser track as well as a double layer coating in their transversal cross sections. The individual laser track height of about 1 mm was achieved when 1750 W laser power, 5 mm/s scanning speed and 150 mg/s powder feeding rate were used as main laser processing parameters during cladding on 60 mm in diameter C45 steel bar. The 30% overlapping of such laser tracks provides 1.3-1.5 mm thickness of single clad layer and 2.4-3.0 mm thick double clad layer. Very low level of internal porosity is evident from microstructural observations on different layer cuts, but also from the surface smoothness after machining, grinding and finishing of the coating surface for final use. We faced small troubles with inter-run porosity [9] between the very first laser tracks when relatively small substrates were used (8 mm in diameter bars) and when laser power has to be reduced due to the fast heating of the whole substrate piece.

3.2 The coating microstructure

Figure 2 shows the microstructure of the laser coating near the coating/substrate interface as well as details of the microstructure in the middle of the coating (perpendicular cross-section).

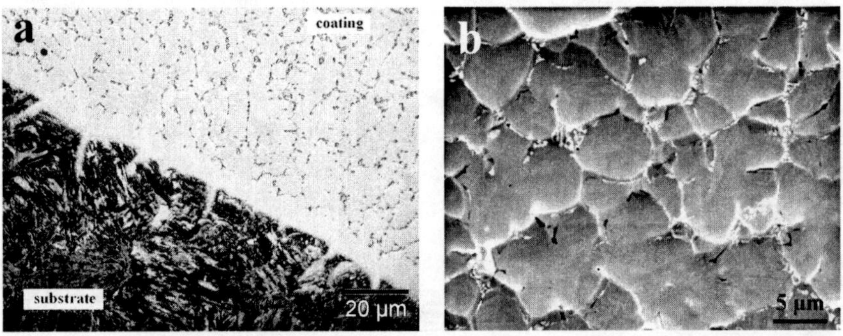

Figure 2: The microstructure of MicroMelt23 laser clad coating on C45 steel substrate. a) Optical microscopy of the microstructure observed near coating/substrate interface (Nital etching); b) SEM secondary electrons image of the coating microstructure.

The microstructure inside the laser track is very homogeneous with fine (5-15 µm) iron based dendrites. Dendrite boundaries and interdendritic space is full of carbide precipitates and hard eutectic interdendritic phase. This microstructure is homogenously distributed through the whole coating, including inside and close to the laser tracks overlapping region, and the overall composition of the clad alloy measured by SEM-EDS is 4Cr-5W-3Mo-85Fe-3V (in wt%).

The interface between coating and substrate is relatively sharp (~ 10µm) and it consists of the thin iron based layer with local perturbations into to the substrate microstructure.

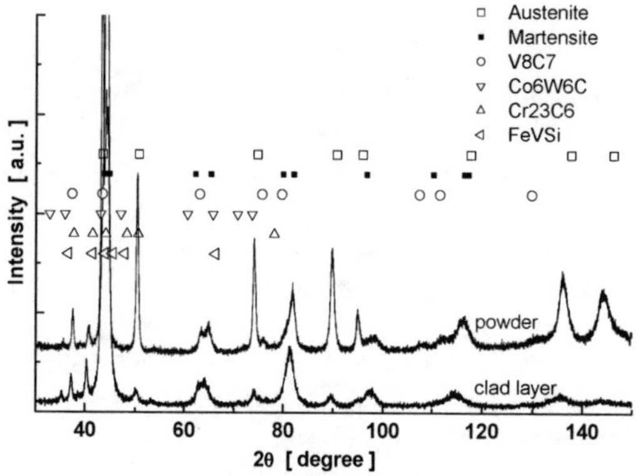

Figure 3: Comparison of X-ray diffraction spectra in θ-2θ set-up measured on initial powder and laser clad coating.

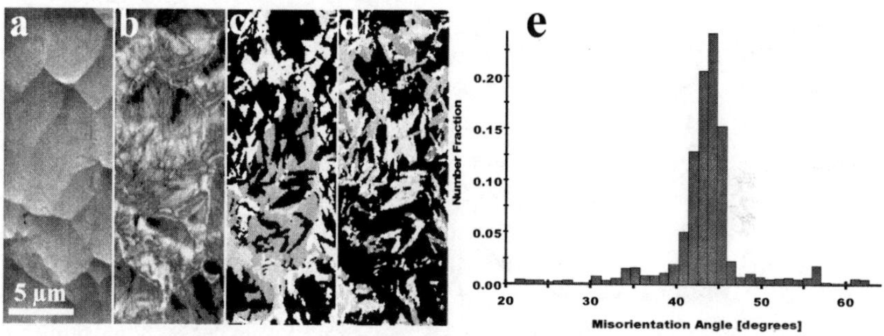

Figure 4: OIM observations of tool steel coating microstructure. a) SE image of the surface after precise ion polishing; b) Image Quality image from the same place, c) OIM image of austenite phase; d) OIM image of martensite phase; e) Misorientation angle between martensite phase inside austenitic grains.

Figure 3 shows the results of phase analysis of the laser clad coating and initial powder via X-ray diffraction of CuKα radiation. The similarity between the X-ray spectra of the powder and the clad layer suggests that the phase composition of the original alloy is conserved, i.e. also inside a clad layer, which means that the change melt composition due to a dilution from the substrate does not influence the phase composition of the coating. However, less content of austenitic iron phase after laser cladding is evident. Analysis of X-ray diffraction experiment and field emission gun electron microscopy observations using both secondary electron and back scatter electron detectors combined with EDS measurements leads to the conclusion, that microstructural dendrites consist of martensite and retained austenite, while hard interdendritic Co_6W_6C and FeVSi phases were formed at the end of solidification process.

Oriented Image Microscopy (OIM) [10] based on analysis of Electron Back Scattered Pattern (EBSP) is a powerful tool for phase and crystallographic analysis of materials by scanning electron microscopy. We used it to quantify the amount of the martensite inside the coating and to visualize austenite and martensite grains orientations. Figure 4 summarizes results of OIM observations from one small place situated at the centre of cross-section from single laser track. Unfortunately, a relative demanding surface polishing procedure [11] requiring Precise Ions Polishing restrains more extensive OIM observations at different coating positions. Figure 4a shows an SEM observation of the surface of mechanically polished tools steel coating followed by 120 min polishing using 4.5 keV Ar^+ ions. The surface profile is characterized by a system of hills and valleys with a local roughness of R_a= 41 nm.

Black areas on the image quality (IQ) map (Fig. 4b) indicate the places, where highly tilted electron beam (70°) is shielded due the local surface roughness and no indexable Kikuchi pattern is generated. OIM maps on Fig. 4c and 4d denote the areas, from which corresponding Kikuchi patterns were recognized as austenite and martensite, respectively. The same colour on these maps belongs to the same grain of corresponding phase. The conclusion concerning a presence of 45% of martensite in retained austenite may be drawn together with the validity of Kurdjumov-Sachs orientation relationship [12] between these two phases, as the distribution of misorientation angles between martensite and austenite grains plotted in Fig. 4e testifies.

3.3 Hardness profiles

Hardness, as an important characteristic that influences the wear performance, was measured to quantify the quality of the coatings.

Figure 5 characterizes the depth profile of Vicker's hardness measured through the cross-section of double layer laser coating on C45 steel substrate bar 60 mm in diameter. It may be concluded that in each clad layer a characteristic decrease in hardness value exists, from the value of about 820 HV0.4 measured at the top of the layer to the value of about 730 HV0.4 measured at its bottom. Consequently quite a sharp discontinuity in hardness value exists at the border of these two layers. Naturally, another sharp hardness change is detected at the coating/substrate interface, because of small hardening effect in Heat Affected Zone (HAZ) of massive C45 steel substrate.

Another type of the hardness profile may be observed on the single layer coating prepared on the relatively small (8.5 mm in diameter) bearing steel bar as Figure 6 clearly indicates. In all four measured vertical directions (Fig. 6b), there is no substantial hardness decrease inside the coating, as it was observed in previous case (Fig. 5). Also a substantial hardening effect in HAZ is present, which moderates the hardness drop inside the substrate material. Due to a small angle between the cutting plane and the laser cladding direction, the Fig. 6a maps a change of coating thickness within one revolution during laser cladding process and it also scans the coating microstructure between two adjacent laser tracks. The graph insert in Fig. 6a shows clearly that a correlation exists between the coating thickness, which varies between 780 and 980 μm, and coating hardness measured at 200 μm depth.

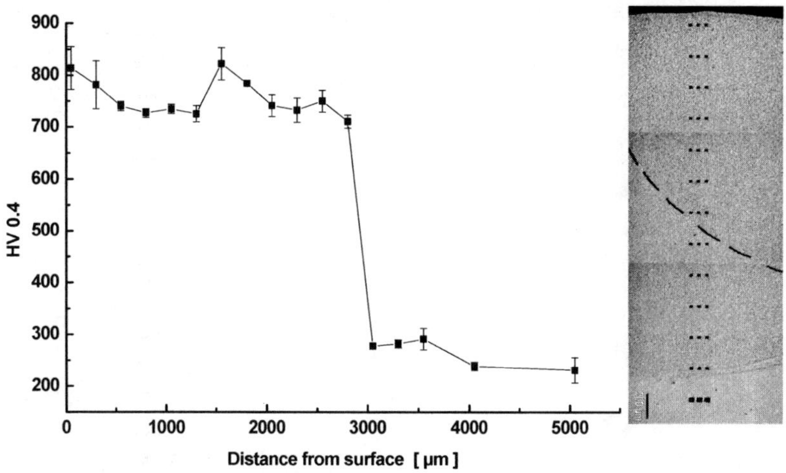

Figure 5: Microhardness profile on transversal coating cross-section plotted as an average value measured from three indents at the same depth of 2.8 mm thick double layer coating. Dashed line denotes microstructural boundary between laser tracks made in the first and the second laser cladding layer.

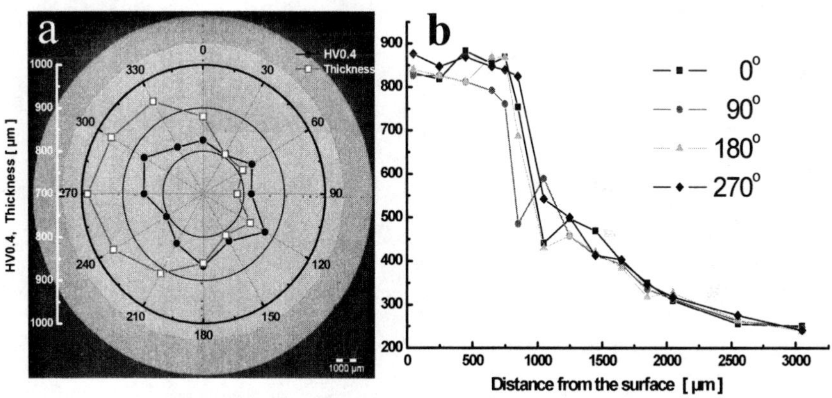

Figure 6: a) Optical micrograph of longitudinal cross section of MicroMelt 23 coating on 8.5 mm bearing steel bar. Insert shows the thickness and the microhardness (in depth of 200 μm from the surface) dependence of the coating on the rotation angle. b) Microhardness profiles of MicroMelt 23 coating and substrate at four different angles.

3.4 Residual stresses

The Vanadis 23 clad layer was ground to eliminate the macro roughness and electrochemically etched. The $Sin^2\Psi$ technique was applied to measure the surface strain and the stress free d_o= 1.08949 Å from the (311) peak of the tool steel alloy powder was used to calculate local strain. An isotropic elastic modulus E = 210 GPa and ν = 0.29 [13] was used to calculate plain stress conditions on the coating surface. The negative slopes of the plots of plane spacing $d_{(311)}$ vs. $Sin^2\Psi$ witness the presence of compressive strains. The biaxial stress tensor was calculated and reveals that the largest principal stress lies along the direction of laser cladding track. The major and minor stress components given in MPa are -1100 and -600, respectively.

4 Discussion

In the Fe-C system the positive effects of vanadium is directly noticed by the increase of strength and wear resistance making these alloys suitable for cutting tools and die applications. The improvement of the mechanical properties is caused by two main mechanisms: the formation of stable V-carbides and the refinement of microstructure [14]. In a steel melt the solubility of the vanadium is quite high [15] (on the order of 6% when the carbon amount is 2% at 1425°C) but it drops drastically in solid austenite (0.23% at 727°C) and it becomes even worse when the ferritic phase is formed (<0.1% on α-Fe at 727°C). The strong temperature dependence of V solubility indicates that the Fe-C-V austenitic system is subjected to a high amount of strengthening by dispersoids where V plays a very important role on the formation of carbides and carbonitrides in interdendritic spaces. The growth rate of the precipitates depends on the supersaturation of vanadium and the ratio of solute concentration in the matrix [16]. A side effect of the enhanced carbide precipitation promoted by vanadium is that the microstructure becomes finer. Further refinement is promoted if a small amount of N is present, which causes the formation of carbonitrides that slow down the austenite-ferrite transformation and promotes nucleation to happen in place of grain coarsening [17].

The primary parameters controlling the size of the grains in austenite/ferrite transformations are the cooling rate and the amount of precipitates formed in the interdendritic spaces. When the melt is rapidly cooled austenite does not have the time to transform in ferrite and due to oversaturation of carbon, a diffusionless transformation takes place resulting in martensite. The resulting microstructure for high cooling rates is then composed of martensite, retained austenite and carbides. Because of the high solidification rates in laser cladding technique the microstructure consists typically of fine grains, supersaturation and non-equilibrium phases. However, different hardness profiles detected for coatings prepared at different laser cladding conditions witness about different content of martensite inside the coating.

Usually, residual stresses detected inside the thick Co based laser cladding coatings on both: macro [8] and micro [18] scale levels are strong tensile stresses, due to a substantial shrinkage after the solidification. On the other hand, the massive martensitic transformation in this tool steel material is associated with expansion and therefore it generates compressive stresses, which in this case overcomes tensile stress components. This behaviour seems to be crucial for the possibility to form the thick coating without cracking, which is often observed when thick and hard coatings are prepared by laser cladding.

Moreover, a full potential of this tool steel coating with an interesting combination of high toughness, hardness and wear resistance at elevated temperatures may be achieved, when appropriate thermal treatment recommended for powder metallurgy products [7] will be applied on the final laser clad coating.

5 Conclusions

Laser cladding of MicroMelt 23 and Vanadis 23 tool steel powder produces homogeneous and continuous coatings, free of defects and with a perfect adherence to ordinary steels substrates in a broad window of processing parameters. The phase analysis of the coating showed the consistent phase composition with the initial powder, indicating a minimal influence of dilution from the substrate steel. The microstructure of the laser deposited tool steel coatings contains fine dendrites, with the presence of subdendritic grains of martensite and retained austenite in approximately equal amounts and hard interdendritic phases. The stress state of the clad layer is compressive near the surface and the larger principal stress is almost parallel to the cladding direction. Large amount of martensite plates formed inside fine dendrites is responsible for the compressive stresses and suppresses the coating cracking often observed on laser clad coating with similar values of hardness and thickness.

Due to the excellent combination of hardness, toughness and wear resistance properties of powder metallurgy products made from this powder, we may expect that laser cladding is a new promising technology to produce coatings with these excellent characteristics.

Acknowledgement

This project is financed by The Netherlands Institute for Metals Research.

References

[1] Steen, W.M., *Laser Material Processing,* Springer-Verlag, London, 408p, 2003

[2] Yellup, J.M., Laser cladding using the powder blowing technique, *Surface and Coatings Technology,* **71,** pp. 121-128, 1995.

[3] Ocelík, V., de Oliveira, U., de Boer, M. & De Hosson, J.Th.M.: Thick Co based coating on cast iron by side laser cladding: Analysis of processing conditions and coating properties, *Surface & Coatings Technology*, in press: http://dx.doi.org/10.1016/j.surfcoat.2006.10.044.

[4] Mazumder, J., Schifferer, A. & Choi, J., Direct materials deposition: designed macro and microstructure, *Mat. Res. Innovat,* **3,** pp.118-131, 1999.

[5] Navas, C., Conde, A., Fernández, B.J., Zubiri, F. & De Damborenea, J., Laser coatings to improve wear resistance of mould steel, *Surface and Coatings Technology,* **194,** pp136-142, 2005.

[6] Wang, S.-H., Chen, J.-Y. & Xue, L. A study of abrasive wear behaviour of laser-clad tool steel coatings, *Surface and Coatings Technology*, **200,** pp.3446-3458, 2006.

[7] Vanadis 23 – Super clean, High performance powder metallurgical cold work tool steel, www.uddeholm.com/files/vanadis_23-english_991019.pdf

[8] de Oliveira, U., Ocelík, V. & De Hosson, J.Th.M., Residual stress analysis in Co-based laser clad layers by lab X-rays and synchrotron diffraction techniques, *Surface and Coatings Technology,* **201,** pp. 533-542, 2006.

[9] Steen, W.M., Weerasinghe, V.M., & Monson, P., Some aspects in formation of laser clad tracks, *SPIE,* **650,** pp. 226-233, 1986.

[10] Electron Backscatter Diffraction in Materials Science, eds. Schwartz, A.J., Kumar, M. & Adams, B.L., Kluwer Academic/Plenum Publishers, Dordrecht, 2000, 339p.

[11] Ocelík, V., Vreeling, J.A. & De Hosson, J.Th.M., EBSP Study of reaction zone in SiC/Al metal matrix composite prepared by laser melt injection, *Journal of Materials Science,* **36,** (20) pp. 4845-4850, 2001.

[12] Kurdjumov, G. & Sachs, G., *Z. Phys.,* **64,** pp. 325-343, 1930.

[13] Carvalho, N., Low friction and wear resistant coatings, Microstructure and mechanical properties, PhD Thesis, Rijksuniversiteit Groningen (2001), 126.

[14] Stacey, A., Macgillivary, H.J., Webster, G.A., Webster, P.J. & Ziebeck, K.R.A., Measurement of residual-stresses by neutron-diffraction, *Strain Analysis for Engineering Design,* **20**, pp. 93-100, 1985.

[15] Sil'man, G.I., Phase diagram of Fe-C-V system and its application to metallography of steels and cast irons, *Metal Science and Heat Treatment* **34,** pp. 665-670, 1992.

[16] Yamasaki, S. & Bhadeshia, H.K.D.H., Modelling and characterisation of V4C3 precipitation and cementite dissolution during tempering of Fe-C-V martensitic steel, *Materials Science and Technology,* **19,** pp. 1335-1343, 2003.

[17] Lagneborg, R., Siwecki, T., Zajac S. & Hutchinson B., The role of vanadium in microalloyed steels *Scandinavian Journal of Metallurgy,* **28,** pp. 186-241, 1999.

[18] de Oliveira, U., Ocelík, V. & De Hosson, J.Th.M., Microstresses and microstructure in thick Co-based laser deposit coatings, *Surface and Coatings Technology,* in press, http://dx.doi.org/10.1016/j.surfcoat.2006.12.013.

Numerical modeling of surface treatment by laser beam

R. Gospavic & V. Popov
Wessex Institute of Technology, Ashurst Lodge, Ashurst, Southampton, UK

Abstract

In this work an analytical approach for analysis of surface treatment by a laser beam is presented. A thermal model of interaction for the case of cylindrical geometry of the material and asymmetric distribution of the laser beam intensity are used. An analytical procedure is developed to analyze the temporal and the spatial distribution of the temperature field inside the bulk of material. This kind of consideration is of practical interest in cases where the excitation by the laser beam is not symmetric in respect to its position or shape, e.g. multi-mode working regimes or asymmetrical distribution of the laser beam intensity. The heating effects were considered in the temperature range up to the melting point. The thermal and the optical parameters of the material were assumed to be independent of the temperature and were given constant values in the temperature range of interest. This approach makes use of the Laplace transform, in order to eliminate dependence on time. The Fourier method of variable separation was used to obtain the temperature field distribution in the Laplace transform domain.
Keywords: surface treatment, laser, thermal model, multi-mode.

1 Introduction

In a general case the analysis of the laser-material interaction, important for practical applications, is very complex and includes analysis of different physical processes such as material removal, material melting, thermal stresses, shock wave, etc. This prevents successful construction of a general analytical solution and different numerical procedures have been used in the past [1]. This work is restricted only to analysis of heating effects of the laser-material interaction.

WIT Transactions on Engineering Sciences, Vol 55, © 2007 WIT Press
www.witpress.com, ISSN 1743-3533 (on-line)
doi:10.2495/SECM070031

In many practical applications of laser beams the finite dimensions of the bulk material and the asymmetric distribution of the laser beam intensity has to be taken into account. The presented analytical approach enables consideration of the heating effects of interactions for the different and complicated laser-beam distributions for cases without axial symmetry. The temperature field distribution in this way can be analyzed for 3D cylindrical geometry allowing monitoring of the time evolution of the temperature distribution. For the arbitrary time dependence, spatial distribution and position of the incoming laser beam intensity the same numerical data could be used, saving in this way memory and CPU time. By using Duhamel's principle [2] the temperature distribution is evaluated by using the convolution integral.

This kind of analyses could be important in many technical applications of laser beams in technology and science [3-5] as well as in the case of laser-material interaction in the multi-mode working regime.

For the multi-mode working regime as well as for the complex laser beam intensity distribution superposition principle could be used. In case of the linear governing partial differential equation (PDE) the final solution could be presented as superposition or sum of the PDE solutions belonging to different parts of the incident loads, i.e. in our cases the incoming laser beam intensity.

2 Mathematical model

Heating of a homogenous cylinder, with a finite or infinite length, by an incident laser beam on the upper surface of the specimen is considered (Fig. 1). The shape of the cross-section, the position of the laser beam on the upper side of the specimen, the distribution and the time dependence of the laser beam intensity can be arbitrary. In the numerical examples presented in this paper, because of simplicity and still without losing generality, only a top head laser beam profile with circular cross section is considered. It was assumed that the laser beam intensity could be approximated by a product of two functions of spatial and time coordinates: $q(r)$ and $\varphi(t)$, respectively [5, 8].

Only the heating effects due to the interaction were considered. The laser-material interaction was modeled by the equivalent thermal flux on the upper side of the specimen. All thermal and optical parameters of the material are considered to be constant and temperature independent, yielding a linear thermal conduction problem. The geometry of the considered problem was represented in a cylindrical reference system.

The temperature distribution inside the bulk material was considered and the convective thermal losses from the lower and the axial surface of the material were taken into account, while the thermal losses from the upper surface of the specimen are neglected. The radiative losses have important contribution to the whole thermal losses at the very high temperatures [8]. Thus, for low temperature of the specimen, the radiative heat losses are smaller than convective ones and could be neglected [8]. Beside this, if the absorption length, for considered laser beam and material, is very short, related to size of the heating affected zone (HAZ), it could be considered that laser beam is absorbed by the surface of the specimen. This is the case for many materials of interest, thus the laser-material interaction could be approximated by the equivalent surface thermal source [8]. Constant and temperature independent value of the coefficient of thermal conductivity were assumed.

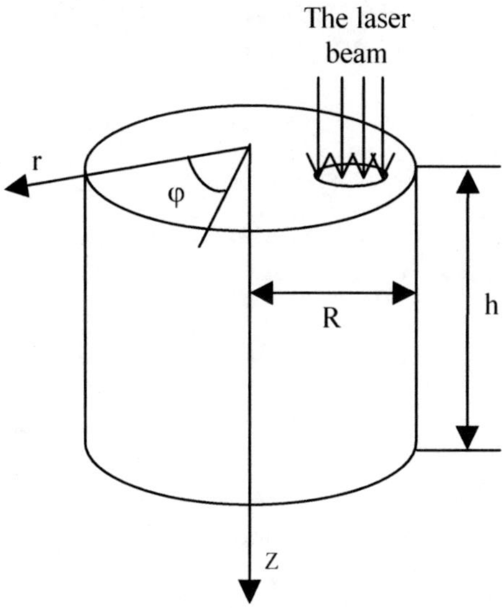

Figure 1: The geometry of considered problem and coordinate system used.

Once these approximations are applied, the heating of the considered cylindrical specimen can be modeled by the following PDE and the corresponding boundary (BC) and initial (IC) conditions [5, 9]:

$$\frac{\partial^2 T}{\partial r^2}+\frac{1}{r}\frac{\partial T}{\partial r}+\frac{\partial^2 T}{\partial z^2}+\frac{\partial^2 T}{\partial \theta^2}=\frac{1}{\alpha}\frac{\partial T}{\partial t};\ t\geq 0;\ 0\leq r\leq R;\ 0\leq z\leq H;\ 0\leq\theta\leq 2\pi \quad \text{(1a)}$$

$$-\lambda\left.\frac{\partial T(r,z,\theta,t)}{\partial z}\right|_{z=0}=Aq(r,\theta)\phi(t);\quad -\lambda\left.\frac{\partial T(r,z,\theta,t)}{\partial r}\right|_{r=R}=h_c\cdot T(r,z,\theta,t)\big|_{r=R}$$

$$-\lambda\left.\frac{\partial T(r,z,\theta,t)}{\partial z}\right|_{z=H}=h_c\cdot T(r,z,\theta,t)\big|_{z=H} \quad \text{(1b)}$$

$$T(r,z,\theta,t=0)=0;\quad 0\leq r\leq R,\quad 0\leq z\leq H,\quad 0\leq\theta\leq 2\pi \quad \text{(1c)}$$

where: λ is the coefficient of thermal conductivity, which is considered to be constant and temperature independent, $\alpha=\lambda/\rho\cdot c$ is the thermal diffusivity, c is the specific heath, ρ is the material density, h_c is the heath transfer coefficient [10], A is the absorption coefficient of the laser radiation by the material [11], R, h are the radius and length of the specimen, respectively, and T is the temperature difference in the interior domain relative to the ambient one. The equation (1b) represents the homogenous BCs while the equation (1c) describes the IC.

The Laplace transform approach was used to eliminate the time dependence and to convert the original problem to the equivalent problem in the Laplace transform domain [12]. Fourier's method of variable separation was used in order to transform the original PDE into three ordinary differential equations as follows [12]:

$$r^2\frac{\partial^2 T_r(r)}{\partial r^2}+r\frac{\partial T_r(r)}{\partial r}+\left(\mu^2\cdot r^2-m^2\right)T_r(r)=0$$
$$\frac{\partial^2 T_z^*(z,s)}{\partial z^2}-\left(\mu^2+\frac{s}{\alpha}\right)\cdot T_z^*(z,s)=0 \qquad (2)$$
$$\frac{\partial^2 T_\theta(\theta)}{\partial \theta^2}+m^2 T_\theta(\theta)=0;\quad T^*(r,z,\theta,s)=T_r(r)\cdot T_\theta(\theta)\cdot T_z^*(z,s)$$

where μ and m are constants, s is a complex parameter, and the asterisk in the superscript denotes functions in Laplace's transform domain. The particular solutions of the governing PDE can be expressed in the following form in the Laplace transform domain [9, 13, 14]:

$$T_{mn}^*(r,z,\theta,s)=T_r(r)\cdot T_\theta(\theta)\cdot T_{zmn}^*(z,s);$$
$$T_r(r)=J_m(\mu_{mn}\cdot r);\quad T_\theta(\theta)=K_{1m}\cos(m\cdot\theta)+K_{2m}\sin(m\cdot\theta);\quad n=1,2,... \qquad (3)$$

where m is integer, $T(\theta)=T(\theta+2m\pi)$, because of the continuity condition, J_m are Bessel functions of the m-th kind, μ_{mn} are positive roots of the characteristic transcendent equations which describe the BC on the axial boundary surface of the specimen, given by the next relation:

$$-\lambda\frac{\mu_{mn}}{2}\left(J_{m+1}(\mu_{mn}R)-J_{m-1}(\mu_{mn}R)\right)=h_c\cdot J_m(\mu_{mn}R) \qquad (4)$$

The next relations hold: $\mu_{m,-n}=-\mu_{mn}$, $(\mu_{mn}\geq 0)$ and $T_{rm-n}^*=(-1)^{-m}T_{rmn}^*$. Due to the boundary conditions on the lower surface of the specimen and according to previous work [9], $T_{zmn}^*(z,s)$ could be expressed in the following way:

$$T_{zmn}^*(z,s)=T_{0mn}^*(s)\cdot\left(\exp\left(\frac{iz\cdot\varepsilon}{h}\right)+\frac{i\lambda\cdot\varepsilon/h+h_c}{i\lambda\cdot\varepsilon/h-h_c}\exp\left(i\frac{2\cdot h-z}{h}\varepsilon\right)\right);\quad i\varepsilon=h\sqrt{\mu_{mn}^2+\frac{s}{\alpha}}. \qquad (5)$$

If the laser beam have rounded cross-section and top head profile, because of symmetry the particular solutions given by (2) have to be odd functions of the angular coordinate θ and accordingly K_{1m} and K_{2m} in (3) can be $K_{1m} = 1$ and $K_{2m} = 0$.

As the particular solutions, for positive values of the constants μ_{mn}, are linearly independent, the solution in the Laplace transform domain could be evaluated by the series of the particular ones:

$$T^*(r,z,\theta,s) = \sum_{n=1}^{+\infty}\sum_{m=0}^{+\infty} a_{mn} \cdot T_{rmn}(r) \cdot T_{\theta m}(\theta) \cdot T^*_{zmn}(z,s) \tag{6}$$

The response to Dirac's pulse induction in time domain $T_\delta(r,z,\theta,t)$ was obtained using inverse Laplace's transformation and Bromwich integral, and it can be expressed in the following form:

$$T_\delta(r,z,\theta,t) = L^{-1}\left\{T^*(r,z,\theta,s)\right\} = \int_{p-i\infty}^{p+i\infty} T^*(r,z,\theta,s) \cdot exp(st) \cdot ds \Rightarrow$$

$$T_\delta(r,z,\theta,t) = \frac{2 \cdot \alpha \cdot A}{h\lambda} \sum_{m,n,j=1}^{N} J_m(\mu_{mn} r) E_j(z) \cdot \left(c_{1mn} \cos(m\theta) + c_{2mn} \sin(m\theta)\right) exp\left(\frac{-t}{\tau_{mnj}}\right) \tag{7}$$

where p is the positive real parameter; L^{-1} denotes the inverse Laplace-s transformation; τ_{mnj} and $E_j(z)$ are given by next relations:

$$\tau_{mnj} = \frac{1}{\alpha\left(\mu_{mn}^2 + \left(\frac{\varepsilon_j}{h}\right)^2\right)}; \quad E_j(z) = \frac{\lambda\varepsilon_j \cos\varepsilon_j (1 - z/h) + h_c \cdot h \sin\varepsilon_j (1 - z/h)}{(\lambda + h_c \cdot h)\sin\varepsilon_j + \lambda\varepsilon_j \cos\varepsilon_j} \tag{8}$$

For arbitrary time dependence of the laser beam intensity, the temperature distribution inside the specimen could be evaluated by a convolution integral as [9, 12]:

$$T(r,z,\theta,t) = \int_0^t \phi(t-\tau) \cdot T_\delta(r,z,\theta,\tau) \cdot d\tau \tag{9}$$

3 Numerical examples

In this section according to the above considerations numerical examples are presented for some characteristic cases. A cylindrical Al specimen is considered and the following characteristic parameters were assumed: $\lambda = 240\,[W/K \cdot m]$; $\rho = 2700$ [kg/m^3]; c = 1021.71 [J/kg K]; h_c = 10 [W/Km^2]; A = 0.64; h=5[mm]; R = 7[mm].

In Figs. 2 the contour lines for the temperature field on the upper surface (X–Y plane) of the specimen are presented. The laser beam with top head profile was assumed to have the following characteristics: power = 500W, radius = 1mm, time duration = 1s, coordinates of the laser beam center: $\theta_0 = 0$; $r_1 = 4$mm.

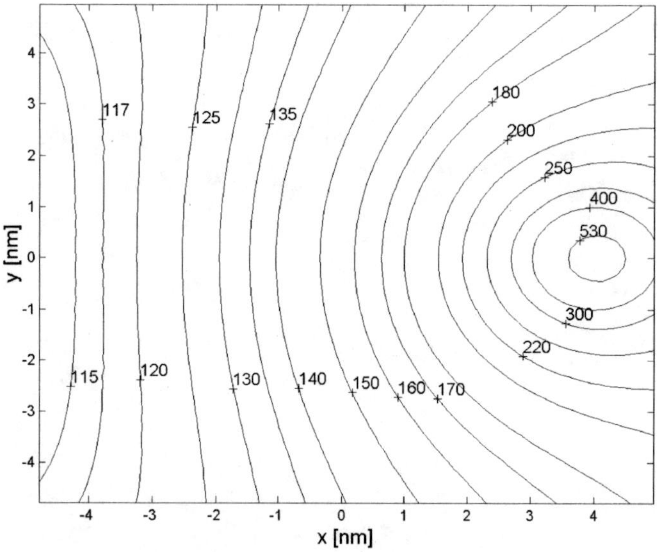

Figure 2: Contour plot of temperature on the upper surface of the specimen. The laser beam properties were: power = 500W, radius = 1mm, time duration = 1s, position $\theta_0 = 0$; $r_1 = 4$mm.

In Fig. 3 the contour plot of the temperature field in x–z plane is presented. The laser beam and the specimen parameters are the same as in the previous case.

In Fig. 4 the contour plot of the temperature difference on the upper surface (x–y plane) is presented, for the case of two laser beams with the same top head profiles. The laser beams have the following properties: power $P_1 = 500$W; $P_2 = 500$W; radii: $r_{01} = r_{02} = 0.5$ mm; time duration 1s, positions of the laser beams center: r_1 = 2mm; θ_{01} = 0 rad; r_2=3.2 mm; θ_{02} = 0 rad.

In Fig. 5 the contour plot in the x–z plane, for the same case, was presented. The dimensions and assumed physical properties of the specimen were the same as in the previous cases.

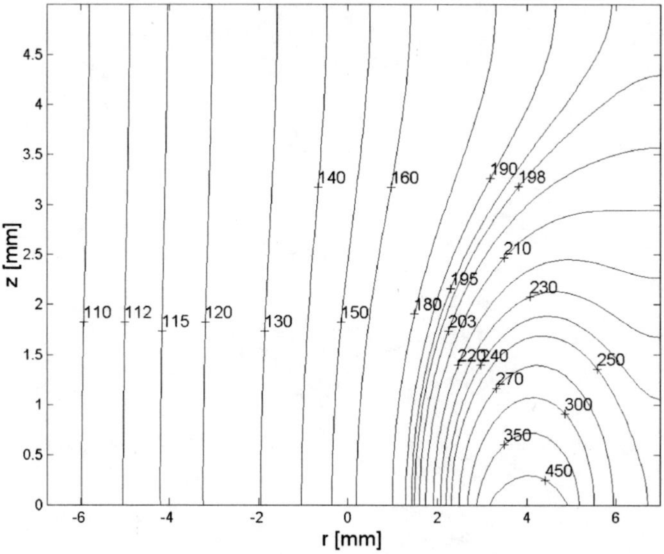

Figure 3: Contour plot in the x–z plane, laser beam properties: power=500W, radius=1mm, time duration=1s, position $\theta_0 = 0;\ \ r_1 = 4\text{mm}$.

Figure 4: Contour plot of temperature on the upper surface of the specimen, for two incident laser beams, positions of the laser beams: r_1 = 2mm; θ_{01} = 0 rad; r_2=3.2 mm; θ_{02} = 0 rad.

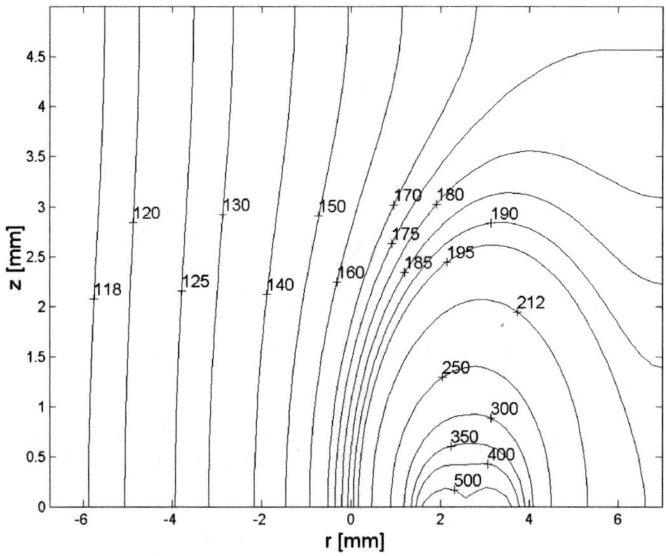

Figure 5: Contour plot in the x–z plane, for two incident laser beams, positions of the laser beams: $r_1 = 2$ mm; $\theta_{01} = 0$ rad; $r_2 = 3.2$ mm; $\theta_{02} = 0$ rad.

4 Conclusions

An analytical approach for solving 3D problems of material heating with laser beam was considered. The heating process was modeled using the linear non-stationary heat equation applied to cylindrical geometry. The spatial and temporal distributions of the temperature fields were considered. Using the method of variable separations and the Laplace transformation, the governing PDE with corresponding BC and IC was solved and the temperature field distributions were presented in closed form.

By using appropriate set of orthogonal functions, the numerical procedure was made more effective, producing saving in CPU time. The obtained numerical results improved as the number of terms used in the series of the solution increased.

Numerical examples were presented for the Al specimen with different characteristic positions of the laser beam. The convolution integral and Duhamel's principle can be used to represent arbitrary time dependence of the laser beam intensity.

For complex profiles of the laser beam the superposition principle was used. The temperature field distribution was considered and presented for two different laser beams targeting the same specimen.

The presented analytical solutions offers further advantage relative to direct monitoring since measurement of the temperature field distribution inside the bulk of the material is difficult to be arranged and can be usually performed in a restricted number of points.

References

[1] T. Thorslund, F.J. Kahlen, A. Kar, Temperatures, pressures and stress during laser shock processing, Optics and Lasers in Engineering 39 (2003) 51-71

[2] S. J. Farlow, Partial Differential Equations for Scientists and Engineers, Dover Publications, INC. New York, 1993.

[3] R. M. Wood, Laser Damage in Optical Materials, Adam Hilger, Bristol and Boston, 1986.

[4] E. M. Bass, Laser Material Processing, North Holland, Amsterdam, 1983.

[5] S. Bojanic, Analyzing of laser-material interaction with condensed matter at viewpoint of physical models, PhD thesis, The Faculty of Electrical Engineering, University of Belgrade, Belgrade, 1997.

[6] C. A. Brebbia, J.C.F. Telles, L.C. Wrobel, Boundary Element Techniques, Springer-Verlag, Berlin, 1984.

[7] L. R. Ram-Moham, Finite Element and Boundary Element Applications in Quantum Mechanics, Oxford University Press, New York, 2002.

[8] N. Rykalin, A. Uglov, A. Kokora, Laser Machining and Welding, MIR Publishers, Moscow, 1978.

[9] R. Gospavic, M. Sreckovic, V. Popov, Modeling of laser-material interaction using semi-analytical approach, Elsevier, Mathematics and Computers in Simulations 65 (2004) 211-219.

[10] R.A. Flinn, P.K. Trojan, Engineering Materials and their Application, Houghton Mifflin Company, Boston, 1975.

[11] Y. Toyozawa, Optical Processes In Solids, Cambridge University Press, New York, 2003.

[12] E. Kreyzig, Advanced Engineering Mathematics, John Wiley & Sons, New York, 1983.

[13] E. Jahnke, F. Emde, F. Lösch, Special functions, Nauka, Moscow, 1968.

[14] M. Abramovic, I.A. Stegun, Handbook of Mathematical Functions, Dover publications, INC., New York, 1972.

Microstructural and tribological observations in metallic glass forming alloy layers produced by high-power lasers

D. T. A. Matthews, V. Ocelík & J. Th. M. De Hosson
Department of Applied Physics and Netherlands Institute for Metals Research, University of Groningen, Groningen, The Netherlands

Abstract

Layers of Cu-based metallic glass forming compositions have been produced using high-power lasers. Laser surface remelting and laser cladding techniques provide sufficient cooling rates to form amorphous individual laser tracks and even coatings. The layers have been characterised by SEM, TEM, confocal and optical microscopy which have shown the layers may be fully amorphous, or (nano)crystalline. Hardness testing reveals that high hardness values are attainable for the layers (> 700 HV). Shear banding is found to be initiated during indentation testing. The processing of laser remelted layers developed from initial cladding has been analyzed and parameters are outlined which govern layer thickness, track width and ultimately the microstructural appearance of the layers.
Keywords: laser cladding, amorphous, hardness, friction.

1 Introduction

The formation of surface layers exhibiting differing mechanical properties to their substrate is now a well documented engineering sphere. Advanced coatings are continually being proposed and explored, which also require new processing routes, such as laser cladding, which is still in its relative infancy, particularly with respect to industrial applications on a large scale. Amorphous, amorphous matrix or nanocomposite coatings are one particularly interesting breed of surface modifications since they have been found to exhibit both extreme hardness and toughness [1]. Their application has somewhat been limited to thin, deposited, layers however, which often means that coating adhesion is limited.

WIT Transactions on Engineering Sciences, Vol 55, © 2007 WIT Press
www.witpress.com, ISSN 1743-3533 (on-line)
doi:10.2495/SECM070041

The processing of thicker layers, with direct metallurgical bonding to a substrate has been proposed some time ago [2], with direction being driven towards high cooling rate processes. Laser processing permits very high, local temperatures and subsequent rapid cooling. Significant progress has also been made in the processing of bulk metallic glasses (BMG) in recent times.

Examples for attractive Glass Forming Alloys (GFA) include the Cu-Ti-Zr-Ni system and some of its derivatives [3,4]. Their application as possible surface layers has been investigated by several research groups and not least by the authors, who show that direct laser remelting of a BMG layer may lead to fully amorphous layers, which exhibit outstanding tribological properties [5,6]. The goal of dual deposition and attainment of amorphous layers has proved somewhat more difficult however, and the reasons for that, progress and properties attainable in layers clad on Ti-(alloy) substrates are outlined in this article.

2 Experimental procedure

Spark-erosion cut and de-greased Ti-alloy (10 cm x 10 cm) have been selected for the application of Ti-containing metallic glass forming alloys. Since, during laser treatments, some of the applied energy may be reflected away from the target, the substrate surface is fine sand blasted to reduce the reflectivity, ergo improving the efficiency of the laser processing. The laser cladding and remelting processes were conducted over a range of processing parameters which will be specified as appropriate with a 2 kW Rofin-Sinar Nd-YAG laser. For all samples deposited on the Ti-alloy, the carrying (delivered at 3 l/min) and shielding gas (10l/min) was argon. The composition of the single track layer was varied by manipulating the feeding rates of a twin hopper powder feeder for a Zr-Ni mix and pure Cu. The powders were purchased commercially and all were at least 99.99% pure. The laser remelted layers were again fed from a twin hopper, with parameters varied to optimise the required composition of BMG production. All resultant fabrications are investigated by optical microscopy, secondary electron microscopy with energy dispersive spectroscopy (SEM with EDS) (Philips XL-30 (The Netherlands)), (high resolution) transmission electron microscopy ((HR)TEM) (FEG Jeol 2010 (Japan)) with in-situ heating and EELS capability. Hardness and scratch test examinations are conducted on a CSM Revetester (Switzerland), using a Standard Vickers geometry indenter for hardness measurements and a 200 μm radius Rockwell C diamond stylus as the scratch counterface. The indenter load was 2 or 4 N for the Ti-containing coatings. Confocal microscopy (μSurf Nanofocus Messtechnik) was also used to characterise the scratch geometries.

3 Results and discussions

The laser cladding technique is well documented and a broader overview, including a schematic overview can be found elsewhere [7]. Various parameters may be manipulated to vary the coating properties. The first stage is the deposition of the layers. In the first instance of this investigation, many layers were deposited at different parameters to gain an understanding of the standard laser cladding technique, and in the second stage a single track is deposited at a relatively slow scanning speed (500 mm/min) to achieve good metallurgical bonding at the coating/substrate interface. The layer may then be remelted at a rate required to develop amorphous layers – previous investigations have shown that 8 mm/min scan speed and +6 mm defocus is sufficient to process fully amorphous layers 300 μm deep in $Cu_{47}Ti_{33}Zr_{11}Ni_6Sn_2Si_1$ BMG alloy [6]. The coatings deposited in the investigation have been achieved in relation to several Ti-containing metallic glass forming alloys.

3.1 Ti-containing layers

A summary of the track dimensions attainable with varying parameters for the Cu-Ti-Zr-Ni alloy are shown in Table 1. The hardness profiles for these alloys are shown in Figure 1 a and b for changes in scanning speed and powder feed rate. The laser power was kept at 1200 W for all examples. Other constants were Ar shielding/carrying gas – this was maintained at 15 and 1 l/min respectively; beam defocus (+ 8 mm) and nozzle angle (58°) and the nozzle was positioned so that the powder was fed “in front” of the laser beam by a displacement of 0.5 mm. The entire Ti proportion in the clad layer was developed through dilution from the substrate. As seen in Table 1, the dilution was around 55-68%. This means that the composition may easily be altered by simple manipulation of the cladding parameters. It is evident from this table that a change in scanning speed has more effective on the track dimensions than a change in powder feeding rate.

Table 1: Table revealing the effect of Cladding Parameters on Track Width, depth and dilution for Cu-Ti-Zr-Ni BMG alloys.

Sample ID	Scanning Speed	Total Powder Feeding Rate	Track Width	Track Depth	Dilution
--	mm/min	g/min	mm	μm	%
A	1000	34.5	2.85	855	62
B	1500	34.5	2.2	650	60.7
C	2000	34.5	1.92	630	55.6
D	2500	34.5	1.73	515	58.3
E	3000	34.5	1.6	440	68.2
F	1500	23	2.1	635	63
G	1500	27.6	2.2	700	57.1
H	1500	31.9	2.2	680	62.5
I	1500	36.3	2.44	735	54.4

The hardness profiles for several of the layers are shown in Figure 1 ((a) and (b)). For the layers produced at varying scan speed, the thickness was too large and hence the layers contained high internal stresses which were released as cracks. Therefore, Figure 1a only reveals results for 2 scanning speeds, 2500 mm/min and 3000 mm/min, which were significantly fast to enable well bound, crack free clad layers. The results show that the hardness is high (700-800 HV0.4), and that, interestingly, the highest hardness is at some depth below the surface in both cases. This is due to compressive (or less tensile) stresses forming in the base of the clad, and is advantageous in terms of creating a functional gradient in the layer. Given that hardness, HV is related to yield stress, σ_y, such that HV ~ 3 σ_y the increase of around 100 HV (1000 MPa) equates to a difference in internal stresses of around 300 MPa, which is an acceptable value for laser clad layers [8]. The discrepancies between the depth noted in Table 1 and the hardness profile 'depth' arises due to the fact that not all profiles were able to be taken through the track centre, since the values are an average of several profiles. Rogue values were removed from the results. The marked increase in dilution at fast (2.5 and 3 m/min scan speed) may also play a role in the better bonding through the layer. The reason for the increased dilution is expected to be a result of the non-equilibrium conditions associated with laser cladding. The total track depth is seen to reduce in accord with the increased scan speed; however the increase in scan speed does not affect the effect of powder feeding rate and laser penetration in the same ways. The inflexion in the dilution values shows this clearly. The beam energy does not penetrate deeper into the substrate, but since the powder feeding rate is constant, the powder amount per unit length is lower. Thus whilst the total clad depth is reduced, the percentage of the clad which comes from the substrate increases. If the beam power (energy density) is kept constant, and the powder feeding rate changed, the effect is not so marked. This is seen clearly in the hardness results which will now be discussed.

No porosity was seen in the layers and the highest hardness, in all cases, was never found at the surface but instead at some depth below the surface. In the case of the change in scan speed, the slightly slower processed alloy actually provides (marginally) the higher hardness. This is a little surprising, but can be explained by the increased dilution at 3 m/min scan speed compared to the dilution at 2.5 m/min, which is described earlier.

If the hardness profile in Figure 1b is viewed, it is seen that the layer depth is not significantly altered by changes in powder feeding rate. The most "stable" deposition appears to result from the layer deposited with a total of 27.6 g/min [Sample G, Table 1] since the hardness is most constant across the full coating depth. The most unstable layer is that with the highest feeding rate – Sample I, Table 1 – 38.3 g/min. This indicates that for this particular system at the prescribed processing parameters, a feeding rate of 23-28 g/min generates the best clad layers.

It is interesting to view the indents since no indents caused cracking. Some indents however, did exhibit apparent shear banding – an example of this is shown in Figure 2 (a), an indent taken from a Cu-Ti-Zr-Ni layer clad at 2000 mm/min [Sample C, Table 1] which was poorly bonded overall and significant clad-cracking was observed, however the hardness (where measurable) of such a clad was almost 1000 HV0.2, and the 4-side shear banding in Figure 2(a) provides a very strong indication that partially amorphous layers are attainable by single pass laser cladding. An indent from a 2500 mm/min scan speed layer [Sample D, Table 1] is shown in Figure 2(b), showing the shear band phenomenon in more detail. At lower cladding speeds evidence of shear banding is again apparent, however if the scan speed and powder feeding rate is reduced (for example 1500 mm/min [Sample F, Table 1] Figure 2(c)), the cooling rate is not so fast and the microstructure is seen to be a fine matrix, containing Ti-rich dendrites and no cracking or shear banding is formed.

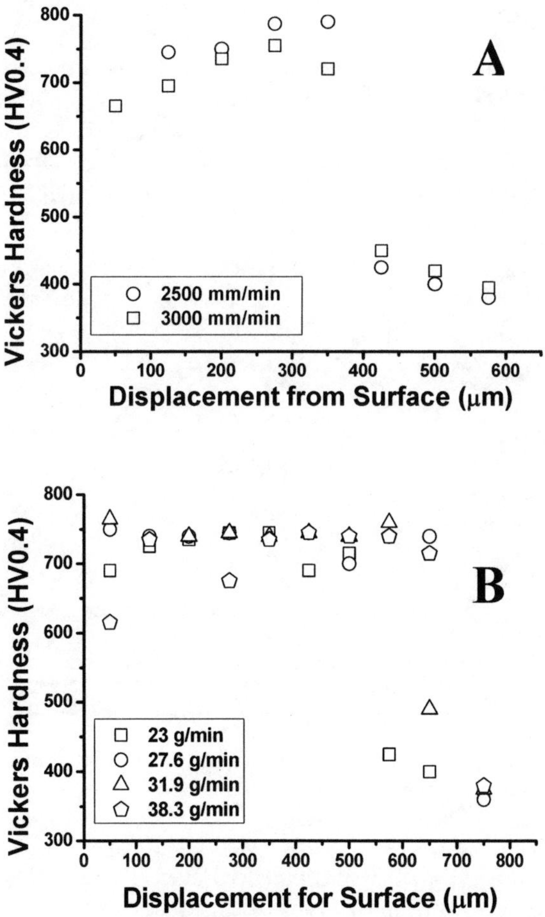

Figure 1: (a) Hardness profiles for laser clad layers produced at 2.5 and 3 m/min scan speed and constant powder feeding rate [Samples D and E, Table 1] and (b) for constant scan speed, but changing powder feeding rates [samples F-I, Table 1].

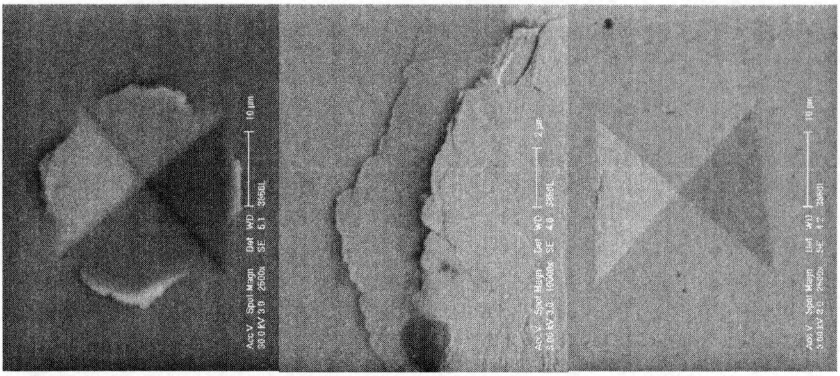

Figure 2: SEM micrographs for Cu-Ti-Ni-Zr laser clad layers having undergone microhardness investigations (a) an indent in Sample D, Table 1 laser clad layer exhibiting 4-side shear banding (b) higher magnification image of the shear band development in sample E, Table 1 and (c) Sample F, Table 1 showing a fine eutectic matrix reinforced with Ti-dendrites which exhibits no peripheral deformation.

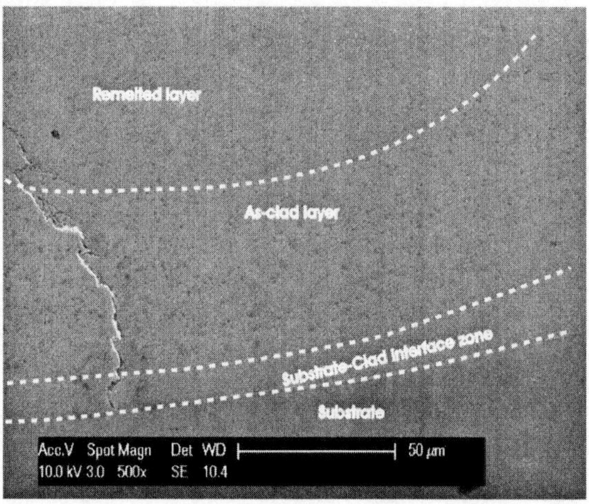

Figure 3: SEM micrograph revealing the zones of a clad and remelted GFA layer on a Ti-alloy substrate.

This relatively slow development of a well-bound layer is highly encouraging and in the instance of a clad and remelted layer (Figure 3), the composition $Cu_{47}Ti_{33}Zr_{11}Ni_6Sn_2Si_1$ was chosen for investigation. The layer forms featureless regions, indicating that rapid cooling is achieved; the cooling rate is too low and therefore prevents a fully amorphous layer forming at the prescribed treatment conditions. The flipside of this is that a thicker layer may be attained, and the mechanical properties between the layer and the substrate are more evenly graded. Upon remelting, however, the faster cooling also leads to higher stresses being developed within the layer, and these are often released by cracking, which may propagate to the as-clad region (Figure 3). The atomic % composition was confirmed at the track centre to be that of $Cu_{47}Ti_{33}Zr_{11}Ni_6Sn_2Si_1$, in accordance with that desired/expected.

The hardness of the as-clad region was found to be slightly lower (670-700 HV0.2) than the remelted area (850-890 HV0.2), as expected (Figure 4(a)). This is due to the enhanced cooling afforded by the rapid scan speed and subsequent refinement of the microstructure. It is also interesting to note, in terms of functionally grading, not only a coating, but also the coating-substrate system, that the hardness of titanium substrate was significantly increased after treatment to a depth of over 200 microns beyond the clad layer. The hardness in this area was 425 Vickers, whilst the hardness of the substrate 1 mm away from the clad layer was only 300 Vickers.

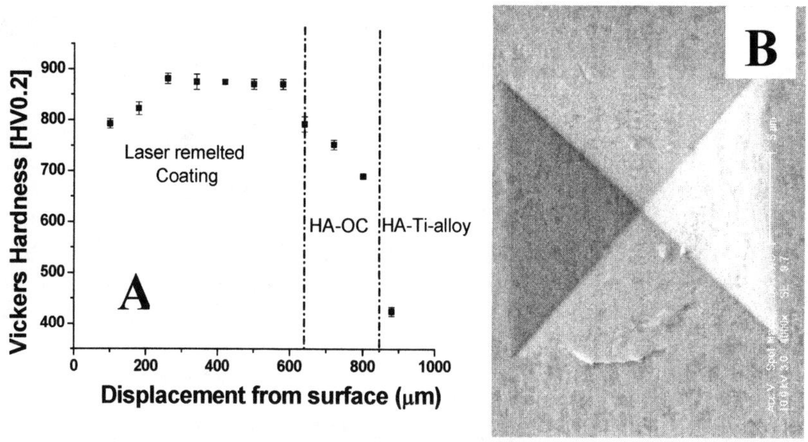

Figure 4: (a) Hardness profile of a laser clad and remelted layer of $Cu_{47}Ti_{33}Zr_{11}Ni_6Sn_2Si_1$ composition and (b) SEM micrograph of a Vickers indent from the centre of that layer revealing shear banding.

The increased 'remelt depth' and heat effects on the titanium substrate are a direct consequence of the low thermal conductivity of titanium. Again, the indentation method appears to induce shear band formation (Figure 4(b)), which indicates that the layer may have amorphous constituent regions. Together with the desired composition achieved, this is very promising, since this procedure involves the deposition of a 5 element powder mix.

TEM observations from the remelted area (an example is shown in Figure 5) show that some areas are amorphous in nature, which may explain the observation in Figure 4(b) that hardness testing induced shear band development in the matrix of the clad layer. There is, however, a greater proportion of crystalline content, with the crystals being for the order of 5 nm up to 2 microns.

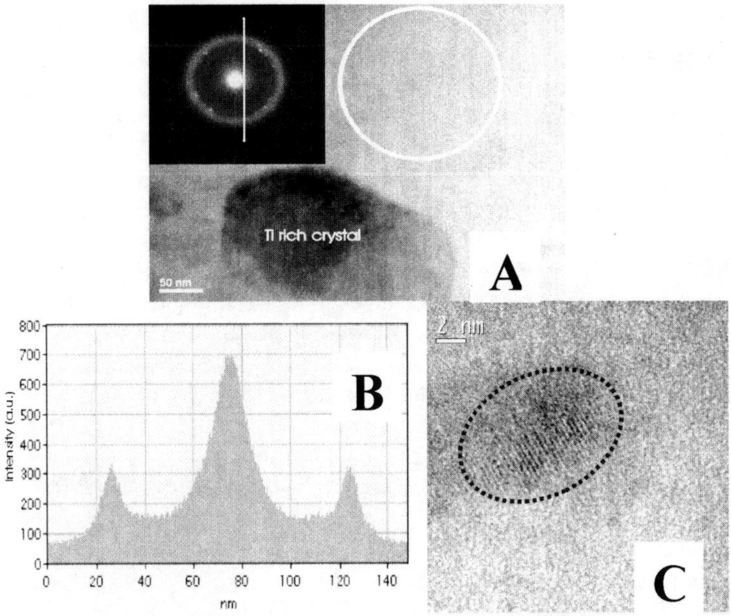

Figure 5: TEM observation revealing an amorphous matrix (see diffraction insert proved by the line scan in (b)) surrounding a Ti-rich crystal around 250 nm in size and (c) High resolution TEM micrograph revealing a 5 nm size Ti-rich crystal surrounded by an amorphous matrix.

3.2 Scratch observations and friction coefficient

For these $Cu_{47}Ti_{33}Zr_{11}Ni_6Sn_2Si_1$ laser remelted layers, the hardness of the layers is promising in terms of mechanical performance. Scratch testing allows information to be gained as regards tribological performance and even hardness characteristics. One parameter attained from scratch testing is the friction coefficient. Single pass friction is very low for the Cu-based layer, (0.032) and repeated scratch testing actually leads to an even lower friction coefficient. This is shown in Figure 6 and signifies that whilst extreme hardness promotes beneficial properties, it can be more advantageous to temper that with a fine microstructure that more readily accepts deformation and provides a ductile counterface as opposed to a brittle one. The dimensions of these tracks are, for 1 pass, depth = 1.3 μm, width = 71.16 μm and for 50 passes depth = 2.1 μm and width = 80.93 μm (including edge pile-up).

4 Summary and conclusions

Properties such as low friction, good wear resistance, high hardness and thermal stability up to 350°C of metallic glasses and layers thereof have been found to be highly attractive. A high-power laser has been proven to be a tool capable of producing such layers. Actual laser cladding, and subsequent remelting of metallic glass compositions have been attempted and proven to be achievable and successful. Laser clad layers have been produced using premixed powders of compositions which are know to promote metallic glass (amorphous structure) formation. Surface coatings of metallic glass forming compositions have been administered to Ti with degrees of success.

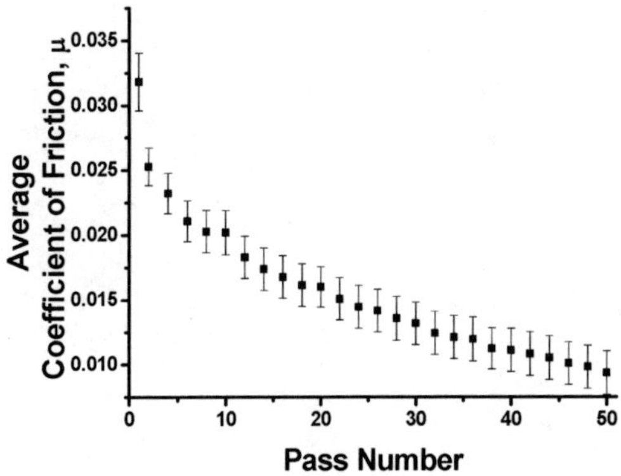

Figure 6: Graph showing the average friction coefficient vs. pass number for a $Cu_{47}Ti_{33}Zr_{11}Ni_6Sn_2Si_1$ laser clad and remelted layer.

Firstly, the surface layers produced on both substrates exhibit excellent substrate-coating metallurgical bonding. The hardness of all Cu-based coatings has been found to be over 700 HV0.4. The indentation experiments exhibit shear band formation in some of the rapidly processed layers. The friction coefficient of all layers was seen to be super-low and after service the friction was seen to reduce to an even lower value of 0.00936 ± 0.00168 for the Cu-based remelted coatings.

References

[1] Y.T. Pei, D. Galvan and J. Th. M. De Hosson, Nanostructure and Properties of TiC/a-C:H composite coatings, *Acta Materialia*, 53, pp. 4505-4521 (2005)

[2] F. Aubert, R. Colaco, R. Villar and H. Sirkin, Production of glassy metallic layers by laser surface treatment, *Scripta Materialia* 48, pp. 281-286 (2003)

[3] X.H. Lin and W.L. Johnson, Formation of Ti-Zr-Cu-Ni bulk metallic glasses, *Journal of Applied Physics* 48 (11), pp. 6514-6519 (1995)

[4] E.S. Park, H.K. Lim, W.T. Kim and D. H. Kim, The effect of Sn addition on the glass-forming ability of Cu-Ti-Zr-Ni-Si metallic glass alloys, *Journal of non-crystalline solids* 298, pp. 15-22, 2002

[5] D.T.A. Matthews, V. Ocelík and J. Th. M. de Hosson, Scratch Test Induced Shear Banding in High Power Laser Remelted Metallic Glass Layers, *Journal of Materials Research* (February 2007), *in press*
[6] D.T.A. Matthews, V. Ocelík, and J. Th. M. de Hosson, Metallic glass layers produced by high power lasers, Bulk Metallic Glasses, *TMS* 2006, pp. 99-108 ISBN 978-0-87339-612-7
[7] W.M. Steen, Laser Surface Treatment An Overview, Laser Processing: Surface Treatment and Film Deposition, (eds. J. Mazumder, O. Conde, R. Villar and W. Steen), NATO ASI Series (1996)
[8] U. Oliveira, V. Ocelik, J. Th. M. De Hosson, Residual stress analysis in Co-based laser clad layers by laboratory X-rays and synchrotron diffraction techniques, *Surface and Coatings Technology*, 201, pp. 533-542 (2006)

Hydrogen permeation behavior in polycrystalline nickel implanted with various elements

R. Nishimura[1], K. Okitsu[1], H. Inoue[1], R. M. Latanision[2]
& G. K. Hubler[3]
[1]*Osaka Prefecture University, Japan*
[2]*Massachusetts Institute of Technology, USA*
[3]*U.S. Naval Research Laboratory, USA*

Abstract

The entry and transport of hydrogen in polycrystalline nickel implanted with various elements (He, Ar, B, P, S, Bi, Ni, Y, Pt, As, Pb and Sb) has been investigated in a fluence range of 1 x 10^{14} to 2 x 10^{17} ions/cm^2 using an electrochemical permeation technique and etching technique, where the elements used are categorized as follows: inert elements (He and Ar), the substrate element (Ni), metalloids (P, S, Sb and As), a catalytic element (Pt) and the other elements (Pb, Bi, Y and B). It was found that the effective diffusion coefficient of hydrogen and the effective solubility were largely dependent upon fluence and elements in comparison to those obtained from non-implanted nickel. The results obtained were qualitatively explained in terms of gas bubbles and defects generated by implantation, compressive stress, catalytic effect, the formation of amorphous phase and so on.
Keywords: hydrogen permeation, ion implantation, fluence, element, hydrogen diffusion coefficient, hydrogen solubility.

1 Introduction

Many researchers have investigated the interrelation between hydrogen embrittlement (HE) and metalloid segregation as metalloids may act as preferential grain boundaries in metals and alloys [1-6]. Segregated sites for the absorption of hydrogen may decrease the cohesive strength of the material.

WIT Transactions on Engineering Sciences, Vol 55, © 2007 WIT Press
www.witpress.com, ISSN 1743-3533 (on-line)
doi:10.2495/SECM070051

Ion implantation may be the best method to prepare surfaces for segregation studies without affecting bulk physical or mechanical properties, because any element can be implanted and the concentration of segregated elements may be controlled over several orders of magnitude. In the present paper, the entry and transport of hydrogen in nickel implanted with various elements (12 elements) were investigated to identify effects caused by the process of ion implantation, a catalytic effect, phase and structure of implanted layers and so on. The results obtained were qualitatively explained in terms of gas bubbles and defects generated by implantation, compressive stress, catalytic effect, the formation of amorphous phase and so on.

2 Experiment

Strips of Ni270 (99.97%) were cold rolled, annealed at 1273K for 10 min and then water quenched. Electropolishing of the specimens was performed in an aqueous 60% H_2SO_4 solution (bath voltage 5V) at room temperature using a platinum electrode as the cathode. The specimens with a thickness of 105 μm were not coated with palladium.

Table 1: Ion implantation parameters and in range statistics.

Ion	Energy (keV)	$Q_{fluence}$ (cm^{-2})	R_p (nm)	ΔR_p (nm)
He	25	$1x10^{16}$, $1x10^{17}$	87	14
B	25	$1x10^{17}$, $3x10^{17}$	35	20
P	50	$1x10^{15}$~$1x10^{17}$	24	12
Ar	60	$1x10^{15}$~$1x10^{17}$	24	12
Ni	90	$1x10^{14}$~$1x10^{17}$	25	12
As	110	$1x10^{16}$, $1x10^{17}$	26	12
Y	125	$1x10^{17}$	25	12
Sn	145	$1x10^{17}$	25	10
Sb	150	$1x10^{15}$~$1x10^{17}$	25	10
Pt	180	$1x10^{15}$~$1x10^{16}$	25	8.6
Pb	180	$1x10^{15}$~$1x10^{17}$	25	8.3
Bi	180	$1x10^{17}$	25	8.2

The electropolished specimens were implanted on the cathodic side (the side where hydrogen evolution occurs) by clamping the edge of a rectangular (5 cm x 2.5 cm) foil along the dimension with razor blade masks. The cryopumped vacuum was kept at pressures between 3 x 10^{-7} and 2 x 10^{-6} Torr and the ion beam current density was held below 1 $\mu A/cm^2$ in order to minimize specimen heating. The ions were implanted at normal incidence to the surface and the beam was raster scanned to produce a uniform fluence measured to better than ±2.0 %. The fluences and energies of implantation are given in table 1, which also shows the calculated projected range R_p of implanted ions and the standard deviation ΔR_p, where R_p was almost constant under the various ion implantation conditions with the exception of helium. The elements used are categorized as follows; inert elements (He and Ar), the substrate element (Ni), metalloids (P, S, Sb and As), a catalytic element (Pt) and the other elements (Pb, Bi, Y and B).

The method developed by Devanathan and Stachurski [7] was used for the permeation measurements. The specimens were mounted between the two half-cells, giving a 0.95 cm^2 area of exposure. The cathodic and anodic compartments contained 0.05 $kmol/m^3$ H_2SO_4 and 0.1 $kmol/m^3$ NaOH solutions respectively. Both solutions were deaerated with pre-purified nitrogen prior to the experiments. Hydrogen charging in the cathodic compartment was controlled by galvanostatic cathodic polarization with a current density of 1.47 mA/cm^2. The anodic side of the specimens was potentiostatically polarized at +0.1 V/SCE. When the anodic current became negligible (less than 10 nA/cm^2), hydrogen charging was started. All experiments were carried out at 303 ± 0.1K. Permeation data were frequently analyzed for D_{lag}, the diffusion coefficient and C_H, the concentration of hydrogen adsorbed just beneath the surface, which are calculated by using the time lag method [7] as follows:

$$D_{lag}=L^2/6t_{lag} \text{ and } P_\infty = D_{lag} \cdot C_H \cdot F/L \tag{1}$$

where L is the thickness of the specimens, t_{lag} the time of a lag, P_∞ the steady state permeation current density and F the Faraday constant. D_{lag} and C_H of the un-implanted nickel (pure nickel) thus obtained were 3.8 x 10^{-10} cm^2/s and 2.5 x 10^{-6} mol/cm^3, respectively. Here, it should be recognized that D_{lag} and C_H for all the implanted nickel samples are taken into consideration to compare with those of the un-implanted nickel samples because of the lack of physical significance of the derived quantities, D_{lag} and C_H. In addition, the implantation depth constitutes only about 1 part in 4000 of the total thickness, which means that the diffusion length of hydrogen is almost equal to that in the bulk and hence D_{lag} and C_H for all the implanted nickels are the effective diffusion coefficient and concentration for the entire sample. As described above, we can obtain three parameters from the hydrogen transients (P_∞, D_{lag} and C_H).

3 Results

3.1 Substrate element (Ni)

Fig. 1 shows the hydrogen permeation transients and three parameters (P_∞, D_{lag} and C_H) of Ni- implanted nickel with a fluence of 1 x 10^{14} to 1 x $10^{17}/cm^2$, where the broken line in the figure indicates a typical permeation transient for un-implanted nickel (pure nickel) as a reference. It was found that the steady state permeation current densities (P_∞) tended to become larger for nickel implanted with fluences of 1 x 10^{16} and 1 x $10^{17}/cm^2$ and smaller for nickel implanted with fluences of 1 x 10^{14} and 1 x $10^{15}/cm^2$ than for un-implanted nickel, even if the difference in membrane thickness shown in the figure is taken into consideration:

the thinner the membrane thickness, the larger P_∞ is. Correspondingly, D_{lag} decreased with increasing fluence, while the fluence dependence of C_H was the same as that of P_∞.

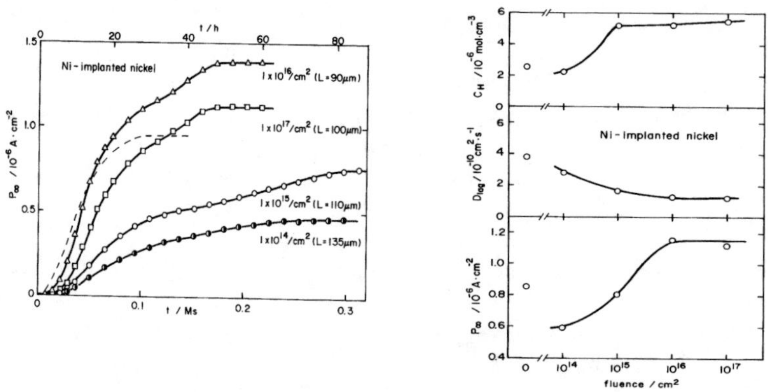

Figure 1: The hydrogen permeation transients and three parameters (P_∞, D_{lag} and C_H) of Ni- implanted nickel with a fluence of 1 x 10^{14} to 1 x 10^{17}/cm^2, where the broken line in the figure indicates a typical permeation transient for un-implanted nickel.

3.2 Inert elements (He and Ar)

Fig. 2 shows the hydrogen permeation transients and three parameters (P_∞, D_{lag} and C_H) of Ar- implanted nickel with a fluence of 1 x 10^{15} to 1 x 10^{17}/cm^2, where the broken line in the figure indicates a typical permeation transient for un-implanted nickel. It was evident that the breakthrough time of the transients was larger than that of un-implanted nickel.

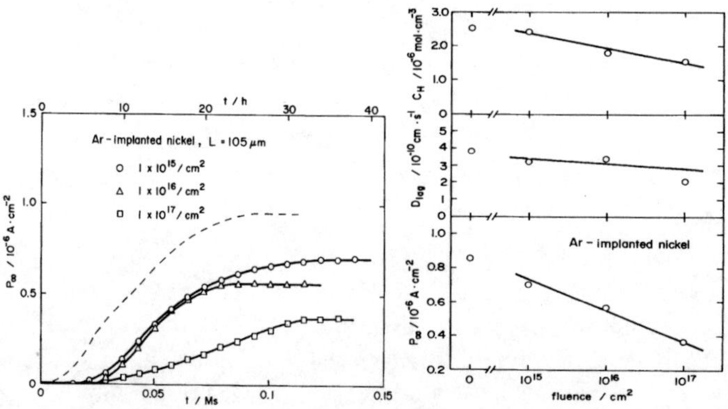

Figure 2: The hydrogen permeation transients and three parameters (P_∞, D_{lag} and C_H) of Ar- nickel implanted with a fluence of 1 x 10^{15} to 1 x 10^{17}/cm^2, where the broken line in the figure indicates a typical permeation transient for un-implanted nickel.

The three parameters of Ar- implanted nickel decreased with increasing fluence and were smaller than those of un-implanted nickel. The fluence dependence of the three parameters was found to be different from that of Ni- implanted nickel. The permeation transients of He- implanted nickel were almost the same behavior as those of Ar- implanted nickel.

3.3 Metalloids (P, S, Sb and As)

Fig. 3 shows the hydrogen permeation transients and three parameters (P_∞, D_{lag} and C_H) of P- implanted nickel with a fluence of 1 x 10^{15} to 1 x 10^{17}/cm^2, where the broken line in the figure indicates a typical permeation transient for un-implanted nickel. Although the breakthrough time and P_∞ tended to increase with the increase in fluence, specifically, for a fluence of 1 x 10^{17}/cm^2 the breakthrough time became almost the same as that of un-implanted nickel and P_∞ was much larger than that of un-implanted nickel. The behavior of the three parameters was also found to depend largely upon fluence and to be different from those in Figs. 1 and 2.

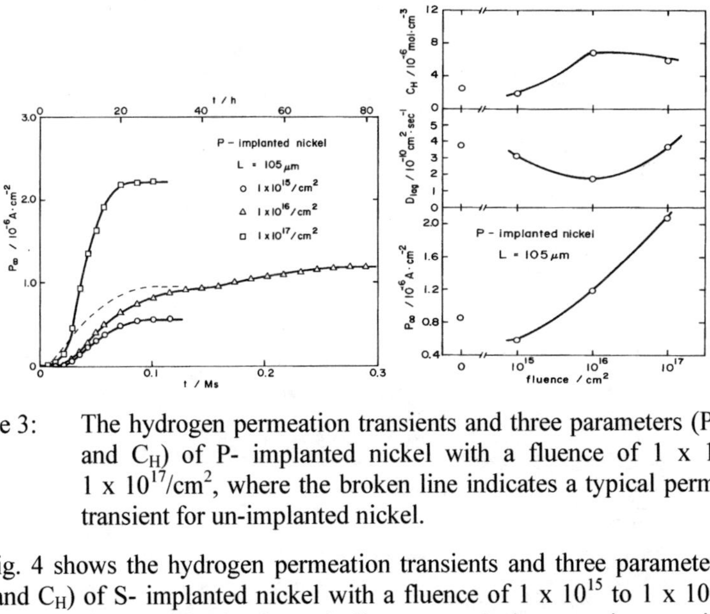

Figure 3: The hydrogen permeation transients and three parameters (P_∞, D_{lag} and C_H) of P- implanted nickel with a fluence of 1 x 10^{15} to 1 x 10^{17}/cm^2, where the broken line indicates a typical permeation transient for un-implanted nickel.

Fig. 4 shows the hydrogen permeation transients and three parameters (P_∞, D_{lag} and C_H) of S- implanted nickel with a fluence of 1 x 10^{15} to 1 x 10^{17}/cm^2, where the broken line in the figure indicates a typical permeation transient for un-implanted nickel. It was found that the breakthrough time decreased with increasing fluence, while P_∞ tended to increase with increasing fluence, but decreased clearly at a fluence of 1 x 10^{17}/cm^2. The fluence dependence of the three parameters was different from those in Figs. 1, 2 and 3.

Fig. 5 shows the hydrogen permeation transients of As- implanted nickel with a fluence of 1 x 10^{16} and 1 x 10^{17}/cm^2, where the broken line in the figure indicates a typical permeation transient for un-implanted nickel. It was found that the permeation transients were almost the same behavior as those of P- implanted nickel with the same fluences.

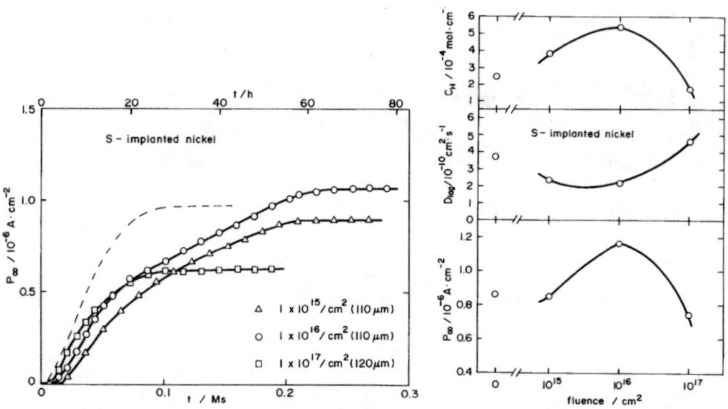

Figure 4: The hydrogen permeation transients and three parameters (P_∞, D_{lag} and C_H) of S- implanted nickel with a fluence of 1 x 10^{15} to 1 x 10^{17}/cm^2, where the broken line indicates a typical permeation transient for un-implanted nickel.

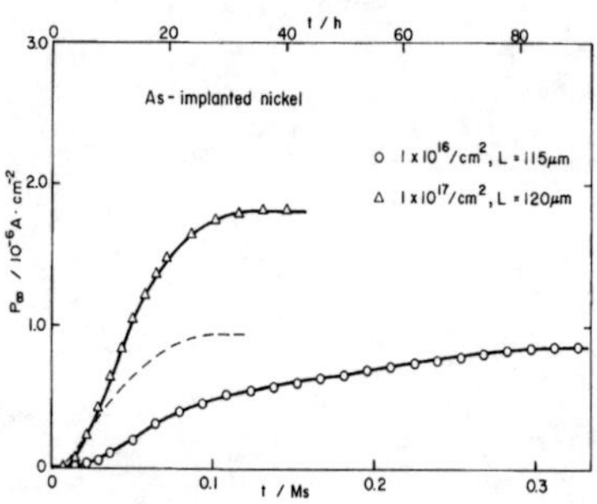

Figure 5: The hydrogen permeation transients of As- implanted nickel with fluences of 1 x 10^{16} and 1 x 10^{17}/cm^2, where the broken line indicates a typical permeation transient for un-implanted nickel.

3.4 Catalytic element (Pt)

Fig. 6 shows the hydrogen permeation transients and three parameters (P_∞, D_{lag} and C_H) of Pt- implanted nickel with a fluence of 1 x 10^{15} to 1 x 10^{16}/cm^2, where the broken line in the figure indicates a typical permeation transient for un-implanted nickel. The range of fluences for Pt- implanted nickel was only 1 x 10^{15} to 1 x 10^{16}/cm^2 because of the difficulty in obtaining high beam currents of platinum ions. The breakthrough time was much longer than for un-implanted nickel and P_∞ was found to increase with increasing fluence. The fluence dependence of the three parameters was similar to that of Ni- implanted nickel.

3.5 Other elements (Pb, Bi, Y and B)

Fig. 7 shows the hydrogen permeation transients of B- and Bi- implanted with a fluence of 1 x 10^{17} or 3 x 10^{17}/cm^2, where the broken line in the figure indicates a typical permeation transient for un-implanted nickel.

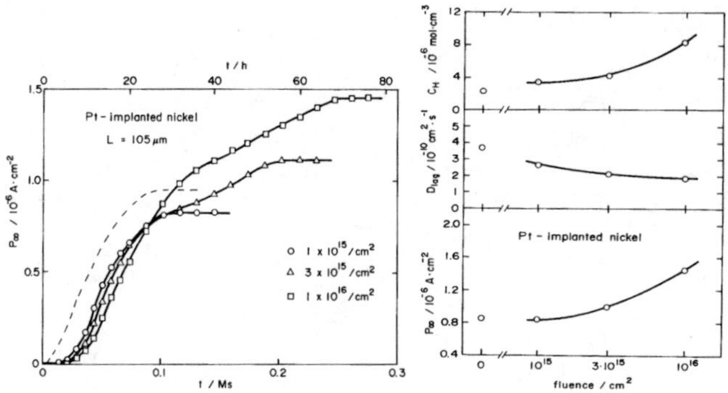

Figure 6: The hydrogen permeation transients and three parameters (P_∞, D_{lag} and C_H) of Pt implanted nickel with a fluence of 1 x 10^{16} to 1 x 10^{17}/cm^2, where the broken line in the figure indicates a typical permeation transient for un-implanted nickel.

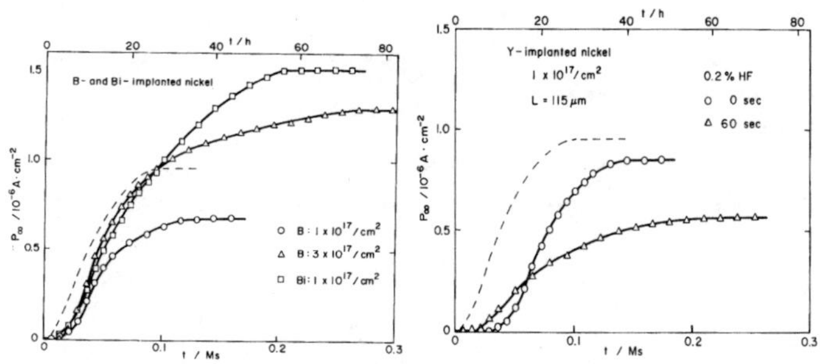

Figure 7: The hydrogen permeation transients of B-, Bi- and Y- implanted nickel with a fluence of 1 x 10^{16} to 3 x 10^{17}/cm^2, where the broken lines in the figure indicate a typical permeation transient for un-implanted nickel.

The breakthrough time of these implanted nickels was almost the same, but smaller than that of un-implanted nickel. P_∞ for Bi- implanted nickel with a fluence of 1 x 10^{17} /cm^2 and for B- implanted nickel with 3 x 10^{17}/cm^2 were larger than that of un-implanted nickel, but that of B- implanted nickel with 1 x 10^{17}/cm^2 was smaller than that of un-implanted nickel. The breakthrough time of Y- implanted nickel is much longer than that for un-implanted nickel, whereas P_∞ was almost the same as that for un-implanted nickel, although it became smaller by the etching treatment of 60s. For Pb- implanted nickel, the breakthrough time was constant and independent of fluence and P_∞ tended to increase with increasing fluence, although P_∞ was larger than that of un-implanted nickel.

4 Discussion

It is well known that shallow and deep traps, compressive stress and occupied interstitial sites are generated by ion implantation 8]. Deep and shallow traps contribute to delay in breakthrough (corresponding to the decrease in D_{lag}), which are observed for almost all the implanted nickels, but lead to the increase in C_H. This is recognized from eq. (1), if P_∞ keeps constant, and is applied to the transient for Y- implanted nickel. In the case that the shallow traps are dominant, P_∞ increases as well as the increase in C_H, because some part of hydrogen trapped in the shallow traps is released, which is observed in the form of the dual rise time transient as shown for Ni- and Pt- implanted nickels with a relatively high fluence in Figs. 1 and 2. On the other hand, the compressive stress leads to the decrease in P_∞ and C_H, since a number of the entry site for hydrogen decreases. This is observed for the transients of almost all implanted nickels with a low fluence of less than $10^{15}/cm^2$, showing the decrease in P_∞ and C_H and little change in D_{lag}. The same effect is caused by the formation of gas bubbles such as He and Ar as shown in Fig. 2 [9].

It is well known that platinum shows a catalytic effect on hydrogen evolution reaction, which means that an amount of the absorbed hydrogen decreases and as a result the decrease in P_∞ and C_H with little change in D_{lag} takes place. However, the transients of Pt- implanted nickels in Fig. 3 cannot support the catalytic effect of platinum. It is confirmed by Rutherford backscattering that a concentration of platinum at the outer layer of the implanted layer in contact with the solution is very low, whereas the maximum platinum concentration becomes about 3% in the implanted layer at a fluence of 3 x $10^{16}/cm^2$. This means that the concentration at the outer layer of the implanted layer is important, but not inside. To evaluate this, after the implanted layer was resolved layer by layer by using 0.2%HF solution, the transients at various etching times were measured. It was found that the catalytic effect of platinum was observed, that is, the decrease in P_∞ and C_H with little change in D_{lag}. On the other hand, arsenic is well known to be a poison for hydrogen evolution reaction, which means that arsenic accelerates hydrogen entry. This appears to apply to the transient with a fluence of 1 x $10^{17}/cm^2$, where P_∞, C_H and D_{lag} are larger than those of the un-implanted nickel, which is also supported from the results obtained by the etching. This implies that arsenic has an inhibiting effect of hydrogen evolution reaction; that is, the increase in the absorbed hydrogen due to the increase in the adsorbed hydrogen.

It is also recognized that the structure of the implanted layer for phosphorous and boron with a high fluence of more than 1 x $10^{17}/cm^2$ moves into the amorphous phase [10], which can apply to P- and B- implanted nickels with fluences of 1 x $10^{17}/cm^2$ and 3 x $10^{17}/cm^2$. In this case, P_∞, C_H and D_{lag} are larger than those of the un-implanted nickel. This behavior is enhanced by the etching of a relatively short time, but disappears with a longer etching time. This suggests that the amorphous phase accelerates hydrogen entry, increases the number of the entry site or enhances hydrogen diffusion coefficient in the implanted layer. In contrast, in the case of S- implanted nickel with a fluence of

1 x $10^{17}/cm^2$, the decrease in P_∞, and C_H, but the increase in D_{lag} was observed, which is more evident by the etching. This may be explained by the structure change from austenite (fcc) to ferrite (bcc), since ferrite was much larger for D_{lag}, but much smaller for P_∞ and C_H than those of austenite.

5 Conclusions

The entry and transport of hydrogen in polycrystalline nickel implanted with various elements (He, Ar, B, P, S, Bi, Ni, Y, Pt, As, Pb and Sb) were investigated in a fluence range of 1 x 10^{14} to 2 x 10^{17} ions/cm^2. The results obtained were summarized as follows:

(1) The effective diffusion coefficient of hydrogen and the effective solubility were dependent upon fluence and element in comparison to those obtained from un-implanted nickel.

(2) The entry of hydrogen was influenced by the elemental composition and structure/phase at the outermost surface of the implanted layer, but not inside the implanted layer; specifically in the case of Pt and As with the catalytic effect.

(3) On the other hand, the transport of hydrogen was mainly affected by gas bubbles and defects generated by implantation, the formation of the amorphous phase and so on in the implanted layer.

References

[1] R.M. Latanision and H. Opperhauser Jr., Metall. Trans. A, 5, pp. 483-490, 1974.

[2] R.D. Kane and B.J. Berkowitz, Corrosion, 36, pp. 29- 36, 1980.

[3] R.H. Jones, S.M. Bruemmer, M.T. Thomas and D.R. Baer, Metall. Trans. A, 13, pp. 241- 249, 1982.

[4] A.W. Funkenbusch, L.A. Heldt and D.F. Stein, Metall. Trans. S, 13, pp. 611- 618, 1982.

[5] B.D. Craig, Metall. Trans. A, 15, pp. 565- 572, 1984.

[6] Y. Obino and T. Yamasaki, Metall. Trans. A, 15, pp. 519- 527, 1984.

[7] M.A Devanathan and Z. Stachurski, Proc. R. Soc. London, Ser. A, 270, pp. 90-102, 1962.

[8] F. Besenbacher, S.M Myers and J.K> Norskov, Nucl. Instrum. Methods B, 7-8, pp. 55-63, 1985.

[9] R. Nishimura, RM. Latanision and G.K. Hubler, Materials Sci. and Engn., 90, pp. 243-251, 1987.

[10] Z.Y.A. Al-Tamimi, W.A. Grant and G. Carter, Nucl. Instr. and mech., 209/210, pp. 363-370, 1983.

Effect of carburizing on fatigue behaviour in a type 316 austenitic stainless steel

K. Tokaji & M. Akita
Department of Mechanical and Systems Engineering, Gifu University, Japan

Abstract

The effect of carburizing on fatigue behaviour of smooth and notched specimens and corrosion fatigue behaviour was studied in a type 316 austenitic stainless steel. The fatigue strength of the smooth specimens was considerably increased by carburizing and the specimens with a thick case exhibited longer fatigue lives than the specimens with a thin case. The fatigue strength of the notched specimens was also increased by carburizing and the extent of increase in fatigue strength decreased with increasing stress concentration factor and then saturated. A slight increase in notch sensitivity by carburizing was seen. In 3%NaCl aqueous solution, the carburized smooth specimens showed no reduction in fatigue strength, indicating the excellent corrosion resistance of the carburized case.
Keywords: fatigue strength, notch effect, corrosion fatigue, carburizing, case depth, austenitic stainless steel.

1 Introduction

In recent years, it has been strongly required to extend the service life of machines and structures due to economic and environmental reasons. To achieve this requirement, various surface engineering techniques have become major interest because they can provide additional properties such as high strength, thermal barrier, and corrosion and wear resistance.

Austenitic stainless steels have excellent corrosion resistance, but they posses relatively low strength and poor wear resistance. Therefore, it is significant to improve those properties by surface treatment. When surface-modified materials are applied to load-bearing components, fatigue properties are critical. Until

WIT Transactions on Engineering Sciences, Vol 55, © 2007 WIT Press
www.witpress.com, ISSN 1743-3533 (on-line)
doi:10.2495/SECM070061

now, the fatigue behaviour of austenitic stainless steels modified by shot peening [1-5], laser [6], dynamic ion mixing [7] and coating [8] has been reported. In addition to these techniques, a carburizing technique has been developed, which can apply to austenitic stainless steels without any loss of their advantages such as corrosion resistance and ductility [9]. Wear resistance and strength can also be improved by this method [9], but fatigue properties have not been studied.

In the present work, rotary bending fatigue tests were performed using carburized smooth and notched specimens of a type 316 austenitic stainless steel in laboratory air and in 3%NaCl aqueous solution. Fatigue behaviour and fracture mechanisms of smooth specimens, notch effect and corrosion fatigue behaviour were discussed.

2 Experimental details

2.1 Material and specimens

The material used is a type 316 austenitic stainless steel of 16 mm diameter whose chemical composition (wt.%) is C 0.05, Si 0.35, Mn 1.35, P 0.033, S 0.025, Ni 10.1, Cr 16.9, Mo 2.11. The material was solution treated at 1353 K for 1 h followed by oil cooling, from which the following fatigue specimens were machined.

Smooth specimens of an hourglass shape with a minimum diameter of 5.5 mm were used. The stress concentration factor, K_t, was 1.03 under bending. In notched specimens, a circumferential notch with a depth of 1 mm and three different notch radii, ρ, of 0.40 mm, 0.10 mm and 0.03 mm was introduced to the gauge section of 8 mm diameter, whose K_t values are 2.08, 3.55 and 6.50, respectively. After machining, the following surface treatment was applied to the fatigue specimens.

2.2 Carburizing

A modified gas-carburizing technique, which is called *pionite* treatment, was performed at a temperature below 773 K in a CO and H_2 gas mixture [9]. During this process, a carbon-diffused zone is formed at the surface region with no Cr-carbides where hardness is remarkably increased. This treatment can improve significantly wear resistance and strength without any loss of ductility and toughness of austenitic stainless steels [9].

In order to produce specimens with two different case depths, the treatment times, t_p, of 15 h and 35 h were applied to the smooth specimens. Hereafter, the smooth specimens treated for 15 h and 35 h are denoted as the 15 h treated specimen and the 35 h treated specimen, respectively, and the specimen not subjected to the surface modification is referred to as the untreated specimen. Only 35 h treatment time was employed for the notched specimens.

2.3 Procedures

Fatigue tests were carried out using cantilever-type rotary bending fatigue testing machines operating at a frequency of 19 Hz in laboratory air and in 3%NaCl aqueous solution. The solution was dropped continually onto the specimen surface by a pump from a reserved tank. Crack initiation and small crack growth were monitored with replication technique. After experiment, fracture surfaces were examined in detail by a scanning electron microscope (SEM).

3 Results and discussion

3.1 Microstructure characterization

Figure 1 shows the microstructures of the untreated and carburized specimens. In the untreated specimen, the microstructure consists of austenitic grains, while in the carburized specimen, a surface region that is clearly distinguished from the core can be recognized, which is the case formed by carburizing. As can be seen in the figure, the carburized case depths are approximately 20 μm and 40 μm for the 15 h and 35 h treated specimens, respectively. It has been indicated that no Cr-carbides were formed in the carburized case and the microstructure underneath the carburized case was the same austenitic microstructure as in the untreated specimens [9].

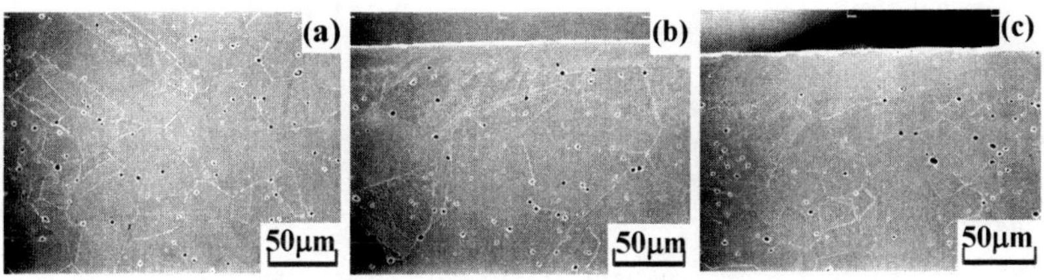

Figure 1: Microstructures: (a) untreated, (b) t_p=15 h, (c) t_p=35 h.

3.2 Hardness profile and mechanical properties

Vickers hardness profiles measured on the minimum cross section in the carburized smooth and notched specimens are represented in fig. 2. The hardness at or near the surface attains to more than 800 *HV* and 1000 *HV* in the 15 h and 35 h treated specimens, respectively. Hardness rapidly decreases with increasing the distance from the surface and then reaches a constant value of approximately 172-220*HV* that is the hardness of the core, i.e. the untreated specimen. The region of the increased hardness is 40-50 μm regardless of treatment time and notch geometry. As can seen in fig. 1, however, the case depth depended on treatment time, thus the case depth, d_p, was defined as the depth established by the microstructure characterization.

The mechanical properties are listed in table 1. In the carburized specimens, tensile strength increases and ductility decreases compared with the untreated specimen and with increasing treatment time, but the differences between the untreated and carburized specimens are small.

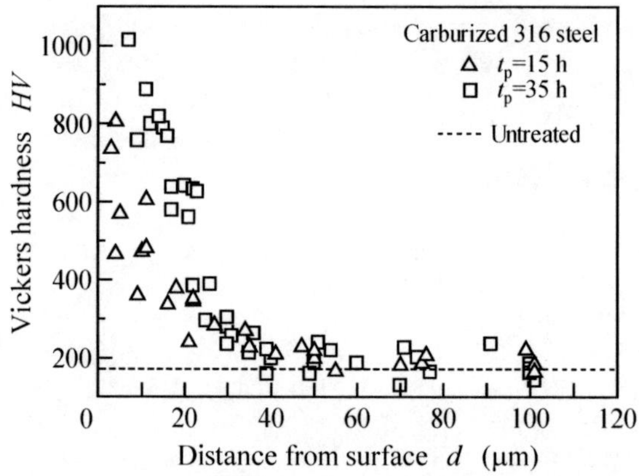

(a) Smooth specimens

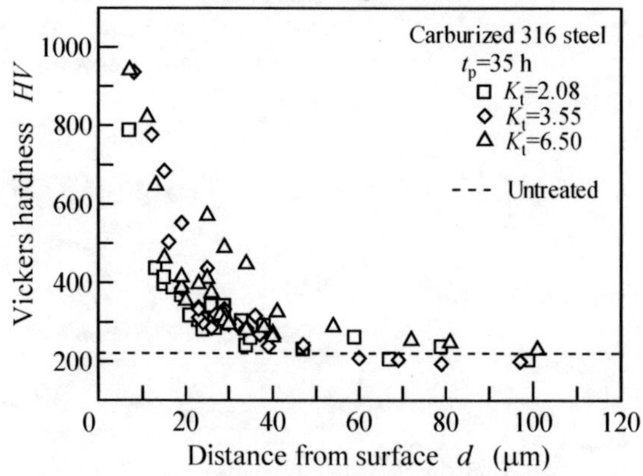

(b) Notched specimens

Figure 2: Vickers hardness profiles.

Table 1: Mechanical properties.

Treatment time t_p	Proof stress $\sigma_{0.2}$ (MPa)	Tensile strength σ_B (MPa)	Elongation ϕ (%)	Reduction of area ψ (%)
Untreated	299	576	67	77
15h		579	61	74
35h		581	57	72

3.3 Fatigue behaviour of smooth specimens

The *S-N* diagram is shown in fig. 3. It can be seen that the fatigue strength is considerably increased by carburizing. The case depth dependence of fatigue strength is slightly seen in the finite life region, where the 35 h treated specimen exhibits slightly longer fatigue lives than the 15 h treated specimen, while there is no discernible difference in the fatigue limit that is 390 MPa for both carburized specimens. The fatigue limit of the untreated specimen is 300 MPa, thus the improvement by 30% is achieved with the modified carburizing employed in the present study. Fatigue tests at the fatigue limit were continued to N=5×10^7 cycles, but no fatigue failure took place.

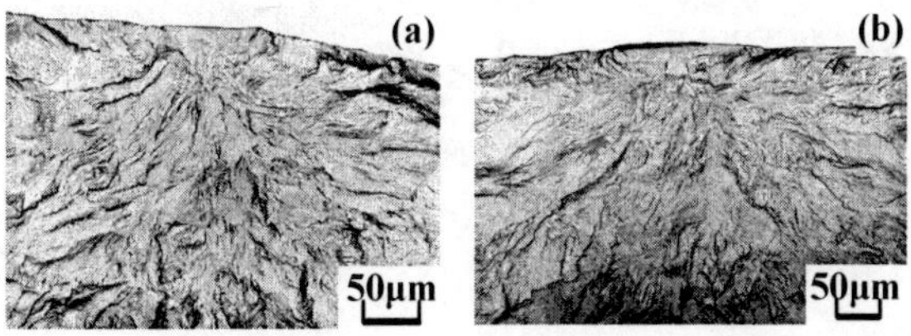

Figure 3: *S-N* diagram for untreated and carburized smooth specimens in laboratory air.

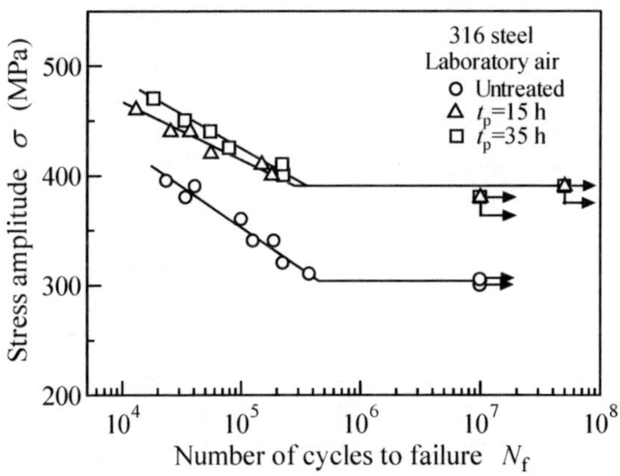

Figure 4: SEM micrographs of fracture surfaces near crack initiation site in laboratory air: (a) t_p=15 h (σ=440 MPa), (b) t_p=35 h (σ=410 MPa).

In the untreated specimen, fatigue cracks generated at the specimen surface due to cyclic slip deformation. Figure 4 reveals typical examples of SEM micrographs of fracture surfaces near the crack initiation site in the carburized specimens. Regardless of applied stress level, cracks initiate underneath the carburized case, i.e. at or near the boundary between the carburized case and the core. Similar subsurface crack initiation was also observed in austenitic stainless steels treated by shot peening [2-4]. A close examination reveals the presence of a smooth facet in the carburized case just above the subsurface crack initiation site, particularly remarkable in the 35 h treated specimens. It is also worth noting that there exists a fish-eye like pattern that extends predominantly into the core and the sizes are approximately 150 μm in the radial direction.

As shown in fig. 3, the fatigue strengths of the carburized specimens were improved considerably compared with the untreated specimens and the fatigue limit increased by 30% by carburizing. This is due to suppression of slip deformation at the specimen surface because of remarkable hardness increase, i.e. the resistance to crack initiation is significantly enhanced in the carburized case. Therefore, crack initiation becomes difficult to occur at the surface and then moves to subsurface between the carburized case and the core. The fatigue limits of the carburized specimens were 390 MPa that is considerably higher than the proof stress (299 MPa) and the fatigue limit (300 MPa) of the untreated specimens. This is because the limiting stress for crack initiation could be enhanced due to constraint of deformation by the hard carburized case.

In the finite life region, the effect of treatment time, i.e. case depth on fatigue strength was slightly seen where the fatigue lives of the specimen with thick case were longer than those of the specimen with thin case. This may also be due to enhancement of the crack initiation resistance and constraint of the carburized case to small crack growth because of increased hardness of the carburized case with increasing treatment time.

3.4 Notch fatigue behaviour

The *S-N* diagram characterized in terms of nominal stress amplitude for the untreated and carburized notched specimens is shown in fig. 5. As commonly observed, fatigue strength decreases with increasing K_t in both the untreated and carburized conditions, but the differences in fatigue strength between the notched specimens with K_t=3.55 and 6.50 become small. In the carburized specimens, the fatigue strengths increase significantly compared with the untreated specimens. As described previously, this is due to the suppression of slip deformation at the notch root surface because of remarkable hardness increase. It should be noted that the extent of increase in fatigue strength is largest in the smooth specimen and decreases with increasing K_t, then tends to saturate at K_t=3.55. It is also worth noting that no non-propagating cracks were seen in all run-out notched specimens in both the untreated and carburized conditions.

In the untreated specimens, it was found that cracks initiated at the notch root surface due to cyclic slip deformation. In the carburized specimens with K_t=2.08 and 3.55, the crack initiation behaviour depended on applied stress level. At high applied stresses, cracks initiated at the notch root surface, while at low applied stresses, underneath the carburized case. Such examples are revealed in fig. 6, where cracks initiated due to cyclic slip deformation underneath the carburized case and then immediately propagated to the surface. On the contrary, in the carburized notched specimens with K_t=6.50, cracks generated at the notch root surface regardless of applied stress level.

The relationship between fatigue strength reduction factor, K_f, and K_t is represented in fig. 7, where K_f is defined as the ratio of the fatigue limit for the smooth specimen, σ_{wo}, to that for the notched specimens, σ_{wk}. The K_f values for the untreated condition are considerably lower than K_t and the difference between both increases with increasing K_t, then tends to saturate at high K_t values [10]. This implies that the present material has very low notch sensitivity.

Similar results have been reported on type 304 and 316 austenitic stainless steels [10-13]. On the other hand, the K_f values for the carburized condition have the same K_t dependence as observed in the untreated condition, but are slightly larger in the entire K_t range studied. This indicates that the notch sensitivity of the present material is only slightly increased by carburizing.

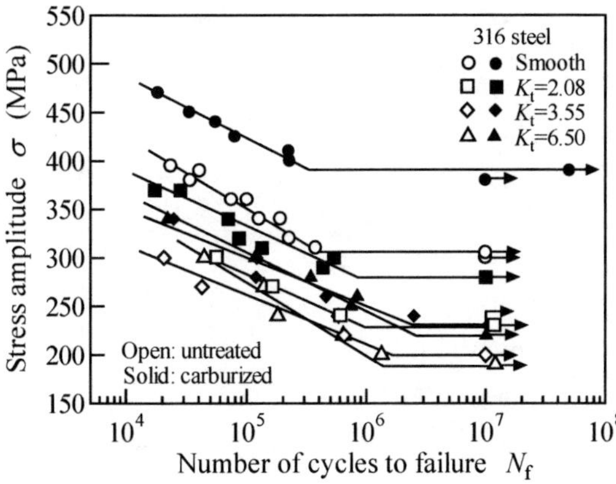

Figure 5: *S-N* diagram for untreated and carburized notched specimens characterized in terms of nominal stress amplitude in laboratory air.

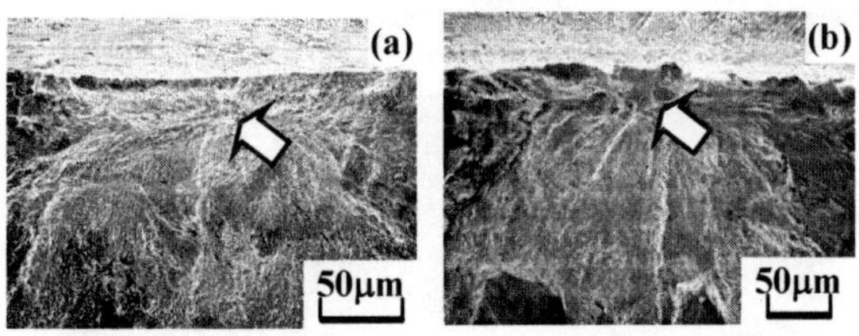

Figure 6: SEM micrographs showing subsurface crack initiation in carburized notched specimens: (a) K_t=2.08 (σ=300 MPa), (b) K_t=3.55 (σ=300 MPa). Arrows indicate the crack origin.

3.5 Corrosion fatigue behaviour

Figure 8 shows the *S-N* diagram for smooth specimens in 3%NaCl aqueous solution. For comparison, the *S-N* curves in laboratory air are also included without experimental data points. In the untreated specimen, the fatigue strength in 3%NaCl aqueous solution is lower than that in laboratory air and the reduction increases gradually with decreasing stress level. This is the well-known corrosion fatigue behaviour. On the contrary, the carburized specimens exhibit longer fatigue lives at high applied stresses than in laboratory air, but tend to exhibit nearly the same fatigue lives as in laboratory air with decreasing applied stress level. In addition, the fatigue limit seems to exist even in the corrosive environment within the range of experiment.

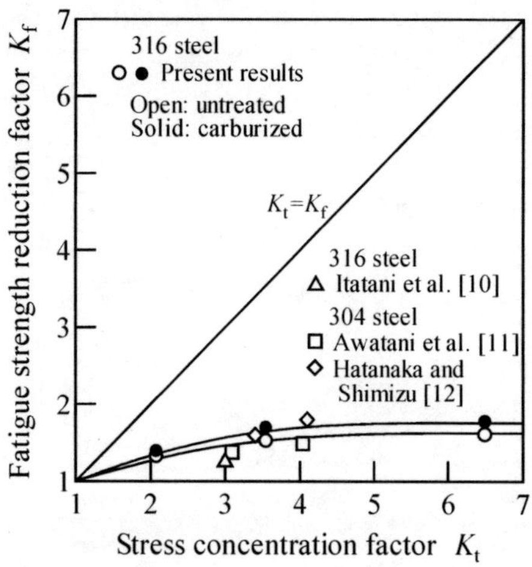

Figure 7: Relationship between fatigue strength reduction factor and stress concentration factor.

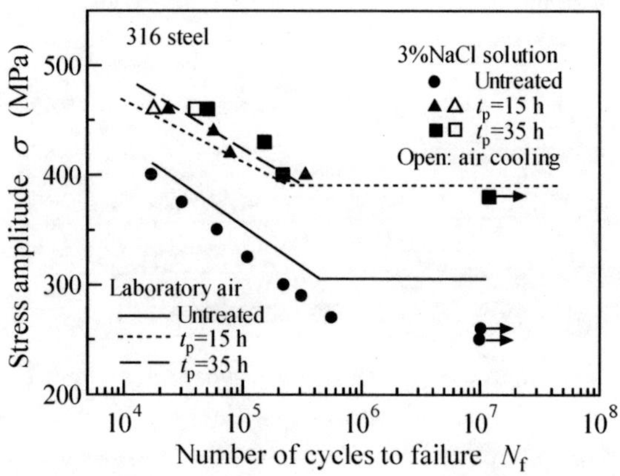

Figure 8: *S-N* diagram for untreated and carburized smooth specimens in 3%NaCl aqueous solution.

In the untreated specimens, cracks initiated at the specimen surface, while in the carburized specimens, always generated underneath the carburized case even in 3%NaCl aqueous solution. A brittle facet was seen in the case just above the subsurface crack initiation site, which has occurred due to crack initiation and subsequent growth into the interior of the specimens.

The increase of fatigue life at high applied stresses in 3%NaCl aqueous solution is believed to be due to suppression of temperature raise resulting from stress cycling. The surface temperature was not measured in the present study, but significant temperature raise has also been indicated in austenitic stainless steel [3, 14, 15]. Thus, a few additional fatigue tests were performed in laboratory air using the carburized specimens that were forced to cool by air. The obtained data are plotted in fig. 8. The fatigue lives are slightly longer than in laboratory air, but shorter than in 3%NaCl aqueous solution. These results seem reasonable, because the cooling effect by the solution is much larger than by air.

As indicated above, it should be emphasized that the specimens hardened by the modified carburizing showed no reduction of fatigue strength in aqueous corrosive environment. Since the fatigue strength of the untreated specimen decreased significantly in the corrosive environment, the modified carburizing can prevent the decrease of corrosion fatigue strength.

4 Conclusions

In the present work, the effect of carburizing on fatigue behaviour of smooth and notched specimens and corrosion fatigue behaviour was studied in a type 316 austenitic stainless steel. The main conclusions can be made as follows.

(1) The fatigue strength of the carburized smooth specimens increased considerably compared with the untreated specimen. The case depth dependence of fatigue strength was seen in the finite life region where the fatigue lives of the specimen with thick case were slightly longer that those of the specimen with thin case, while there was no discernible difference in the fatigue limit.

(2) The fatigue strength of the notched specimens was increased by carburizing and the extent of increase decreased with increasing stress concentration factor and then saturated. Both the untreated and carburized specimens indicated significantly low notch sensitivity, with a slight increase by carburizing.

(3) In 3%NaCl aqueous solution, the carburized specimens exhibited no reduction of fatigue strength, indicating excellent corrosion resistance of the carburized case.

(4) In the carburized smooth specimens, cracks initiated at subsurface underneath the carburized case regardless of applied stress level and environment. In the carburized notched specimens, the crack initiation behaviour was dependent on applied stress level and stress concentration factor. In the specimens with moderate stress concentration factors, crack initiation occurred at the notch root surface at high applied stresses, while underneath the carburized case at low applied stresses. In the specimens with s severe stress concentration factor, cracks initiated at the notch root surface regardless applied stress level.

References

[1] Hayashi, M. & Enomoto, K., Effect of preliminary surface working on fatigue strength of type 304 stainless steel in air and pure water at 288°C. *J. Soc. Mat. Sci. Jpn*, **45**, pp.1107-1112, 1996.

[2] Masaki, K., Ochi, Y. & Ishii, A., Behaviors of hardness distribution, residual stress distribution and fatigue cracks during the fatigue process. *Mater. Sci. Research Int.*, **4**, pp.200-205, 1998.

[3] Ochi, Y. & Masaki, K., Improvement of fatigue strength and fracture surface morphology of hard shot-peened type 316L steel. *Proc. of the 12th Bienniel Conference on Fracture*, eds. M.W. Brown, E.R. de los Rios & K.J. Miller, EMAS: West Midlands, **I**, pp.127-132, 1998.

[4] Masaki, K., Ochi, Y. & Matsumura, T., Estimation of high cycle fatigue limit of hard shot peened austenitic stainless steel. *Proc. of the 10th Int. Congress on Fracture*, Elsevier Science: Oxford, CD-ROM, 2001.

[5] Masaki, K., Ochi, Y. & Matsumura, T., The effects of hard shot-peening on high cycle fatigue properties of SUS316L steel. *Proc. of the 7th Int. Fatigue Congress*, eds. X.R. Wu & Z.G. Wang, Higher Education Press: Beijing, EMAS: West Midlands, **III**, pp.1219-1224, 1999.

[6] Stamm, H., Holzwarth, U., Boerman, D.J., Dos Santos Marques, F., Olchini, A. & Zausch, R., Effect of laser surface treatment on high cycle fatigue of AISI316 stainless steel. *Fatigue Fract. Eng. Mater. Struct.*, **19**, pp.985-995, 1996.

[7] Villechaise, P., Mendez, J. & Delafond, J., Improvement in fatigue resistance of 316L stainless steel and copper by aluminum coating with dynamic ion mixing. *Surf. Modification Technol.* **IV**, pp.335-347, 1991.

[8] Berrios, J.A., Teer, D.G. & Puchi-Cabrera, E.S., Fatigue properties of a 316L stainless steel coated with different TiNx deposites. *Surf. Coat. Technol.*, **148**, pp.179-190, 2001.

[9] Aoki, K. & Kitano, K., Surface hardening for austenitic stainless steels based on carbon solid solution. *Surface Eng.*, **18**, pp.462-463, 2002.

[10] Itatani, M., Asano, K. & Iida, K., Fatigue strength of notched austenitic stainless steel for nuclear power component. *ASME PVP* 374, pp.145-152, 1998.

[11] Awatani, J., Katagiri, K., Shiraishi, T. & Matsuyama, T., Fatigue character of stainless steel related to nonpropagating cracks. *J. Soc. Mater. Sci., Jpn*, **25**, pp.151-156, 1976.

[12] Hatanaka, K. & Shimizu, S., Fatigue strength in long life range and non-propagating crack in SUS304 type stainless steel. *Bulletin of the JSME*, **25**, pp.1859-1866, 1982.

[13] Linder, J. & Larsson, M., Notch sensitivity of austenitic and duplex stainless sheet steels. *Internal report Swedish Institute for Metal Research*, IM-3491, 1997.

[14] Tokaji, K, Ando, Z. & Mizutani, H., Fatigue strength of austenitic stainless steel in various environments: initiation, density and distribution, and growth of small fatigue cracks. *J. Soc. Mater. Sci., Jpn*, **34**, pp.816-822, 1985.

[15] Ogawa, T., Tokaji, K. & Kinpara, D., High cycle fatigue properties of SUS304 stainless steel under load-increasing conditions. *Trans. Japan Soc. Mech. Engrs*, **65**, pp.1684-1689, 1999.

Section 2
Thin coatings

Microstructural evolution of TiC/a-C nanocomposite coatings with pulsed magnetron sputtering

Y. T. Pei[1], K. P. Shaha[1], C. Q. Chen[1], J. Th. M. De Hosson[1], J. W. Bradley[2], S. Voronin[2] & M. Cada[2]

[1]Department of Applied Physics, The Netherlands Institute for Metals Research, University of Groningen, Groningen, The Netherlands

[2]Department of Electrical Engineering and Electronics, University of Liverpool, Liverpool, UK

Abstract

The microstructure and property of magnetron sputtered coatings are strongly affected by the intensity of concurrent ion impingement, in particular, by the energy distribution of impinging ions and the flux ratio between impinging ions and depositing atoms. In this paper, we report some striking results in the microstructure manipulation and residual stress control of TiC/a-C nanocomposite coatings with pulsed-DC magnetron sputtering. Ion mass/energy spectrometry of plasma diagnostics reveals that, depending on the waveform, frequency and width of pulses, pulsing the magnetrons can control the flux and energy distribution of Ar^+ ions over a very broad range, in comparison with DC sputtering. The latter delivers only low energy Ar+ ions and also less flux. With increasing pulse frequency, the nanocomposite coatings exhibit evolutions in the morphology of growing interface from rough to smooth and in the microstructure from strongly columnar to non-columnar. AFM, SEM, HR-TEM and nanoindentation are employed to characterize the deposited coatings, supported with plasma diagnostic experiments for a better understanding of the pulsed sputtering process.

Keywords: pulsed magnetron sputtering, ion energy distribution, plasma diagnostics, microstructural evolution, nanocomposite coating.

WIT Transactions on Engineering Sciences, Vol 55, © 2007 WIT Press
www.witpress.com, ISSN 1743-3533 (on-line)
doi:10.2495/SECM070071

1 Introduction

Pulsed DC magnetron sputtering in combination with the unbalanced magnetron configuration has become a major technique in the deposition of advanced coatings during the last decade [1, 2, 3]. It has the significant advantage over DC magnetron sputtering in suppressing the formation of arcs on the cathodes/targets. In particular it improves the microstructure and properties of dielectric films, which strongly depend on the intensity of the concurrent ion impingement on the growing interface of a deposited film. The crucial parameters determining the intensity of ion impingement are the energy distribution of the impinging ions and the flux ratio between impinging ions and depositing atoms. It has been observed that pulsing magnetrons in mid frequency (up to 350 kHz) leads to a much extended energy distribution of impinging ions and rather high ion current density towards the substrate [4, 5]. As a result, dense and well-structured dielectric and metallic coatings can be grown in pulsed mode [6]. However, a thorough understanding of the mechanism of plasma controlling and microstructure manipulation with pulsed DC sputtering is still lacking.

Our recent work shows that the column boundaries (CBs) of TiC/a-C:H nanocomposite coatings are a potential source of failure under loading and contact sliding [7, 8]. The CBs are harmful as initiation site of cracks and preferential cracking path, attributed to the fact that the homogeneity of nanocomposite is interrupted by the CBs that are enriched in carbon and voids. In reactive sputtering, the CBs can be readily restrained by employing a high voltage substrate bias or increasing the carbon content in the coatings. The mechanism of column restraint in this case results in a smooth growing interface of deposited coatings via intensive ion impingement and highly mobile carbon adatoms. However, it is hardly applicable to DC non-reactive sputtering of TiC/a-C nanocomposite coatings, where the landing of sputtered atoms interrupts and ion impingement may fluctuate to a large extent when the substrates pass from one target to another. The purpose of this work is to study the effects of pulsed DC sputtering on the depositing process and microstructure evolution of TiC/a-C nanocomposite coating, in particular, on the restraint of column growth. The results are striking in the sense of microstructure optimization and plasma manipulation.

2 Experiments

TiC/a-C nanocomposite coatings were deposited with non-reactive sputtering in a TEER UDP400/4 closed-field unbalanced magnetron sputtering (CFUMS) system. The system was configured of one Ti target (99.7%), one Cr target (99.95%) and two graphite targets (99.999%) opposite to each other. The two magnetrons with a metallic target were powered by a Pinnacle 6/6 kW double channel DC power supply (Advanced Energy) and the other two magnetrons with graphite targets were powered by a Pinnacle Plus 5/5 kW double channel pulsed DC power supply (Advanced Energy). The substrates were biased by a Pinnacle Plus 5 kW single channel pulsed DC power supply (Advanced Energy). All the power supplies for sputtering were operated at current control mode via a computer controlling system. The UDP 400/4 system was installed with an oil-free pumping system (a turbo molecular pump plus a diaphragm backing pump). The base pressure before deposition was $3\sim4\times10^{-6}$ mbar and the deposition pressure 2.6×10^{-3} mbar controlled by a constant flow rate of argon gas. No intended heating on the substrates was used during deposition.

The substrates used for each coating were ø30×6 mm discs of hardened M2 steel for tribological tests and ø100 mm Si wafer for microscopic observation of coating fracture cross-sections and for residual stress measurements by monitoring the curvature change. A 200 nm thick ductile CrTi interlayer of optimized composition and structure was employed to enhance the interfacial adhesion of TiC/a-C nanocomposite coatings. The interfacial adhesion was quantified by scratch tests with a CSM Revetester. The hardness and indentation modulus of the coatings were measured by nanoindentation with an MTS Nanoindenter XP®. The microstructural evolutions of the coatings were characterized with high resolution scanning electron microscopy (HR-SEM) on fracture cross sections and atomic force microscopy was used to image the surface morphology and to measure the surface roughness. The nanostructure of the coatings was revealed with high resolution transmission electron microscopy.

An EQP300 quadrupole mass spectrometer/ion energy analyser (Hiden Analytical Ltd) was used to measure the energy distribution of impinging ions in a single unbalanced magnetron sputtering system, GEN*LAB* (GENCOA Ltd), installed with a ø150 mm circular magnetron. It had a typical unbalanced configuration of adjustable magnetic field strength, which was set to resemble the field strength of the magnetrons installed in the TEER UDP400/4 rig. The single magnetron was powered with a 5kW Pinnacle Plus unit and operated at the same sputtering parameters used for coating deposition. The extractor head of the EQP300 instrument was pointed to the racetrack of the target and fixed directly behind a metallic substrate, which had a large opening hole that was covered with a fine nickel grid and aligned with the entrance orifice (ø300 μm) of the extractor. For the detailed setup see [4]. The ion energies were effectively measured with reference to ground potential and averaged during a measure time of one second. Therefore, the ion counts of different energies were a direct measure of the ion flux and can be confidently compared between measurements, provided the instrument settings remained unchanged.

3 Results and discussions

3.1 DC magnetron sputtering of TiC/a-C nanocomposite coatings

In our recent work on the deposition of TiC/a-C:H nanocomposite coatings with reactive DC sputtering in an argon/acetylene atmosphere, it has been revealed that the undesired columnar microstructure can be fully restrained by applying a higher substrate bias voltage (up to 150V) or/and a higher flow rate of acetylene gas that corresponds to a maximum carbon content of 88 at.% (excluding hydrogen) [7]. The columnar growth is directly related to the interface structure of the growing coatings. Typical cauliflower-like patterns characterize the surface topography of the nanocomposite coatings such that the cauliflower branches are separated by groove networks with a maximum depth of about 10 nm. These deep groove networks are likely the origin of the columnar boundaries (CBs), which is supported by the fact that the CBs intersect with the growing interfaces at the deep groove networks [9]. One can imagine that restraining the formation of groove networks will lead to the destruction of CBs and thus a column-free coating. In fact, the mechanisms involved in manipulating the two deposition parameters (bias voltage and flow rate of acetylene) are different. That is to say, the former breaks down the groove structure through intensive ion impingement and the latter directly fills the grooves with hydrocarbon adatoms of high mobility, respectively.

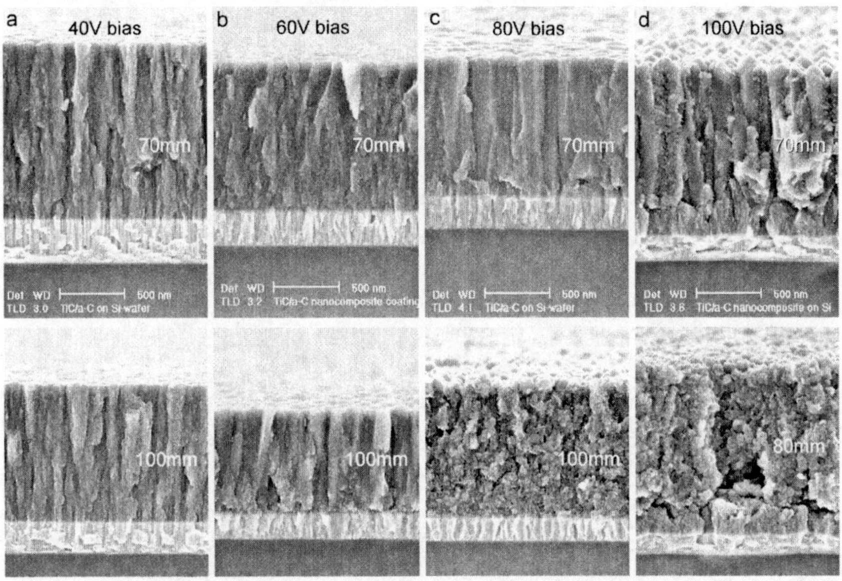

Figure 1: SEM micrographs showing the fracture cross sections of TiC/a-C nano-composite coatings deposited with DC sputtering at different substrate bias voltage (two hours deposition time): (a) 40V, (b) 60V, (c) 80V and (d) 100V with indicated distance from the targets.

Following this approach, hydrogen-free TiC/a-C nanocomposite coatings have been deposited by non-reactive magnetron sputtering graphite and titanium targets. Figure 1 shows the microstructural evolution of DC sputtered TiC/a-C coatings of nearly the same composition with increasing substrate bias voltage. It is obvious that the columnar microstructure is always present in the coatings, even though the columns become thicker and the coatings get denser with increasing the bias voltage up to 80V. Further increase of the bias voltage to 100V makes the columnar character stronger and the coating looser in reverse, due to the formation of facets on the head of growing columns and thus a much rougher interface of the growing coating. Finally, the coating deposited at 120V bias voltage flakes off. There is no way to fully suppress the columnar microstructure of TiC/a-C nanocomposite coatings by only increasing substrate bias voltage under DC magnetron non-reactive sputtering. In addition, it is worth to note that an increase in substrate bias voltage narrows the space suitable for depositing the coatings, in terms of the distance from targets. When applying a lower negative bias voltage (40 V or 60 V), it is possible to deposit dense coatings at the distance of 100 mm from the targets though the deposition rate is lower than that at the distance of 70 mm (Figs. 1(a) and 1(b)). However, the coating deposited at the same distance of 100 mm becomes porous and very loose once the bias voltage is increased to 80 V. An even worse situation is at 100V bias where the coating deposited at 80 mm distance is already porous (Fig. 1(d)). In this case, only the coating deposited at 60 mm distance from the targets is fully dense but still with a columnar microstructure (not shown). Such a reduction of suitable space for deposition is attributed to the over impingement of highly energetic but low flux ions, as to be discussed in the next session.

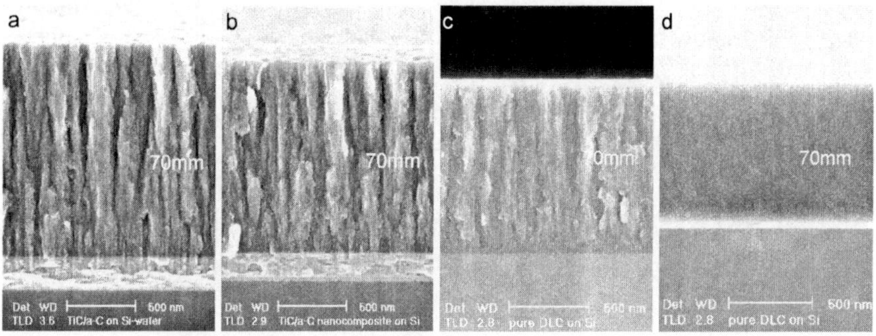

Figure 2: SEM micrographs showing the fracture cross sections of TiC/a-C nano-composite coatings deposited with DC sputtering at 60V substrate bias and of different carbon content: (a) 82 at.%, (b) 94 at.%, (c) 100 at.% and (d) 100 at.% on a stationary substrate.

The next step in our study of DC sputtering deposition parameters is to vary the carbon content of the coatings and to check the effect on the restraint of columnar growth. The results are shown in Fig. 2. By increasing the carbon content from 82 at.% to 94 at.%, the columnar character of the coatings does not show an essential difference, except for the fact that the coating of higher carbon content becomes slightly denser. Even the coating of pure carbon exhibits columnar microstructure (Fig. 2(c)). Apparently, the effect of increasing carbon content is minor in restraining the columnar growth, which is rather different from the situation of reactively sputtered TiC/a-C:H nanocomposite coatings [7]. In conclusion, it is almost impossible to deposit column-free TiC/a-C coatings with DC magnetron sputtering. The only exception is the pure carbon coating deposited on the substrate that is kept stationary in front of the graphite target (Fig. 2(d)), where a continuous impingement of high flux ions is available. In this sense, it provides a clue how to prohibit the columnar microstructure in TiC/a-C coatings. That is to enhance the flux of the impinging ions especially during the travel of substrates passing from one target to another.

3.2 Pulsed-DC magnetron sputtering of TiC/a-C nanocomposite coatings

Figure 3 shows the fracture cross section of the TiC/a-C coatings deposited with pulsed DC magnetron sputtering. Switching from DC to pulsed DC (*p*-DC) sputtering of 100 kHz frequency, the columnar microstructure does not change much, but becomes more diffuse CBs (Fig. 3(a)). By increasing the frequency of sputtering DC pulses to 250 kHz and keeping all the other parameters unchanged, the coating surface becomes much smoother and the columnar feature is almost invisible. Up to 350 kHz pulsed DC sputtering, the deposited coating is fully columnar free, with a surface (actually the growing interface) so smooth that it hardly shows some contrast under the magnification used for SEM observation. It should be pointed out that the restraint of columnar microstructure with pulsed DC sputtering is achieved at the expense of a reduced deposition rate. The deposition rate of TiC/a-C coatings at 50% duty cycle and *p*-DC frequency 100 kHz, 250 kHz and 350 kHz is 80%, 60% and 58%, respectively, of the deposition rate with DC sputtering. Obviously, the reduction extent of deposition rate is dependent on the sputtering yield of different kinds of targets at different frequencies and the duty factor. Once the *p*-DC frequency is above 250 kHz, the deposition rate does not significantly decrease any further.

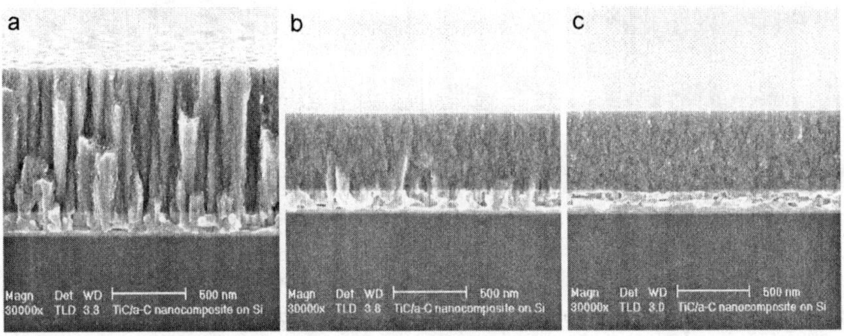

Figure 3: SEM micrographs showing the fracture cross sections of TiC/a-C nano-composite coatings deposited with pulsed-DC sputtering at 40V substrate bias and different pulse frequency (50% duty cycle): (a) 100 kHz, (b) 250 kHz and (c) 350 kHz (deposition time: two hours).

It is interesting to note that, under *p*-DC sputtering, the space suitable for coating deposition is wide enough. As shown in Fig. 4(a), the thickness of the coatings changes inversely as a function of the distance from the targets but its negative slope is smaller than that with DC sputtering. Moreover, the coatings deposited in the distance range of 60-100 mm exhibit continuously a dense and columnar free microstructure. This feature is of particular importance for industrial applications where workpieces of large sizes or large quantity of workpieces need to be coated in one batch. The optimized microstructure of TiC/a-C coating deposited with *p*-DC magnetron sputtering is shown in Fig.

4(b), consisting of a ductile interlayer, smooth and mirror-like surface, and fine and column-free microstructure.

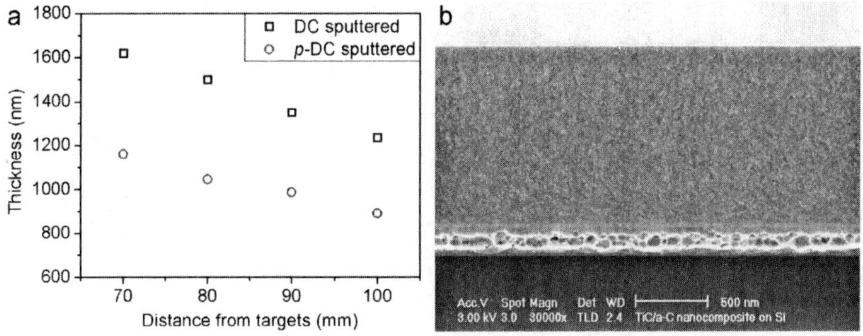

Figure 4: (a) Thickness of deposited coatings versus the distance from the targets and (b) optimized microstructure of TiC/a-C nanocomposite coatings.

3.3 Ion energy distribution and plasma diagnostics of *p*-DC sputtering

To understand the mechanism of column restraint under *p*-DC magnetron sputtering, ion mass/energy spectrometry has been employed to diagnose the sputtering plasma, in particular the energy distribution and flux of the impinging Ar^+ ions pulled onto the growing interface of the coatings. Figure 5 presents the *p*-DC voltage waveform applied onto the targets, time resolved energy distribution and the flux of the impinging Ar^+ ions. The voltage waveform of the asymmetric bipolar pulsed DC exhibits three characteristic periods: the pulse on period A, the reversing period B and the pulse off period C. The significant features of the asymmetric bipolar *p*-DC power supplies used in this research include the overshoot peak of around 200 V and the adjacent fluctuations in the reversing period B as well as the low positive voltage remaining during the pulse off period C. The time averaged energy distribution curves of the impinging Ar^+ ions under *p*-DC magnetron sputtering present three distinct populations of the ions that reflect the target voltage waveforms. That is to say, impingements of Ar^+ ions of low energy (< 20 eV, peak A in Fig. 5(b)) occur during the pulse-on period, supported by the fact that DC magnetron sputtering produces only the low energy ions of the same category. The ions of intermediate energy (20~50 eV, peak C in Fig. 5(b)) are generated during the pulse-off period, which is evident because the population of this category ions diminishes with decreasing pulse-off period (i.e. increasing the duty cycle at a chosen *p*-DC frequency) as shown in Fig. 5(c). Those ions of high energy extending over 200 eV are created during the reversing period. Because of the detection limit of the energy spectrometer, the energy distribution function curves at a *p*-DC frequency above 200 kHz are partly cut off beyond 195 eV. It is clear that the energy distribution function of the impinging Ar^+ ions is governed by the sputtering mode, *p*-DC frequency and the duty cycle.

More important in control of the microstructure of the coatings is the flux of the impinging ions generated in the sputtering plasma. As shown in Fig. 5(d), the flux of the total impinging Ar^+ ions at *p*-DC sputtering mode of low frequencies (e.g. 100 kHz) is comparable with that under DC magnetron sputtering. However, the total ions flux dramatically increases with increasing the frequency beyond 100 kHz. One should keep it in mind that the total ions flux measured at a frequency above 200 kHz is slightly lower than the true value, due to the cut off of the measurement of the high energy ions beyond 195 eV. This explains the smaller increase of the measured ion flux between the high frequencies (Fig. 5(d)). It is clear that the flux of total impinging ions may differ by one order of magnitude between DC magnetron sputtering and *p*-DC magnetron sputtering at high frequencies. In particular, the impinging ions in the intermediate and high energy bands become more and more dominant with increasing *p*-DC frequency, delivering much more energy for impingement on the growing interface of deposited coatings.

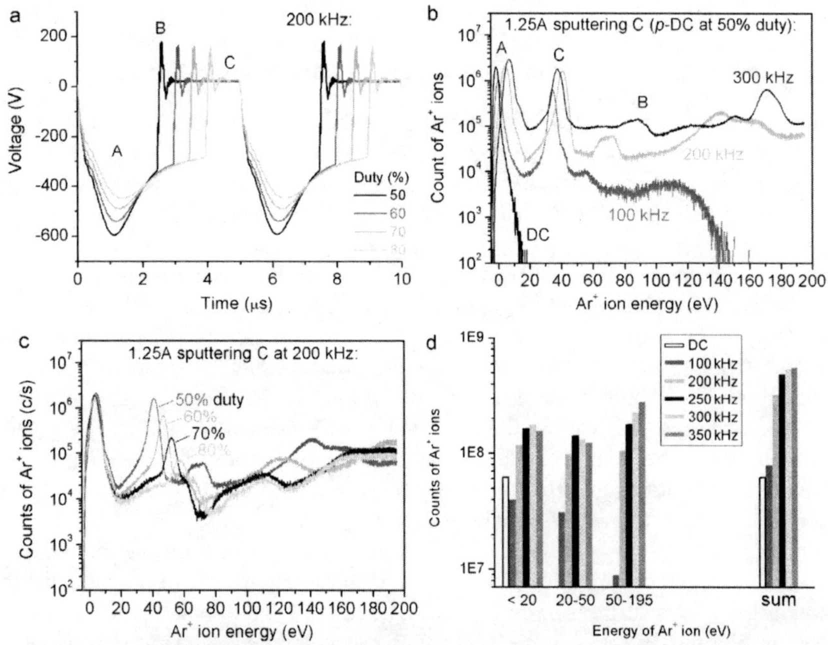

Figure 5: (a) Voltage waveform of *p*-DC at 200 kHz; energy distribution of impinging Ar^+ ions incident on the growing coatings under DC sputtering or pulsed DC sputtering of different frequencies (b) and different duty cycles at 200 kHz (c); (d) flux of the Ar^+ ions in three energy ranges as well as the sum at different *p*-DC frequencies.

Besides the enhancement on the flux and energy distribution of impinging ions, another significant feature of *p*-DC magnetron sputtering is the expansion of the plasma torch in between the unbalanced magnetrons. Figure 6 compares the confine of the plasma torch incident onto the substrate at different sputtering modes and frequencies. Under DC sputtering, the dense plasma torch confined in front of the target can cover only the center part of the substrate (Fig. 6(a)). In contrast, the plasma torch at 100 kHz *p*-DC sputtering covers nearly the whole substrate, but with a large intensity difference from the center to the outer fringe (Fig. 6(b)). Up to 250 kHz, the plasma torch gets much denser and homogeneously covers the entire substrate (Fig. 6(c)). It even expands to the surrounding area but the intensity is still very low. At the highest frequency 350 kHz, the extensive plasma not only covers the entire substrate but also fills in the whole chamber (Fig. 6(d)). It is this expanded plasma that keeps the growing coating under intensive and in particular continuous impingement when passing by from one target to another in a closed-field configuration of multi-targets sputtering system.

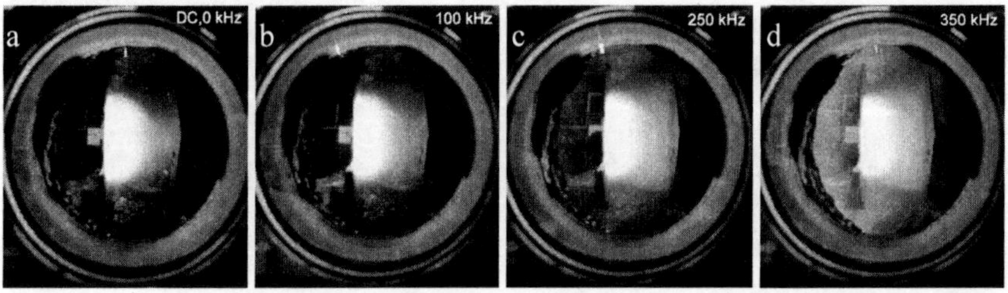

Figure 6: Photos showing the shape and density of the plasma incident at the rectangular substrate positioned to the right of viewport. A much denser and extended plasma is obtained at higher frequency pulsed DC sputtering, compared with DC sputtering.

In the so called closed-filed unbalanced magnetron sputtering system such as the Teer UDP 400/4 rig used in this work, targets stand vertically along the circular chamber wall and the substrates mounted on the sample carousel rotate around the central axis of the chamber (single spindle rotation) and pass at a chosen distance the targets one by one. For reactive sputtering, the decomposition of the reactive gases occurs in the plasma everywhere inside the chamber and provides some of the species for deposition that continuously reach the growing film from various angles and at any position, besides the sputtered atoms from the targets. Even in non-reactive sputtering with a co-planar configuration of magnetrons facing towards the same area or a substrate, such a continuous landing of different kinds of depositing atoms can be readily realized as well. Therefore, there is less challenge in control of the microstructure of DLC based composite coatings in these two cases. Different from the preceding situation, during non-reactive sputtering with a closed-filed unbalanced magnetron sputtering system the landing of sputtered atoms interrupts and ion impingement may fluctuate to a large extent in between the targets. This may readily lead to either multilayered coatings due to composition variation or undesired microstructures because of unstable ion impingement. This work focuses on the influence of pulsed magnetron sputtering on the depositing process and consequently the microstructure evolution of the deposited coatings, rather than the sputtering process itself. It is clearly demonstrated that pulsing the unbalanced magnetrons may dramatically change the property of the plasma and consequently enhance the intensity of the concurrent ion impingement on the growing interface of deposited coating, in comparison with DC sputtering. As a result, a desired microstructure may be readily tailored to achieve the superior performances of advanced coatings.

4 Conclusions

Pulsed DC magnetron sputtering can control the flux and energy distribution of Ar^+ impinging ions over a very broad range, in comparison with DC sputtering that delivers only low energy Ar^+ ions and also much smaller flux. The most striking result observed is the capability in the microstructure manipulation of TiC/a-C nanocomposite coatings with pulsed-DC magnetron sputtering, especially constraining the formation of columnar microstructure. With increasing the pulse frequency, the nanocomposite coatings exhibit evolutions in morphology of the growing interface from rough to smooth, in the microstructure from strongly columnar to non-columnar and become fully dense.

Acknowledgements

The authors acknowledge financial support from the Netherlands Institute for Metals Research (NIMR) and the Foundation for Fundamental Research on Matter (FOM-Utrecht).

References

[1] S. Schiller, K. Goedicke, J. Reschke, V. Kirchhoff, S. Schneider and F. Milde. *Surf. Coat. Technol.*, 61 (1993), 331-337.
[2] P.J. Kelly and R.D. Arnell. Vacuum, 56 (2000), 159-172.
[3] R.D. Arnell, P.J. Kelly and J.W. Bradley. *Surf. Coat. Technol.*, 188 (2004), 158-163.
[4] J.W. Bradley, H. Bäcker, Y. Aranda-Gonzalvo, P.J. Kelly and R.D. Arnell. *Plasma Source Sci Technol.*, 11 (2002), 165-174.
[5] H. Bartzsch, P. Frach and K. Goedicke. *Surf. Coat. Technol.*, 132 (2000), 244-250
[6] P.J. Kelly and R.D. Arnell. *J. Vac. Sci. Technol.* A17 (1999), 945-953
[7] Y.T. Pei, D. Galvan and J.Th.M. De Hosson. *Acta Mater.*, 53 (2005), 4505-4521.
[8] D. Galvan, Y.T. Pei and J.Th.M. De Hosson. *Surf. Coat. Technol.*, 200 (2006), 6718-6726.
[9] J.Th.M. De Hosson, Y.T. Pei and D. Galvan. Surfaces and Interfaces in Nanostructured Materials II. Eds: *S.M. Mukhopadhyay, N.B. Dahotre, S. Seal, and A. Agarwal.* TMS, 2006. 59-68.

Ionic migration behavior in minute wiring on flexible substrate

Y. Kimura[1], S. Isawa[2], M. Chino[2], H. Hara[2], K. Tamayama[2] & A. Suzuki[1]
[1]*Department of Materials Science and Technology, Kogakuin University, Shinjuku-ku, Tokyo, Japan*
[2]*MISUZU Industrial Corporation, Nakaminowa, Minowacho, Kamiinagunn, Nagano, Japan*

Abstract

Recently, miniaturization and high-density mounting have been demanded in various electronic devices. Above all, flexible print circuit (FPC) technology has been a focal point, especially in the field of notebook computers and cellular phones, for obtaining a higher degree of freedom in design. In order to actualize the fine pitch of wiring in FPC, the insulation degradation due to the migration becomes a problem that cannot be ignored. Because of differences in its structure and wiring from the former printed circuit board, the analysis of migration behavior in FPC becomes important for evaluating the reliability of an electronic component based on FPC. In this research, in order to grasp the migration behavior of FPC, a 45μm pitched COF specimen was used. In addition, the paragraphed 30μm pitch tooth profile wiring specimen was designated as the test material. Then, a reliability test for evaluating the migration behavior was conducted. As a result of a Steady-State Temperature Humidity Bias Life (THB) test, the migration occurred progressively. Various investigations were conducted for explaining about the mechanism of ionic migration. Also in this paper, the adhesive strength between the polyimide film and copper (Cu) wiring pattern, and that between the polyimide film and the under-filling material, was measured in order to investigate the influence of the surface state of polyimide film upon the migration behavior.
Keywords: ionic migration, FPC, reliability, THB testing.

www.witpress.com, ISSN 1743-3533 (on-line)
doi:10.2495/SECM070081

1 Introduction

Recently, miniaturization and high-density mounting have been demanded in various electronic devices. Above all, flexible print circuit has been a focal point especially in the field of notebook computers and cellular phones for realizing high density mounting and also obtaining a higher degree of freedom in design. The insulation failure, due to an ionic migration damage [1,2] with an advanced fine pitch wire mounting, becomes a serious problem [3]. Ionic migration is an electrochemical phenomenon related to chemical solutions under electric potential. The reaction mechanisms of ionic migration can be classified into three processes: anodic dissolution, mass transfer, and cathodic deposition [4]. Migration in electronic materials is caused by an electrochemical phenomena related to chemical solutions and electric potential [5,6]. Deposition reactions and metal ion transfer reactions are involved in migration, induced by such factors as metal dissolution reactions and diffusion [7,8]. Migration is especially likely to occur when electronic devices using high-density mounting are affected by both materials and environmental conditions [9]. When migration occurs, it causes changes in electrical characteristics and results in equipment failure.

"Migration" refers to a phenomenon in which metal ions are transferred from one metal electrode to the opposite metal electrode, this process results in metal or alloy deposits. The transfer occurs between the electrodes of devices such as printed circuit boards (PCBs) when an electric field is impressed in the presence of moisture such as dew condensation adhering between the electrodes. Migration is classified as either dendrite or conductive anodic filament depending on the shape of the deposits and the conditions leading to the occurrence [1,10] "Dendrite" refers to dendritic-shaped metal or metal oxide deposits on the surface of PCB insulation.

The analysis of migration behaviour in FPC becomes important for evaluating the reliability of electronic devices based on FPC. In this research, THB tests under the conditions of 85°C and 85% RH, with applying electric potentials of 5, 25 and 45v, were conducted using COF as a sample to understand the ionic migration behaviour in FPC. Various investigations concerning generated ionic migration behaviour were conducted employing SEM and EDX. Some trials were conducted for establishing the evaluation method of ionic migration behaviour. In addition, improvement of ionic migration character was conducted employing alternative under fill resin in FPC materials system.

2 Experimental procedures

2.1 Specimen and THB accelerated migration test

COF structure of the test specimen is shown in detail in fig. 1. The specimen used in this study is that composed of polyimide substrate and consisted of an 8μm thick Cu wiring that was covered with 0.2 μm thick plated tin (Sn), coating as shown in fig. 1 (c). IC, whose size is 1.96 mm x 19 mm, is mounted on Sn plated Cu wiring on FPC using gold (Au) bump with a thickness of 14 μm.

Epoxy resin was used for an under fill material and polyimide resin was used as a solder resistant of FPC. The COF specimen used in this study was prepared using Cu wiring plated on a Cr/Ni seed layer sputtered on polyimide resin substrate.

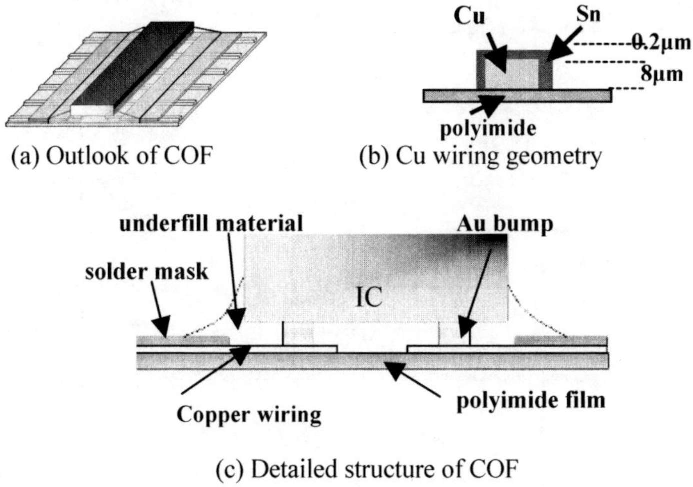

Figure 1: COF structure of test specimen: 45μm pitch.

Employing this COF sample with a 45μm pitch, an accelerated migration test was conducted up to 168 hrs long employing a THB test. The conditions of the THB test were 85℃ and 85% RH whilst applying electric potentials of 5, 25 and 45v.

2.2 Observation of migration morphology and leak test result

After conducting THB tests of 24, 70, 96 and 168 hrs, morphology change concerning migration generation in test specimens was examined by using an optical microscope. At the same time, a leak test was conducted for each specimen under a constant condition.

2.3 SEM and EDX evaluation of migration behaviour

Morphologies of migration damage and the generated dendrite structures were investigated by SEM. Characterization of generated materials observed on a cross section of a specimen, with THB accelerated migration test, was conducted employing EDX.

3 Experimental results and discussions

3.1 Result of THB accelerated migration test

THB tested specimens under the condition of 85℃ and 85% RH showed insulation failure for the first time after 168 hrs. Color change was recognized in degraded specimens on the interface between the Cu line and polyimide substrate after 24 hrs of THB testing. Also dendrite was generated from the side part of Sn coated Cu wiring, which was anode. These degradations (shown in fig. 3) were mainly recognized in the areas from A to C originally indicated in fig. 2, which were covered with under fill material. Therefore, under fill material may affect migration behaviour of this specimen.

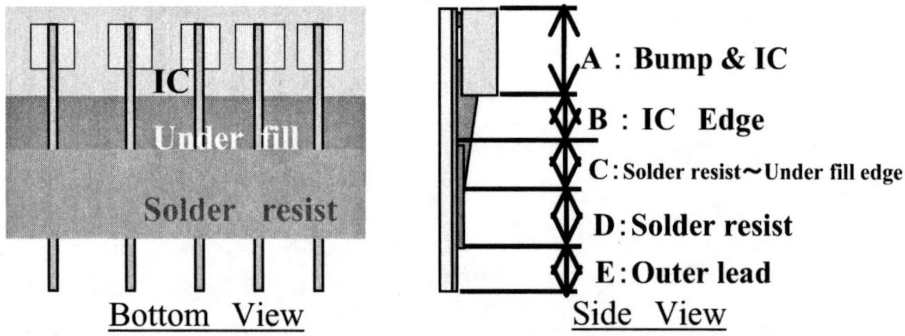

Figure 2: Observed sample parts.

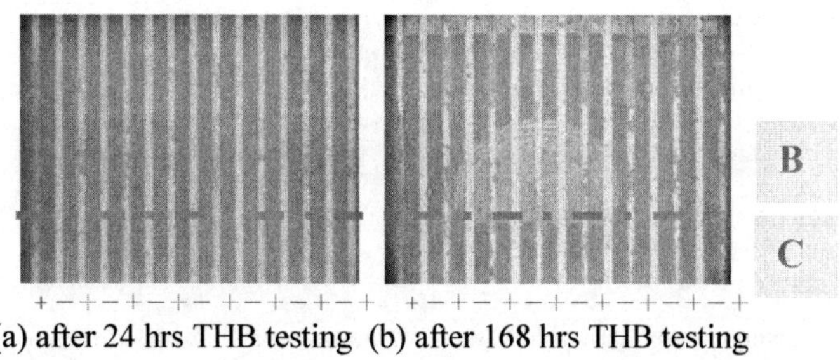

Figure 3: Bottom view of migration damage observed by an optical microscope mainly generated in B and C regions shown in fig. 2.

Migration behaviour generated in COF can be characterized by the color change due to the dissolution of the Cu lines and dendritic precipitation of Cu. These electrochemical processes may result in the insulation failure of FPC. The generation of migration with time, under THB testing conditions of 85℃ and 85% RH was investigated and summarized in Table 1. As a result, color change occurred after 24 hrs testing due to the dissolution of Cu lines and dendritic precipitation of Cu. On the contrary, insulation failure was detected for the first time only after 168 hrs THB testing. Therefore, the migration process was proceeded from the initial stage of THB testing at which insulation failure was not detected.

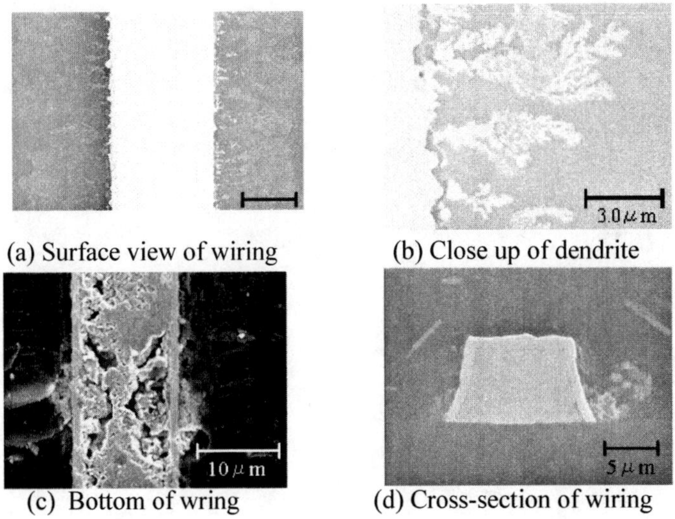

(a) Surface view of wiring (b) Close up of dendrite

(c) Bottom of wring (d) Cross-section of wiring

Figure 4: Migration observed by SEM.

Table 1: Generation of migration with time.

Time	Outlook		Impedance
	Color change	Dendrite	
24	Yes	Yes	Good
70	Yes	Yes	Good
96	Yes	Yes	Good
168	Yes	Yes	Degraded

3.2 Migration behaviour detected by SEM and EDX

The generated morphology of migration was indicated in fig. 4. Surface morphology of wiring, after generating migration and close up features of dendrite, were examined by SEM and indicated in fig. 4 (a) and (b). Damaged morphologies on the bottom and the cross-section of wiring are shown in fig. 4 (c) and (d). Therefore dendrite formation was essential for the migration process. Dissolved morphology was recognized on the bottom of Cu wiring; however, Sn plated coating was left undamaged. Cross-sectional view of wiring shown in fig. 4(d) indicates that migration damage was initiated from interfacial region between Sn plated coating and polyimide substrate. Then the central parts of the Cu line bottom were dissolved out with the progress of migration damage.

The results of the EDX analysis for the cross-sectional area of wiring are shown in fig. 5. This figure indicates that Cu was dissolved from the line via an interface between plated Sn coating and polyimide substrate. From these observations, elements of dendrite were composed of Cu dissolved from lines and precipitated in the under fill resin. Also, in EDX mapping shown in fig. 5, the chlorine's (Cl) presence peaked in surrounding areas around Sn plated Cu wiring, which was included in under fill resin. Then, morphology of dendrite was investigated in detail and shown in fig. 6. As a result, initiation of dendrite formation was not recognized in the vicinity of Sn plating, however, dendrite formation was initiated from some distant area from Sn plating. Judging from this figure and observed results obtained from specimens before THB testing, the initiation site of dendrite was understood to be the Cr/Ni seed layer which is an essential element for making specimens by sputtering and plating process.

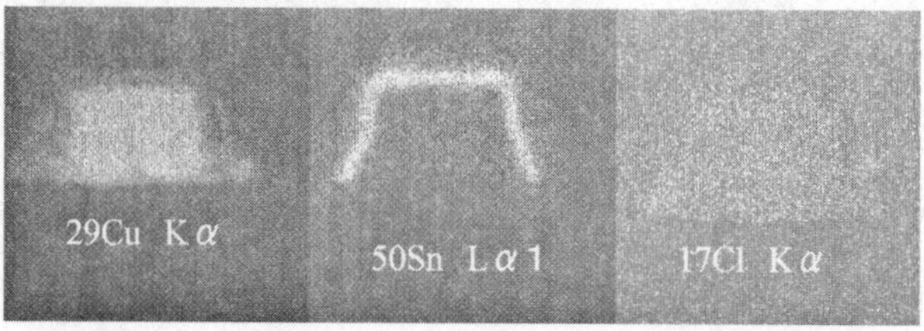

Figure 5: Analysis of migrated element by EDX.

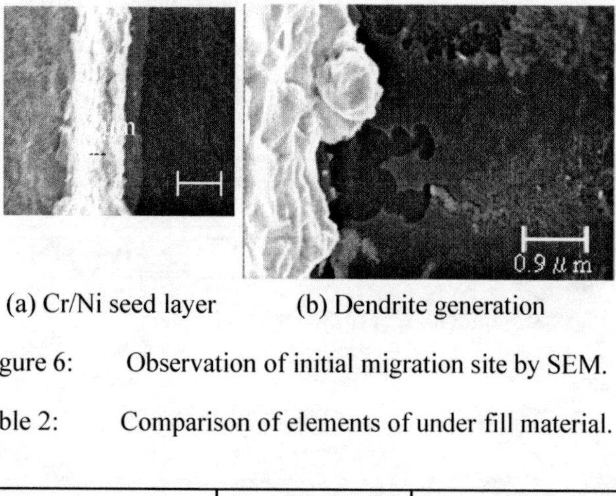

(a) Cr/Ni seed layer (b) Dendrite generation

Figure 6: Observation of initial migration site by SEM.

Table 2: Comparison of elements of under fill material.

	Usual under fill	Improved under fill
Main element	Epoxy	Epoxy
Cl^- ion (ppm)	0.41	0.12
Na^+ion (ppm)	0.32	0.15
pH	5	4.1
Conductivity (μS/cm)	21.2	27.7
Water absorption(wt.%)	1~2	1.4
Viscosity (Pa・s)	0.45	19

3.3 Improvement of migration resistance by applying alternative under fill resin

The EDX result shown in fig. 5 indicated that Cl content of under fill resin might significantly affect migration behaviour. Therefore, some trials for improving migration resistance were conducted employing alternative resin whose content of Cl was suppressed. Chemical compositions of both usual and alternative under fill resin were indicated in Table 2. Another type of COF specimen whose dimensions were the same as shown in fig. 1, except for under fill material, was prepared. THB testing under the same condition was then conducted employing this type of specimen. After 168 hrs THB testing, good results were obtained, both in outlook evaluation and impedance testing, for this type of specimen with alternative under fill material system.

3.4 Adhesive strength degradation after THB testing

SEM/EDX analysis indicated that migration was initiated from an interface between the substrate and plated Sn coating. Therefore, the adhesive strength of the interface between the substrate and coating may affect the migration behaviour of COF. Evaluation of degradation characteristics of adhesive strength due to THB testing is extremely important. In this chapter, comparison of adhesive strength before and after THB testing was conducted. Also in this research, the difference in under fill resin was taken into consideration. The peeling strength of under fill resin from substrate was measured and the test results are shown in fig. 7. The initial peeling strength was different depending upon the type of resin. Alternative under fill material that suppressed Cl content showed improved adhesive strength compared with the usual type of resin.

Degradation in peeling strength was detected in both usual and improved under fill resin; however, degraded ratio in peeling strength is obtained to be better for improved resin. The value of the peeling load for improved resin, after THB testing, remained at about 80% of the usual resin before THB testing. Therefore, the advantage of improved resin was indicated from the viewpoint of peeling strength degradation by THB testing.

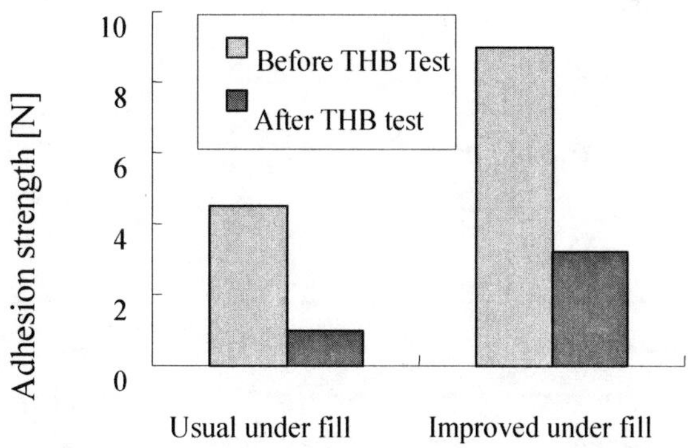

Figure 7: Comparison of adhesion strength before and after 168 hrs THB test.

3.5 Dependency of migration behavior of FPC upon manufacturing process and under fill resin

In this section, effects of manufacturing methods and under fill resin upon the migration behavior of FPC were evaluated employing a 30 μm pitched tooth profile wiring specimen shown in fig. 8.

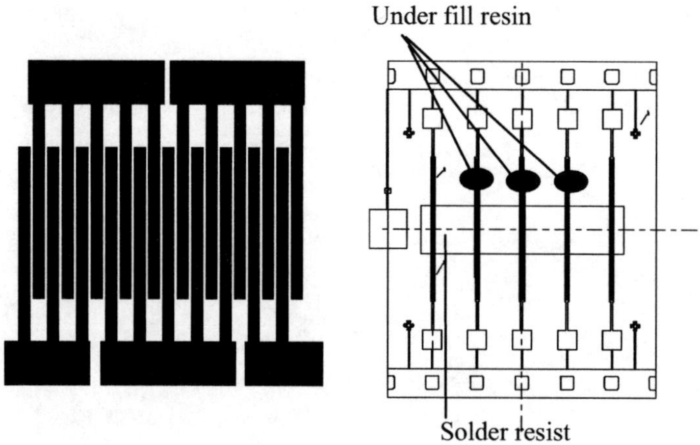

(a) Enlarged view of tooth profile wiring   (b) Outlook of tooth profile wiring specimen

Figure 8: FPC tooth profile wiring specimen with 30 μm pitch.

Table 3: Combination of under fill resin and substrate.

Under fill resin type	Substrate: a [Casting method]	Substrate: b [Cr/Ni Sputtering & Cu electroplating]
Resin A(Epoxy)	Sample A-a	Sample A-b
Resin B(Epoxy)	Sample B-a	Sample B-b
Resin C(Epoxy)	Sample C-a	Sample C-b
Resin D(Epoxy)	Sample D-a	Sample D-b
Resin E(Epoxy)	Sample E-a	Sample E-b
Non	Sample 0-a	Sample 0-b

In this paper, both the sputtering & electroplating (substrate: b) and casting method (substrate: b) were employed for manufacturing FPC. On polyimide substrate, about 40 μm thick, 8 μm of Cu line, with 0.2 μm of Sn plating, was formed. Polyimide resin was also used for solder resistant and 5 under fill resin types (A, B, C, D and E) of epoxy resins were applied. A specimen without under fill resin was also used. On some part of the tooth profile wiring specimen shown in fig. 8 (b), under fill resin was applied and hardened under adequate curing conditions. Specimens used in this study are summarized in Table 3. THB testing was employed in these specimens by measuring the impedance of them with time. Test conditions were the same as the previous test, that is, 85°C, 85% RH and a bias potential of DC45V.

After 200 hrs of THB testing had been conducted, an optical microscope detected the outlook of the specimen. As a result, some color change was detected in some part of specimens using substrate:a, however no dendrite formation was recognized in these types of specimens. On the contrary, in the case of specimens using substrate:b, color change on wiring and also dendrite formation were clearly recognized. Color change and dendrite formation were initiated from the boundary layer between wiring and polyimide substrate. These degradations were generated only from the anode part of wiring as shown in fig. 9. These results almost corresponded with the results indicated in previous sections that were obtained from the test specimen shown in fig. 1.

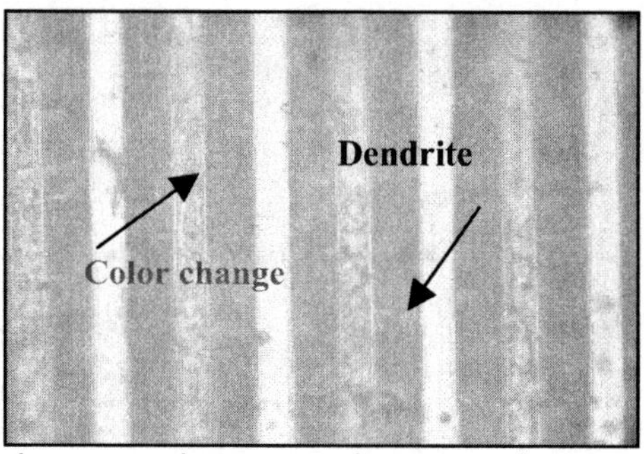

Figure 9: Dendrite formation after 200hrs THB testing. Sample E-b.

From the above mentioned result, color change and dendrite formation behavior have a dependency on under fill resin and manufacturing processes. Therefore in the following process, dependency of migration behavior on FPC upon under fill resin and substrate material was examined through evaluating impedance change with THB testing time. Insulation degradation with THB testing, detected by impedance change, is summarized in Table 4.

Judging from the impedance value of the specimen, using substrate:a showed better ionic migration resistance. Impedance values have been kept greater than 100 MΩ for three kinds of under fill resins: A, B & E, and the value was kept greater than 1 MΩ for two kinds of under fill resin: C & D in this case. In the case of specimens using substrate:b, no specimen kept its impedance value grater than 100 MΩ after 200hrs of THB testing. Only one kind of specimen with under fill resin D and the non-resin coated specimen managed to keep a good impedance condition (I≧1 MΩ) after 200hrs THB testing. The other four kinds of under fill resin coated specimens showed insulation degradation within 10 hrs. From these test results, it was clarified that specimens using substrate:a showed better insulation degradation resistance compared with those using substrate:b.

Table 4: Insulation degradation detected by impedance change.

	Substrate a	Substrate b
Resin A	Sample A-a	Sample A-b
Time to insulation degradation	◎	× : 4.4hrs
Resin B	Sample B-a	Sample B-b ×
Time to insulation degradation	◎	8.6hrs
Resin C	Sample C-a	Sample C-b
Time to insulation degradation	○	× : 5.8hrs
Resin D	Sample D-a	Sample D-b
Time to insulation degradation	○	○
Resin E	Sample 0-a	Sample 0-b
Time to insulation degradation	◎	× : 5.8hrs
Non resin	21	23
Time to insulation degradation	× : 161.8hrs	○

◎ : Fairly good (I≧100MΩ)
○ : Good(I≧1MΩ)
× : Degraded (I≦1kΩ)

For evaluating long-term reliability of specimens an additional 800hrs THB testing was conducted till 1,000 hrs was achieved through employing four kinds of specimens, that is, B-a, C-a. D-a and E-a, which showed no color change and kept good impedance value larger than 1 MΩ after 200 hrs THB testing as shown in Table 4. Results obtained in this test are summarized in Table 5. No impedance degradation was detected in specimens B-a and C-a, however two other kinds of specimens were degraded at the stage of more than 800 hrs THB testing as shown in Table 5.

In the specimens B-a and C-a, which showed no impedance degradation, some color change was detected as shown in fig. 10. In the specimens D-a and E-a, which showed impedance degradation before 1,000 hrs, remarkable color change was detected and is also shown in fig.10 (c) and (d).

Table 5: Impedance change with THB test time till 1,000 hrs.

Sample	I : Impedance
B-a	Fairly good : ◎(I≧100MΩ)
C-a	Good : ○(I≧1MΩ)
D-a	Degraded : ×(I≦1kΩ) after 997.6hrs
E-a	Degraded : ×(I≦1kΩ) after 810.8hrs

4 Conclusions

In this research, Steady-State Temperature Humidity Bias Life Test (THB Test) was conducted using COF as a sample to understand the ionic migration behaviour in FPC. Then, investigation concerning generated ionic migration behaviour was conducted employing SEM and EDX. In addition, some trials were conducted for establishing the evaluation method of ionic migration behaviour.

Results obtained are summarized as follows:

1. Migration behaviour such as color change that corresponded with dendrite formation was recognized in the specimen after 24 hrs THB testing, however, insulation failure of specimen was detected after 168 hrs THB testing.

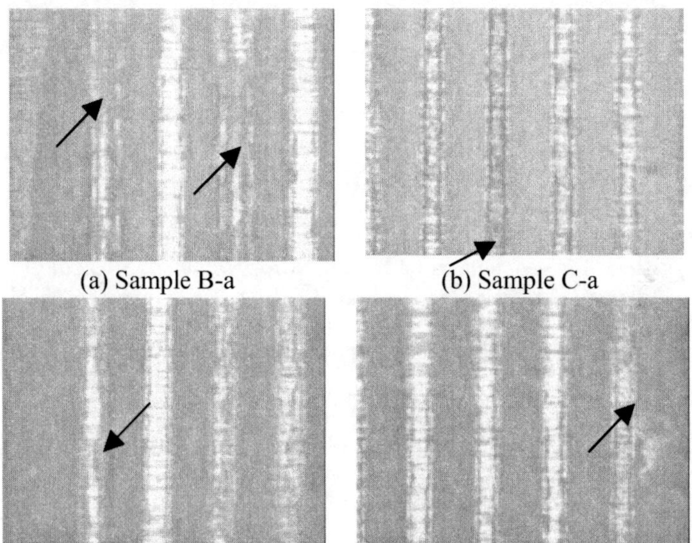

(a) Sample B-a (b) Sample C-a

(c) Sample D-a (d) Sample E-a

Figure 10: Outlook of specimen after additional 800 hrs (total 1,000 hrs) THB test.

2. Dendrite generated from wiring on FPC was composed of Cu, and this dendrite was initiated from the interface between plated Sn coating and polyimide substrate.
3. Cl included in epoxy resin used for under fill material is dominant factor affecting migration behaviour. Absorbed moisture in under fill resin, and the bias potential between anode and cathode electrodes, accelerated migration behaviour.
4. Generation of migration was initiated from some distant area due to Sn plating. The initiation site of dendrite was understood to be the Cr/Ni seed layer, which is an essential element for making specimens by the sputtering and plating process.
5. Results of THB testing till 1,000 hrs, employing 30 μm pitched tooth profile wiring specimen, showed that manufacturing method and types of under fill resin clearly affect migration behavior of FPC.

References

[1] G.T. Kohman, H.W. Hermance and G.H. Dowenes, "Silver Migration in Electrical Insulation" Bell System Tech. Journal, Vol.34, No.6, pp.1115-1147, 1955.

[2] S. Krumbein, "Metallic Electromigration Phenomena," IEEE Transactions on Components, Hybrids, and Manufacturing Technology, Vol.11, No.1, pp.5-15, 1988.

[3] Tsutomu Tsukui, "Insulation Deterioration and the Prevention Method by Electrochemical Migration of Electronic Equipment (Part 1)", The Journal of Japan Institute of Electronics Packaging, Vol.8, No.4, pp.339-345, 2005.

[4] Hirokazu Tanaka, "Factors leading to ionic migration in lead-free solder", ESPEC Technology Report No. 14, pp. 1-9, 2002.

[5] M. Pourbaix, "Atlas of Electrochemical Equilibria in Aqueous Solutions", NACE, 1966.

[6] J.A. Augis, D.G. DeNure, M.J. LuValle, J.P. Mitchell, M.R. Pinnel and Welsher, T.L. "A Humidity Threshold for Conductive Anodic Filaments in Epoxy Glass Printed Wiring Boards", 3rd International SAMPE Electronics Conference, pp. 1023-1030, 1989.

[7] T.L. Welsher, J.P. Mitchell and D.J. Lando, "CAF in Composite Printed-Circuit Substrates: Characterization, Modeling, and a Resistant Material," Reliability Physics, 18th Annual Proceeding, pp 235–237, 1980.

[8] J.P. Mitchell and T.L. Welsher, "Conductive Anodic Filament Growth in Printed Circuit Materials," Proceedings of the Printed Circuit World Convention II, pp. 80-93, 1981.

[9] G. W. Warren, P. Wynblatt and M. Zamanzadeh, "The role of electrochemical migration and moisture adsorption on the reliability of metallized ceramic substrates", *J. Electron. Mater.* Vol.18, No.2, 339-353, 1989.

[10] B. S. Rudra and M. G. Pecht, "Assessing Time-to-Failure Due to Conductive Filament Formation in Multi-Layer Organic Laminates," Packaging and Manufacturing Techniques, Part B, Vol. 17, No. 3, pp. 269-276, 1994.

Multilayer contacts in electrical connectors: experimental results and modelling

F. Ossart[1], S. Noel[1], D. Alamarguy[1], S. Correia[2] & P. Gendre[2]
[1]*Laboratoire de Génie Électrique de Paris, Supélec,*
Universités Paris VI et Paris XI, Gif sur Yvette, France
[2]*PEM, PEM Siaugues Saint Romain, Siaugues Sainte Marie, France*

Abstract

Electrical contacts are an essential part of electrical circuits and many reliability problems are related to their failure. The present work uses numerical simulation in view of a better analysis of the electromechanical phenomena, in the case of multilayer electrical contacts. We study a ball/plane contact made of bulk CuZn alloy, protected by a thin Sn surface layer. A coupled finite element analysis is performed in order to calculate the contact resistance of the device: an elasto-plastic model is used to determine the geometry of the contact area, then an electrical model gives the resulting constriction resistance. Results of the simulation are compared to experimental data. The respective contributions of the mechanical and electrical phenomena are analysed.
Keywords: multilayer contact, finite element modelling, constriction resistance, tin coating, electric contact.

1 Introduction

Electrical connector performances are closely linked to the mechanical behaviour of the contact. Much work has been devoted to understand the mechanisms involved and their complexity has been shown [1, 2]. Many contacts are made of cuprous alloys with various plating layers. Tin finishes are commonly used because they are cheap and can protect fairly well the copper substrates. This study is part of a larger one aimed at investigating and improving the properties of electroplated tin layers [3]. The contacts under study are of the ball plane type; their geometry simulates elementary contacts in connectors. They are submitted to a normal load and to the various failure tests representative of their conditions

WIT Transactions on Engineering Sciences, Vol 55, © 2007 WIT Press
www.witpress.com, ISSN 1743-3533 (on-line)
doi:10.2495/SECM070091

of use. Various platings are shown to display different properties; the influence of the tin thickness is particularly important. When brass (CuZn) is used for the substrate, a nickel underlayer is often deposited in order to prevent diffusion of zinc at the surface. Thus we have investigated the influence of such an underlayer on the contact behaviour.

Electrical contacts are characterised by their contact resistance. This global parameter takes into account both mechanical and electrical phenomena: its calculation requires a coupled analysis. We describe here the first results.

2 Experimental results

Samples are brass (Cu 70 Zn 30) coupons with a layer of tin electrodeposited from a bright or matte bath. The properties of the materials used for samples are described in table 1. Tin baths are based on MSA (methane sulfonic acid) chemistry. An underlayer of matte nickel was deposited on some of the coupons. Series of brass coupons were coated with tin layers of various thicknesses from 0.1 µm to 10 µm. Particular care was taken in order to have reproducible and constant values of the tin thickness. Micro-hardness measurements were performed with a Leitz Miniload 2 apparatus on the cuprous substrates either bare or with especially thick (50 µm to 100 µm) layers of the two types of tin electrodeposits. The bright finish tin layer has a higher value of hardness than the matte tin finish. The experimental yielding stress values σ_Y of table 1 are deduced from these microhardness measurements. A sphere of 1.3 mm of radius was stamped in some of the flats.

Several special devices (described in [4] for example) allowed the simulation of the degradation mechanisms leading to connector failure. The effect of mating and unmating connectors was simulated with a reciprocating wear test (constant speed 0.5 mm/s, amplitude 2mm; normal load 250 gf). The contact resistance *Rc* values were measured in static and then during the friction test, at the end of each wear track with a four wire method with +/- 20 mA in order to eliminate thermal fem. A dedicated device was used to study the evolution of the ball/plane contact resistance during a fretting test (small amplitude oscillatory movements simulating the effect of vibrations): with the following parameters: normal load 250 gf, amplitude 50 µm and frequency 1 Hz.

These tests allow the characterisation of the properties of platings for the application. Figure 1 shows the evolution of *Rc* (in static) for substrates coated with matte tin and bright tin layers of increasing thickness. It also shows the resistance values obtained with a nickel underlayer between CuZn and tin. *Rc* values lay between 0.8 mΩ and 2 mΩ. They decrease with increasing tin thickness. The nickel underlayer has no effect for bright coatings but causes the resistance values to increase for matte tin ones.

When the contacts are submitted to mechanical tests (wear or fretting) very different behaviours can be recorded for the various platings. An example is shown in Fig. 2 where we have plotted the number of cycles (of amplitude 50 µm) after which the electrical properties are degraded (when the contact resistance values reach 10 mΩ).

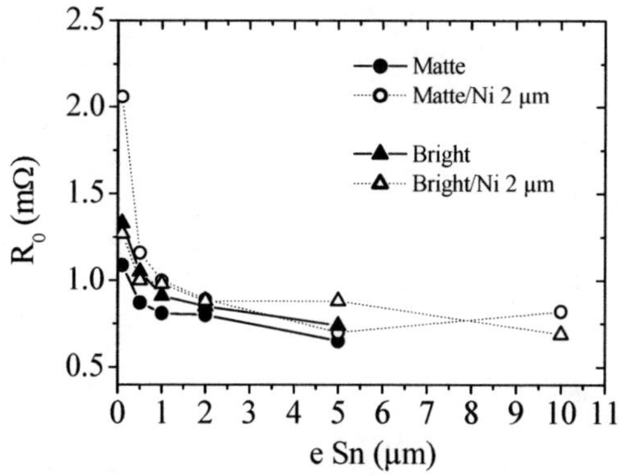

Figure 1: Experimental contact resistance values (F=250 gf) for bright and matte tin layers on CuZn substrate or CuZn with a nickel underlayer, for various tin coating thicknesses.

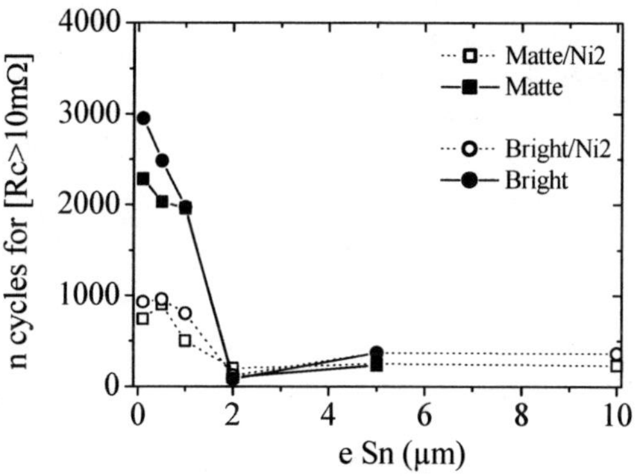

Figure 2: Number of “fretting” cycles (F=250 gf, a=+/-25 mm, f=1 Hz) at which Rc > 10 mΩ for different tin platings and different tin layer thicknesses.

3 Finite element model

The contact resistance which characterises electrical contacts is a global parameter resulting from both mechanical and electrical phenomena: hence, its calculation requires a coupled analysis. In the case of multilayered contacts submitted to plastic deformations, simple analytical models are not available and a numerical approach is needed. The finite element method is well suited to account for the geometry of the device and for the non-uniform mechanical and electrical properties of the layers. Only a weak coupled analysis is required: the contact resistance depends on the mechanical behaviour via the geometry of the contact area, but conversely the mechanical state of the device is not affected by its electrical behaviour. This corresponds to the case of low level electrical contacts where no thermal effects occur. Hence, we do a mechanical analysis, simply followed by an electrical one for a static contact. The commercially available finite element software ANSYS Multiphysics [5] was used.

3.1 Mechanical analysis

The ball/plane contact is an axi-symmetric problem and a 2d-model can be used. An elasto-plastic finite element analysis accounting for large deformations and contact is performed. The geometry, boundary conditions and loading are shown on Fig. 3. The contact between the upper and lower parts of the contact is managed by ANSYS, which determines which face-to-face elements come into contact for a given load. The radius of the contact area is computed for increasing loads up to 300 gf, and for Sn thickness between 0 and 10 μm. 6-node triangular elements are used and special care was taken to properly mesh the layers and the contact area. Due to the meshing, the contact radius increases discontinuously with F, whenever two new elements come into contact. Two assumptions are made. First we suppose that there is no friction between the two parts of the contact and secondly that there is perfect adhesion of the tin layer on the substrate. The material 'stress–strain' behaviours are represented by bilinear curves, characterised by the elastic Young modulus E for stresses below the yielding stress σ_Y, and a hardening coefficient $\gamma = d\sigma / d\varepsilon$ above. The bulk part of the contact is made of a CuZn alloy. We model two types of Sn-layers with different hardness values, referred to as "soft" and "hard" Sn hereafter. The influence of a 2 μm-thick Ni underlayer was also studied.

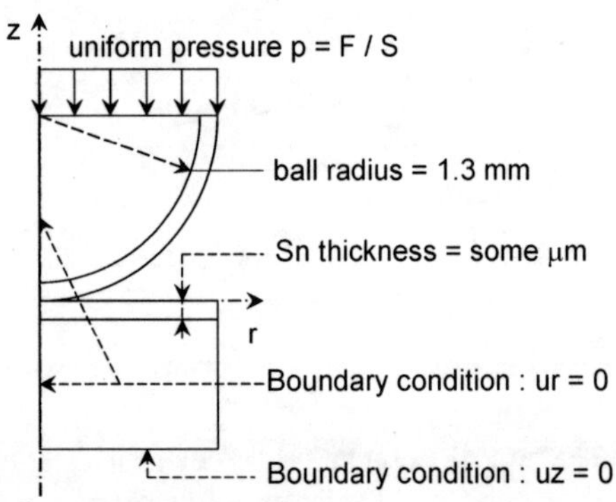

Figure 3: Geometry and boundary conditions of the 2d-axisymmetric model.

The characteristics of the materials are summarised in table 1.

In a first stage, the validity of the model is checked, by comparing with the Hertz model which provides an analytical solution for perfect contacts (no friction and homogeneous elastic materials). Formula (1) gives the Hertz contact radius a. Parameters R, F, E and ν are respectively for the ball radius, the applied force, the Young modulus and the Poisson coefficient of the material [7].

Table 1: Mechanical and electrical characteristics of the materials used. (a) manufacturer data, (b) measured data, (c) Hammam [6].

	E (GPa)	σ_Y (MPa)	γ (GPa)	ν	ρ (Ωm)
CuZn	$114^{(a)}$	$370^{(b)}$	$0.55^{(a)}$	$0.34^{(a)}$	$6.2.10^{-8(a)}$
Sn matte	50	$25^{(b)}$	0.55	0.34	$26.5.10^{(c)}$
Sn bright	50	$70^{(b)}$	0.55	0.34	$26.5.10^{-8(c)}$
Ni	220	$> 600^{(b)}$	0.55	0.34	$6.9.10^{-8(c)}$

$$a = \sqrt[3]{\frac{3}{4} \cdot \frac{FR}{E^*}} \qquad \text{with } E^* = \frac{E}{2(1-\nu^2)} \tag{1}$$

The finite element model should give the same results as Hertz model at low forces (elastic regime). But because the finite element model accounts for plasticity, it will give larger contact area after some critical load has been reached (elasto-plastic regime).

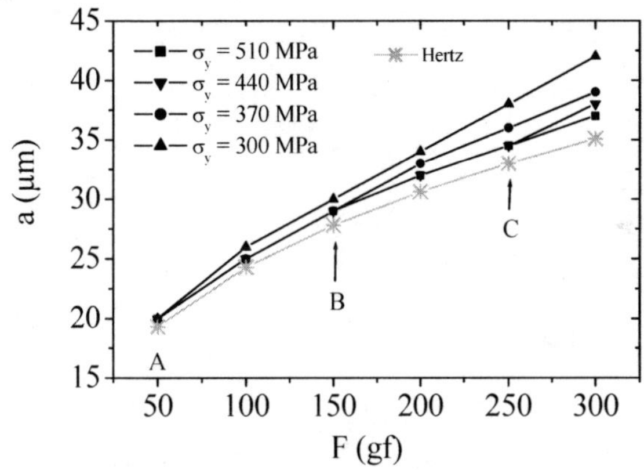

Figure 4: Contact radius a (µm) vs. applied force F (gf), for different values of σ_Y: 510 MPa, 440 MPa, 370 MPa (exp. value) and 300 MPa.

Different yielding stresses σ_Y corresponding to various usual cuprous substrates are tested, so as to see how the critical force between the elastic and the elasto-plastic regime depends on this parameter (Fig. 4). For σ_Y =510 MPa, the global behaviour remains elastic up to 3 N. We note that the finite element model overestimates the contact radius. This systematic error (around 1 µm) is due to the mesh. Using smaller elements allows this error to diminish, but the computational cost is high and no significant changes of the global curves are observed. For lower values of σ_Y, the elastic regime ends for lower value of F (point A : F=50 gf for σ_Y=300 MPa, point B : F=150 gf for σ_Y=370 MPa and point C : F=250 gf for σ_Y=300 MPa). We obtain critical values of F which are consistent with the 'rule of the thumb' given by formula (2).

$$3F_y \leq F_{critical} \leq 5F_y \qquad \text{where } F_y = 21\frac{R^2 \sigma_Y{}^3}{E^{*2}} \tag{2}$$

The next step is to study the influence of the Sn-layer on the contact area. Figure 5 shows the contact radius as a function of the Sn-thickness, for applied forces ranging between 50 gf and 300 gf, in the case of hard and soft Sn. The curves give expected continuous trends. The contact radius increases with the load. It also increases with the Sn thickness because this material is softer than the CuZn substrate. Larger contact areas are obtained with the softer Sn layer. An example of deformed Sn layer is shown on Fig. 6 (soft Sn, e=2μm).

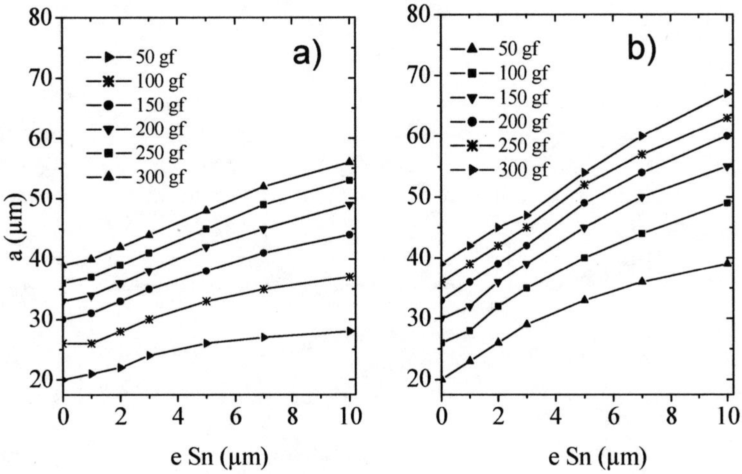

Figure 5: Contact radius a, versus Sn thickness e, for different applied forces. a): hard Sn (σ_Y=70 MPa) – b): soft Sn (σ_Y=25 MPa).

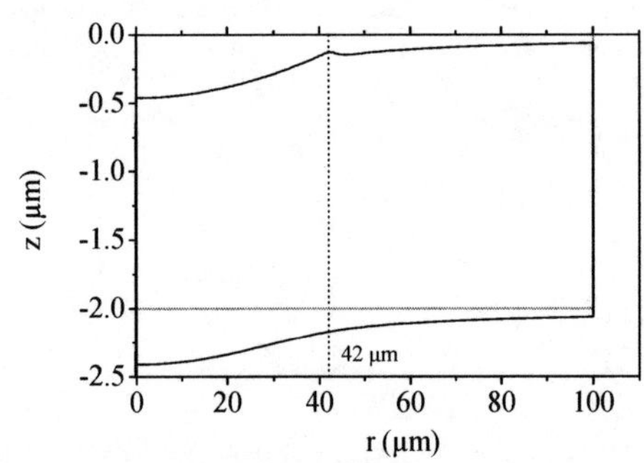

Figure 6: Geometry of the bottom Sn layer in the contact area (2 μm).

The contact area spreads between the vertical axis and the dashed line (*r*=42 μm). It is worth noticing that the geometry of the layer does not change significantly. There is a slight bending and, although the tin is very soft, its thickness remains the same. This behaviour is due to the perfect adhesion of the Sn film on the CuZn substrate, which prevents the radial displacement of Sn caused by the contact pressure.

3.2 Electrical analysis:

The electrical model allows the global resistance of the contact to be calculated, for given geometry and electrical properties. Figure 7 shows the current lines in the contact area, when a difference of potential ΔV is applied between the top and bottom surfaces of the contact. Let us denote I the resulting intensity. The contact resistance R_c is defined as $R_c = \Delta V / I$. The small section of the contact area leads to a strong constriction of the current in this region and a high local resistance.

When two infinite electrodes touch at a single circular spot of radius a, the formula for the contact resistance is $R_C = \rho / 2a$ [1, 7]. The contact radius is the only geometric parameter, because constriction is the dominant mechanism in the contact.

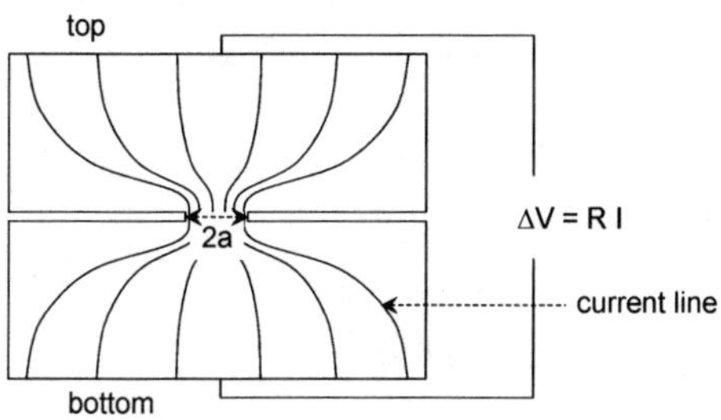

Figure 7: Constriction of the current lines in the contact area.

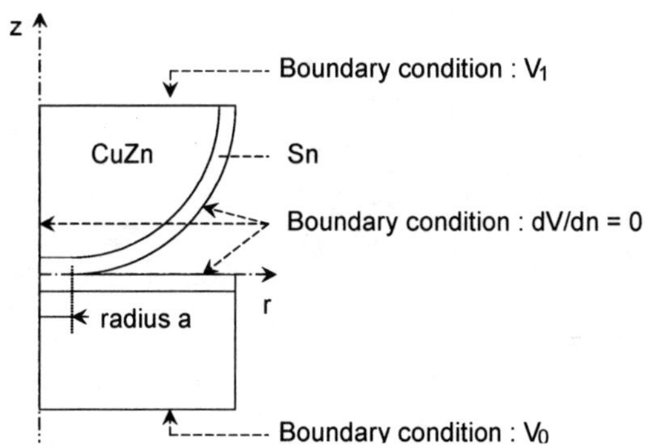

Figure 8: Geometry and boundary conditions of the finite element model.

In the case of a multilayer contact $R_C = \rho / 2a$ does not hold and a numerical analysis is needed. Again, we use the finite element method to deal with the non-uniform properties of multilayer contacts. As we have seen, the results of the mechanical analysis indicate that the change of geometry of the Sn layer under the load is very small. Hence, we neglect it. We assume a perfect electrical contact between both sides of the contact. The radius of the contact area is the result of the mechanical analysis. Figure 8 summarises the problem solved.

A difference of electric potential is applied between the top and the bottom of the contact, and the resulting current distribution is calculated. The total current I is obtained by integration of the current density, and the global resistance of the contact Rc is given by: $R_c = (V_1 - V_0)/I$.

Rc was calculated as a function of the Sn-layer thicknesses for all the applied forces. Since the value of the resistivity of thin tin layers found in literature could vary from 6 to 26 10^{-8} Ω.m, calculations were performed for these various values. Results are plotted for hard tin Fig. 9 a) and for soft tin Fig. 9 b). These figures also display the calculation of $R_c = \rho / 2a$ (with ρ the resistivity of the substrate CuZn and a the contact radius calculated from the mechanical finite element FEM model) and the experimental values measured for the bright (hard) and matte (soft) tin layers on the CuZn substrate.

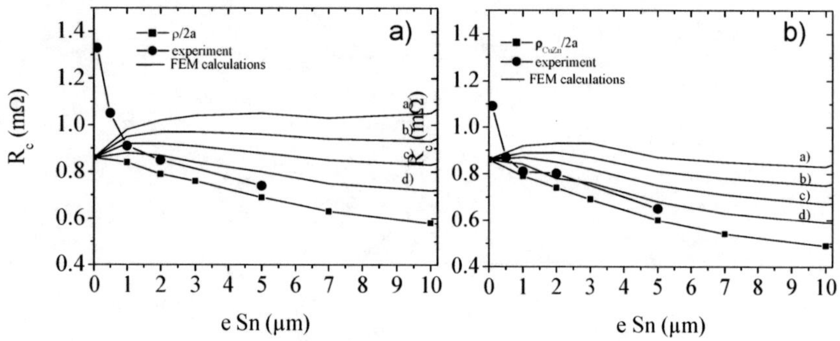

Figure 9: Contact resistance vs. Sn layer thickness, for different values of ρ [a) 26.5.10^{-8} Ωm – b) 21.10^{-8} Ωm – c) 16.10^{-8} Ωm – d) 11.10^{-8} Ωm], compared to ρ/2a and to experimental values. a) hard tin (calculation) and bright tin (experimental) b) soft tin (calculation) and matte tin (experimental).

4 Discussion and conclusion

The values calculated from the electro-mechanical model are compared to those measured experimentally. Several observations can be made. The calculated values have the same order of magnitude as the experimental ones. For thicknesses comprised between 1 and 10 µm a fairly good agreement is found between the experiment and the model when taking low values of resistivity for the tin layer. The agreement between the experiment and the result of the calculations is better for the soft and matte tin layers. For very thin tin layers (0.1 µm to 1 µm) experimental values are much higher than the calculated ones. The calculations were done for no tin layer (e=0). No experimental value could be measured for this case since CuZn without any tin layer is prone to severe oxidation and thus Rc values measured would not correspond to the bare CuZn case.

No calculation was done for the very thin layers (0.1 to 1 µm) but the calculated variations must be continuous. The measured high values of Rc are thus difficult to account for. Several questions can be raised. We have supposed that the resistivity of the Sn layers was constant for a given finish, whatever the film thickness. Surface analysis has shown the presence of an oxide film on the Sn surface. The ratio of the tin oxide thickness over the tin total thickness can have an effect on the resistance value. The assumption of perfect adhesion of the tin layer on the substrate could also be too simple. In any case it seems that the geometric effect (increase of the contact surface) is stronger than the electrical effect of having a less conducting layer in the interface.

Experimental data show that a nickel underlayer changes the contact behaviour (Fig. 1). To understand the mechanism involved, the influence of such an underlayer was simulated. Results not reported here show that, despite a high yielding stress, the nickel has no visible influence on the contact area. A closer look at some results obtained without the nickel layer explains why this lack of effect is consistent with the model used. Figure 10 displays the vertical plastic deformation ε_{pzz} (compression) along the z-axis, for increasing values of the applied force, in the case of a 2 µm thick Sn-layer.

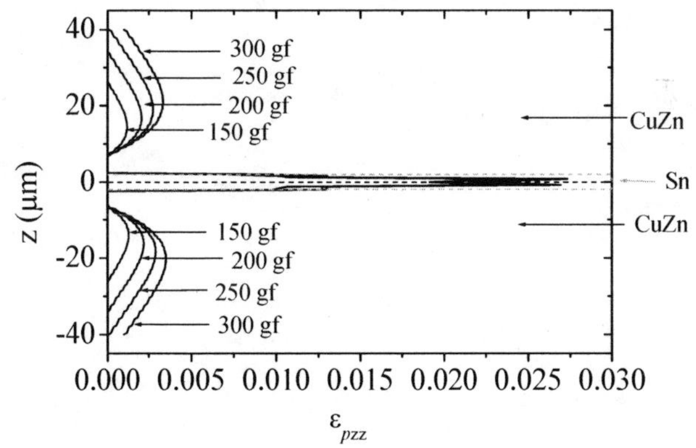

Figure 10: Compression plastic deformation ε_{pzz} along the symmetry axis (Sn thickness = 2µm).

The plastic deformation appears first in the Sn layer and later in the substrate, about 5 µm below the interface. For the considered range of applied force, there is no plastic deformation just under the Sn-layer, and thus adding a material with a high yielding stress at this place has no mechanical influence. The Ni resistivity being about the same as the CuZn one, the presence of this material has also no electrical influence.

These results lead us to question an important point of the mechanical model: the adhesion of the Sn layer on the CuZn substrate was assumed to be perfect, but since strong shearing was calculated to occur at the interface, this assumption might not be satisfied. If the model includes a possible relative sliding at the interface, the mechanical behaviour of the contact changes, leading to larger contact area, thinner Sn layer after deformation, and hence lower contact resistance by geometric effect. A nickel underlayer is bound to strongly affect the behaviour of the interface, and this may be the mechanism which explains its influence on the contact. Calculations have been started to explore this explanation, but results are not available yet.

References

[1] Slade, P., *Electrical contacts: principle and applications*, M. Dekker, 1999.

[2] Tangena, A.G., The correlation between mechanical stresses and wear in a layered system, *Wear*, **121,** pp. 27-35, 1988.

[3] Noël, S., Lécaudé, N., Correia, S., Gendre, P., Grosjean, A., Electrical and tribological properties of tin plated copper alloy for electrical contact in relation to intermetallic growth, *Proc. 52ième IEEE Holm Conference,* Montreal, Canada, pp.274-280, pp. 1-10, Sept. 2006.

[4] Noël, S., Lecaudé, N., Alamarguy, D., Boyer, L., Friction properties of perfluorinated polyethers for hot-dipped tin low level separable electrical contacts, *Synthetic Lubrication*, (**18**), pp. 179-189, oct.2002.

[5] Ansys website : www.ansys.com

[6] Hammam, T., Friction, wear and electric properties of tin-coated tin bronze for separable electric connectors, *Proc 42nd IEEE Holm Conf. On Elec. Contacts,* Chicago, USA, pp. 321-330, 1996.

[7] Holm, R., *Electric contacts theory and application*, Berlin, Germany: Springer-Verlag, 1976.

Section 3
Surface problems in contact mechanics

Contact problems between optical lenses and shrink fitter for a new type of laser microscope with a wide field of view

I. Nitta[1] & A. Kanno[2]
[1]*Department of Mechanical and Production Engineering, Niigata University, Japan*
[2]*Sanjo Tsubame area Research Core, Japan*

Abstract

We have studied contact stiffness between randomly rough surfaces. This research led to the development of a shrink fitter consisting of a new ring-shaped machine element to join mechanical components with different coefficients of thermal expansion, such as a ceramic shaft and a metal cylinder. As one application of the shrink fitter, several optical lenses were shrink-fitted in a lens-barrel. Such a scanning lens, assembled using the shrink fitter, can focus laser light well over a wide scanning width because the optical axes of the several lenses in the lens-barrel compliment each other well regardless of changes in room temperature. Thus, a laser microscope with such a scanning lens can observe object surfaces over a relatively wide area. However, contact pressures between the shrink fitter and the optical lenses will change the curvature of the lenses resulting in deterioration of the scanning properties of the scanning lens. Therefore, it is very important to calculate the contact pressures acting on optical lenses to examine the scanning properties of the scanning lens. In this study, we have developed a new type of laser microscope with a field of view of 10 × 8 mm. The pixel number in the laser scanning direction of 10 mm was 20,000 and 16,000 in the perpendicular direction of 8 mm. Thus, one field of view of this laser microscope had 320,000,000 pixels. The observation results of some surfaces by this laser microscope are reported.

Keywords: contact pressure, shrink fitter, laser microscope, field of view, lens.

WIT Transactions on Engineering Sciences, Vol 55, © 2007 WIT Press
www.witpress.com, ISSN 1743-3533 (on-line)
doi:10.2495/SECM070101

1 Introduction

An optical or scanning electron microscope is usually used to observe object surfaces. However, only a very small fraction of the surface can be observed at a time because of limitations of the field of view of such microscopes, Nitta et al [1]. To obtain an image of the whole object surface to be investigated, large numbers of pictures of adjacent positions must be photographed using a CCD camera through the microscope with the specimen moved in a step-by-step feed operation. These pictures are combined with each other to make an image of the whole object surface. Thus, such observations have hardly been performed. However, observation of the whole surface will provide better information than that of only part of the surface. We have been engaged in attempts to make a new type of laser microscope with a resolution almost as high as that of optical microscopes and with a much wider field of view.

Figure 1 shows a schematic diagram of a new type of laser microscope, the key technology in which is a scanning lens unit, fθ lens. It is necessary to converge a laser beam bundle on the object surface, *i.e.*, the focal plane, to a few micrometers over a wide scanning width. Such a fine laser spot can be focused easily on the object surface only at the centre of the scanning width, but is difficult at the ends. The precise location of each scanning lens relative to the housing can adversely affect the size and shape of the laser spot on the focal plane, especially near the ends of the scanning width. The optical axes of each lens must be coincident with each other and should not be moved even with changes in room temperature, Yoder [2].

Here, we report that a shrink fitter, a new machine element developed by one of the authors, Nitta et al [3-9], greatly improves performance of the fθ scanning lens. The fitting pressure between the optical lenses and the metallic housing decreases with temperature because their coefficients of thermal expansion differ from each other. However, the shrink fitter could maintain the fitting pressure of the shrinkage fit of such a combination despite changes in room temperature.

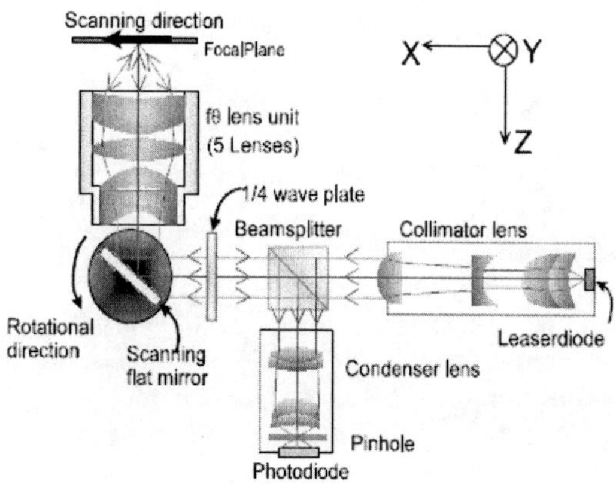

Figure 1: Schematic diagram of the new type of laser microscope.

With this shrink fitter technology, we have developed a new type of laser microscope with a field of view of 10 × 8 mm. The pixel number in the laser scanning direction of 10 mm is 20,000 and that in the perpendicular direction of 8 mm is 16,000. The observation results of some surfaces using this laser microscope are described.

2 Shrink fitter

The shrink fitter is a new machine element of cylindrical geometry. When two machine elements with different thermal expansion coefficients are shrink-fitted, the fitting pressure will vary with temperature. However, it is possible to keep the fitting pressure constant using the shrink fitter, even if the circumferential temperature changes. The shrink fitter has been applied to the connection of polygon mirror and self-acting air bearing, Nitta et al [7]. In addition, an assembly of an all-ceramic bearing and a metal housing has also been examined to improve the performance of the ceramic bearing using the shrink fitter. In the present study, the shrink fitter was applied to accurately locate the lenses in the housing of a new type of laser microscope with a wide field of view. The cylindrical shrink fitter is made of a plastic material, the Young's modulus of which is lower than that of the metal by about two orders of magnitude. This means there will be a larger degree of interference for the shrinkage fit of such a combination. If the room temperature increases, the housing will expand more than the lenses. The interference of the shrinkage fit, if the shrink fitter is not used, will decrease at elevated temperatures. However, the coefficient of thermal expansion of the plastic material is larger than that of the metal. Thus, if the thickness of the shrink fitter is designed appropriately, the interference will not change even at elevated temperature. Consequently, the contact pressure acting on the lens rim due to the shrinkage fit will be kept constant regardless of changes in temperature.

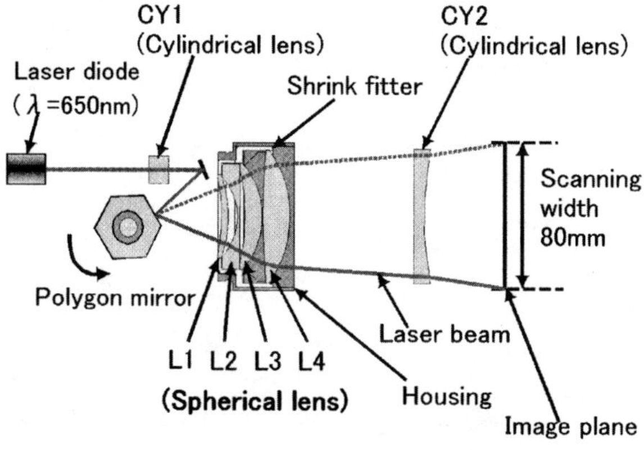

Figure 2: Laser scanner for description of lens deformations.

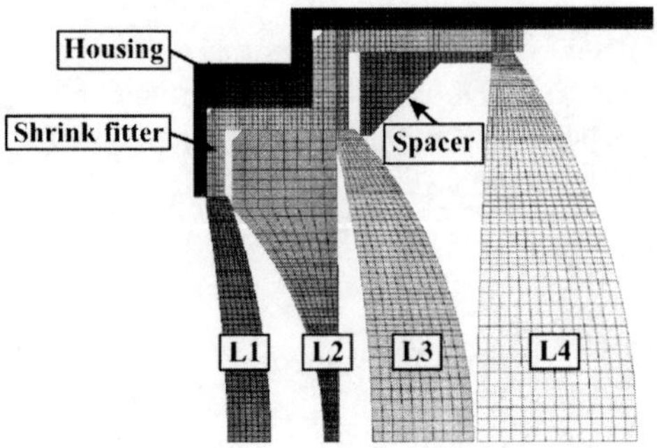

Figure 3: FEM mesh of the assembly.

3 Shrink fitter method for a new laser microscope: procedure for critical interference

In the shrink fit method, excessive interference will cause large-scale deformation of fitted lenses, resulting in deterioration of the focusing performance, Nitta et al [3]. Therefore, the critical interference under which the focusing performance does not become worse must be determined for appropriate use of the shrink fit method. A procedure to obtain the critical interferences is described below along with the laser scanner system shown in fig. 2. The fθ lens shown in fig. 2 consists of four lenses, named L1, L2, L3 and L4, respectively. These four lenses were shrink-fitted into an aluminium housing by the shrink fitter. A laser beam with a wavelength of 650 nm is irradiated from a laser diode. The fθ lens was designed so that the laser spot may be kept at a constant diameter of 17 μm over a wide scanning width of 80 mm on the image plane. Figure 3 shows two-dimensional FEM meshes of the fθ lens assembly. The mechanical properties of each component are shown in table 1. Deformations of each lens corresponding to various interferences were calculated in an axisymmetric analysis using commercial FEM software, MARC, developed by MSC Software Corporation. The calculated lens deformations of both the entrance plane of L1 and the exit plane of L4 are shown together with the measured values in fig. 4. The calculated deformations were in excellent agreement with the measured values. The laser beam diameters were then simulated using commercial optical design software, CodeV, developed by Optical Research Associates, taking all lens deformations into consideration. Table 2 shows the critical interference of each lens under which the laser spot diameter on the image plane does not become larger than the designed value of 17 μm. The allowable interference is 60 μm. Thus, we can assemble the fθ lens by the shrink fitter method without reducing the focusing performance if the interference can be kept below the critical value of 60 μm. In measurement of the sizes of the laser beam spots on the image plane shown in fig. 2, we confirmed that the laser beam bundles could be kept constant at about 17 μm over the wide scanning width of 80mm when the interferences were below 60 μm.

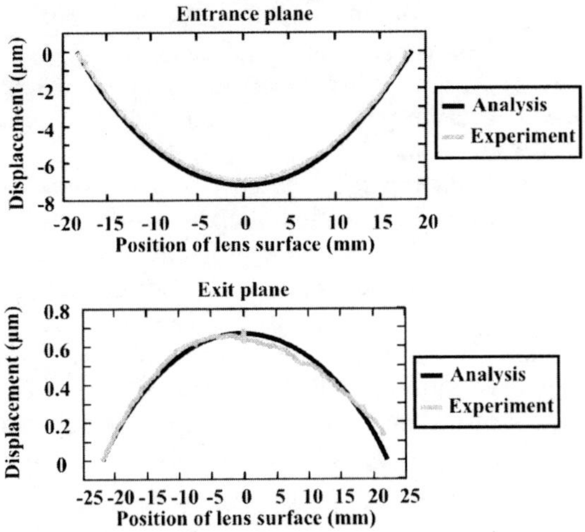

Figure 4: Profile of the deformed lens surface.

Table 1: Mechanical properties of each element.

Elements	Materials	Young's Modulus [GPa]	Poisson's Ratio
L1	Glass	83.2	0.25
L2	Glass	55.2	0.23
L3	Glass	111.8	0.28
L4	Glass	106.1	0.28
Shrink fitter	POM	2.8	0.47
Housing	Al(A5056B)	69.0	0.34

Table 2: Limited interference of each lens simulated by FEM and CODEV [μm].

Lens No.	L1	L2	L3	L4
Allowable value	64	67	60	521

4 Outline of the developed laser microscope

Figure 1 shows a schematic diagram of the new type of laser microscope developed in this study. First, a laser beam bundle emitted by a laser diode was made as parallel as possible through a collimating lens. In addition, a collimated laser beam, which was linearly polarised, was transformed into a circularly polarised beam by a quarter wave plate. The laser beam was scanned by a rotating flat mirror at a rotational speed of 9,000 rpm and passed through the fθ lens unit to focus the laser beam bundle on the focal plane at a laser spot diameter of about 3 μm over a scanning width of 10 mm. Then, *via* a polarising beam splitter and the quarter wave plate, the outgoing laser beam and the reflected laser beam became polarised at right angles to each other, which allowed good separation at the polarising beam splitter. Finally, the reflected laser beam passed to a photo-detector through a pinhole. The intensity of the reflected laser beam was transformed into digital data by a 12-bit A/D converter at a conversion rate of 100 MHz. The rotating flat mirror scanned the laser beam in the horizontal direction and the specimen attached to the motor-driven stage could be moved in the vertical direction at a constant speed controlled by a microcomputer. Thus, the surface of the specimen being observed could be scanned by the fine laser beam and the image of the surface be made by arranging the signals of the reflected laser light in the horizontal and vertical directions.

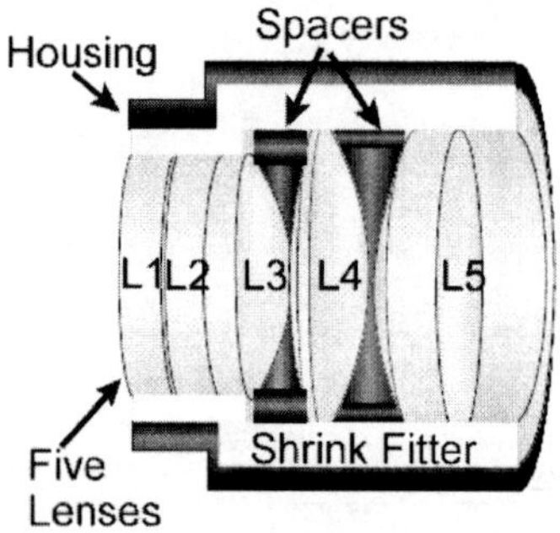

Figure 5: fθ lens for the laser microscope.

Figure 6: Developed laser microscope.

This fθ lens unit has a telecentric property such that the outgoing laser beams from the fθ lens to the specimen to be inspected are parallel to the optical axis of the fθ lens unit. Telecentric lenses yield constant magnification over a range of working distances, virtually eliminating viewing angle error. With telecentric lenses, the image size remains almost unchanged when the object distance changes, provided the object to be inspected stays within the given field depth/telecentric range. In addition, the fθ lens unit must be designed such that the image height is proportional to the scan angle (Theta), not the tangent of that angle: converting the equiangular motion of the laser beams to the constant speed motion for the scanning operation.

Figure 5 shows the fθ lens unit used in this study, consisting of 5 lenses: named L1–L5 from the left side. The diameters of the lenses are 25 mm and 28 mm. Figure 6 shows a photograph of the new type of laser microscope.

5 Basic performance of the laser microscope

The laser spot diameters on the focal plane were simulated using the optical design software, Code V. The fθ lens was designed so that the laser spot diameters could be kept at almost a constant value of 2 μm over the whole scanning width. Figure 7 shows the laser spot diameters as a function of scanning position over the scanning width of 10 mm, measured using a SpotScan model 0390, Photon Inc. The laser spot sizes at several scanning positions were measured in two different directions: one was in the scanning direction by the rotating flat mirror, main scanning direction, and the other was in the orthogonal direction, sub- (or vice-) scanning direction, as the laser spot emitted from the laser diode is not perfectly circular but slightly elliptical although the laser beam from the laser diode is reshaped into an almost circular spot through two cylindrical lenses before the collimator lens. The laser spots measured at the middle of the scanning width in the main scanning direction were about 2.7 μm in diameter. The laser spots became larger as the laser beam was scanned toward both the scanning ends. If the diameters of the laser spot in two different directions are compared, the spot diameter in the main scanning direction is smaller than that in the sub-scanning direction.

The resolution of the laser microscope was evaluated using a USAF resolution target, consisting of bars organised into groups and elements, Edmund Optics [10]. A positive target has only the pattern chrome deposited on the glass substrate, so it had a black pattern with a clear field. Each group consists of six elements (*i.e.*, elements 1–6), each of which is composed of three horizontal and three vertical equally spaced bars. Each element within a group corresponds to an associated resolution, based on the bar width/space, in the range from 0.8 μm to 500 μm. The group and element define the resolution of an imaging system just before the black and white bars begin to blend together. The vertical bars are used to calculate horizontal resolution and the horizontal bars are used to calculate vertical resolution. One line pair equals one black and one white bar.

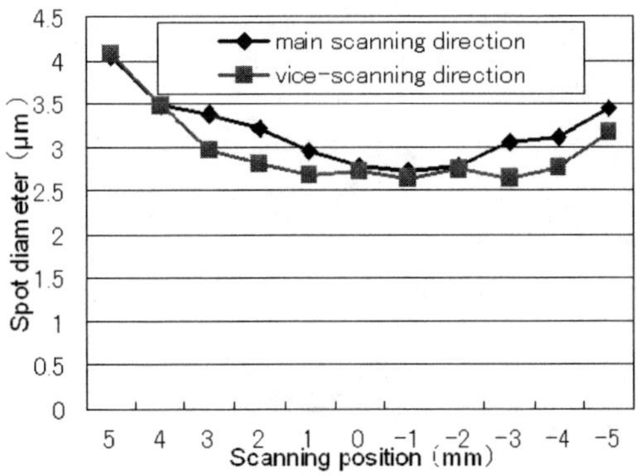

Figure 7: Measured laser spot sizes.

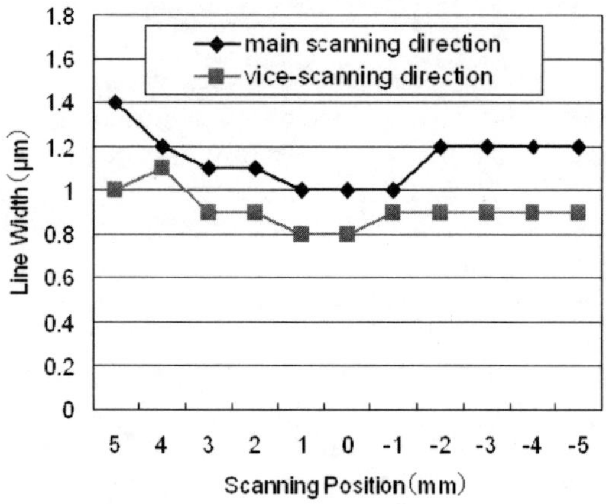

Figure 8: Measured resolution, MTF.

Figure 8 shows the line widths of the resolution targets that can be resolved by the laser microscope as a function of the scanning position. The resolution in the main scanning direction was about 1 μm around the middle of the scanning width. It decreased towards both the scanning ends corresponding to the tendency of the laser spot sizes. The resolutions in the sub-scanning direction were slightly better than those in the main scanning direction corresponding to the laser spot sizes on the focal plane shown in fig. 7.

On the other hand, the same resolution target was also observed with an ordinary optical microscope. Using an objective with a magnification of 10 the resolution was about 1.0 μm. Thus, the magnification of the laser microscope was about 10 by conversion of the objective of the optical microscope.

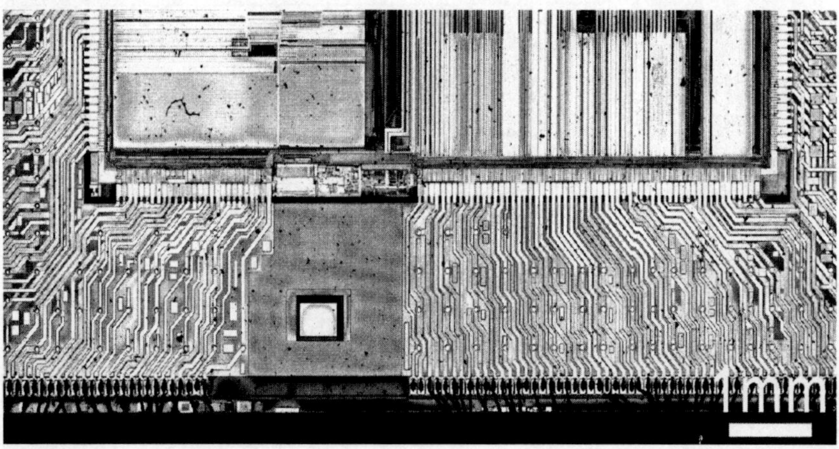

Figure 9: Image of a CPU observed using the developed laser microscope.

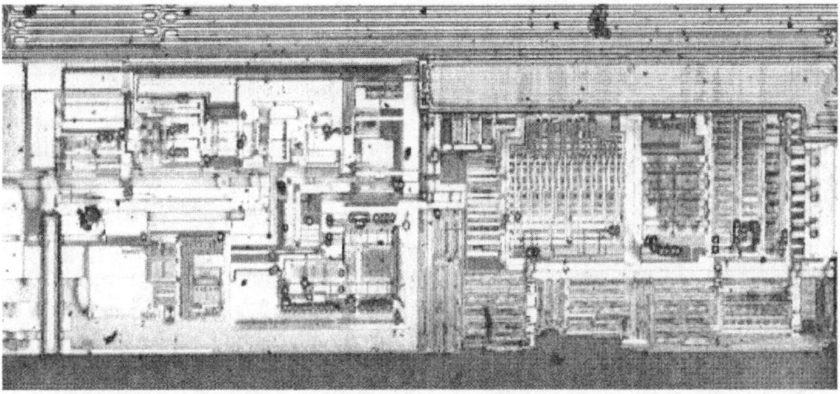

Figure 10: Detailed image of the white rectangular portion in fig. 9.

6 Images observed with the laser microscope

Figure 9 shows an image of a CPU observed with the laser microscope. The cover of the CPU package for a personal computer was removed and set on the stage of the laser microscope to observe its circuit pattern. The measured area was 10 mm in width and 4 mm in height. The scanning width of 10 mm is a limitation of the fθ lens unit. However, the vertical scanning width of 4 mm is not a limitation of the laser microscope but is dependent on the limitation of the mechanical stage used in the laser microscope. Thus, longer surfaces than this can also be observed.

It is obvious that a huge area can be observed at a time, compared with the conventional optical microscope. This laser microscope also has a deep depth-of-focus so that the image was very clear despite the rough surface.

The pixel number in the laser scanning direction of 10 mm was 20,000, and thus the distance from pixel to pixel is 0.5 μm. Figure 9 has a total of 320,000,000 pixels. Thus, any portion in fig. 9 can be seen at higher magnification. Figure 10 shows a portion of fig. 9 surrounded by the white rectangle on a finer scale; very fine patterns that could not be seen in fig. 9 were visible.

7 Conclusions

The shrink fitter technology has made it possible to shrink fit optical lenses into the housing without serious reduction of the focusing performance. The critical interferences under which the focusing performance does not become worse were obtained. The procedure to obtain the critical interferences was described. With the shrink fitter technology, a new type of laser microscope with a huge field of view was developed. Several images obtained using this laser microscope were presented.

References

[1] Nitta, I., Kanno, A., Komata,K. & Iguchi, S., "New joining method for laser scanner lenses by using a shrink fitter", Proc. 5th int. conf. on computational methods in contact mechanics, ed. Dominguez & Brebbia C.A., WIT press, pp.31-40, 2001.

[2] Yoder, P.R., "Opto-mechanical systems design , third edition", CRC Press, pp.202-204, 2005.

[3] Nitta, I., Kanno, A. & Komata,K., "Effect of Interference on Scanning Performance of fθ Lens Fixed by a Shrink Fitter", Optical review, Vol.10, No.4, pp.321-324, 2003.

[4] Nitta, I., Kigoshi, K., & Kato, K., "Study of the fitting strength between ceramic and metal elements with the use of a shrink fitter at elevated temperature", JSME international Journal, Series III, 32, pp.632-639, 1989,.

[5] Nitta, I., Nakashizuka, K. & Hara, T., "The fitting strength between ceramic and metal with the use of a bimetal shrink fitter at elevated temperature", JSME international Journal, Series I, 34, pp.249-256, 1991.

[6] Nitta, I., Kusama, K. & Hara, T., " Shrink fit between a ceramic and a metal element using a hybrid shrink fitter, JSME international Journal, Series C, 38, pp.617-624, 1995.

[7] Nitta, I., Furukawa, H., Komata, K. & Konno, D., "New method of joining a polygon mirror using a shrink fitter (in Japanese), Trans. Jpn. Soc. Mech. Eng. 62, pp.2785-2791, 1996.

[8] Nitta, I., Kanno, A., Komata,K. & Iguchi, S., "New joining method for laser scanner lenses by using a shrink fitter", Proc. 5th int. conf. on computational methods in contact mechanics, ed. Dominguez & Brebbia C.A., WIT press, pp.31-40, 2001.

[9] Nonaka, S., Nitta, I., Kanno, A. & Nishimura, M., "Study of a laser material processing system with fine optical setup using a shrink fitter", Proc. of the Int. Conf. Leading Edge Manufacturing in 21st Century, pp.873-877, 2003.

[10] EdmundOptics, http://www.edmundoptics.com/techSupport/Display Article.cfm/articleid= 248.

C-sphere strength as an indicator of rolling contact performance of silicon nitride

W. Wang[1], A. A. Wereszczak[2] & M. Hadfield[1]
[1]*School of Design, Engineering and Computing, Bournemouth University, Poole, UK*
[2]*Material Science and Technology Division, Oak Ridge National Laboratory, Oak Ridge, Tennessee, USA*

Abstract

Silicon nitride material has been used as a bearing material due to its superior performance against bearing steel. Its successful application as a bearing element leads to the development of other rolling contact applications in the automotive industry, especially the engine manufacturing industry. Its excellent rolling contact performance can make significant savings on warranty costs for engine manufactures. However, the remaining difficulty for the broader application is the high component machining cost. Further understanding of the rolling contact performance of silicon nitride material in relation to its surface integrity will enable engine manufactures to produce components that meet the design requirements while at the same time reduce the machining cost. In the present study, the relationship between the C-sphere strengths of silicon nitride specimens and their rolling contact fatigue life is investigated. The C-sphere test is designed to compare the strengths of three batches of Sintered and Reaction-Bonded Silicon Nitride (SRBSN) specimens with different subsurface quality induced by varying the machining parameters. The rolling contact fatigue (RCF) performance of three batches of SRBSN ball specimens are studied on a modified four ball tester. The results show that the most aggressively machined specimens have the weakest C-sphere strength and the shortest RCF life. This positive relationship can give component manufactures a valuable reference when they make selections of candidate material and finishing standards.
Keywords: ceramics, silicon nitride, flexure strength, rolling contact.

WIT Transactions on Engineering Sciences, Vol 55, © 2007 WIT Press
www.witpress.com, ISSN 1743-3533 (on-line)
doi:10.2495/SECM070111

1 Introduction

Silicon nitride (Si_3N_4) has been used in rolling contact applications in various industries such as turbomachinery, power and automotive industries. Compared with steel traditionally used in these applications, it has significant advantages due to its low density, low friction, corrosion resistance and excellent performance under extreme conditions. However, a major limitation of its wider application is its high material and machining cost, especially the cost associated with the component finishing stage. The high material cost is partly due to the high cost of raw silicon nitride powder, and another contributor is the high energy consumed and the demanding environment involved in the sintering process of high strength Si_3N_4, such as Hot Isostatically Pressed (HIPed) Si_3N_4 [1]. For different types of components with different degrees of machining complicacy, the machining cost can contribute to 55%-70% of the overall component cost [2]. For a Si_3N_4 bearing element, the cost of surface finishing contributes to around 70% of the overall cost. A Sintered and Reaction Bonded Silicon Nitride (SRBSN) has been developed to reduce the material cost without significant compromise on strength. It uses silicon power as the raw material instead of Si_3N_4 power, and the silicon power is nitrodized through reaction bonding. High density Si_3N_4 is achieved by sintering process after reaction bonding, however, the sintering temperature and pressure is lower than that of HIPed Si_3N_4. In order to materialise the benefit of low cost SRBSN, better understanding of the relationship between the finishing process and the contact reliability is necessary to optimise the machining process and reduce the cost in this stage.

The machining effect on the surface quality of SRBSN has been studied by researchers [3, 4]. It was found that by changing the grit size of diamond dressed on the grinding wheel, different depth of subsurface damage can be generated on SRBSN rod and bar specimen. It is reported that the size of micro-cracks on the rod specimen generated in grinding process varies from 10-50 microns from 1200 grit to 180 grit diamond dressed grinding wheel. It is important to quantify the surface integrity of SRBSN in relation to its surface strength, and linking the surface strength with rolling contact fatigue performance of the specimen. In the present study, the machining effect on the surface strength and rolling contact fatigue performance of a Ceralloy 147-31N (Ceradyne Inc, United States) SRBSN ball is studied. The surface strength is measured by the compressive C-sphere flexure strength.

2 Ceralloy 147-31N SRBSN and specimen machining

The micrographs of a Ceralloy 147-31N Si_3N_4 specimen are shown in Figure 1. The microstructure is polished using diamond suspension, and the size of the diamond paste used in the suspension is gradually reduced from 15 micron to ¼ micron. The polished surface is plasma etched using a mixture of carbon fluoride and oxygen for 6.5 minutes before it is coated for SEM examination. The darker grey phase in the microstructure is β-Si_3N_4, which is the major constituent phase in the microstructure. The lighter grey phase, which can be found between the grains boundary is a glassy phase, which is formed by sintering aides (rare earth oxides). The needle-like grains interlock with each other, which are thought to be responsible of improving fracture toughness. However, the relatively long length of the needle-like grains can limit the strength of the material, so a compromise must be struck between desired fracture toughness and strength.

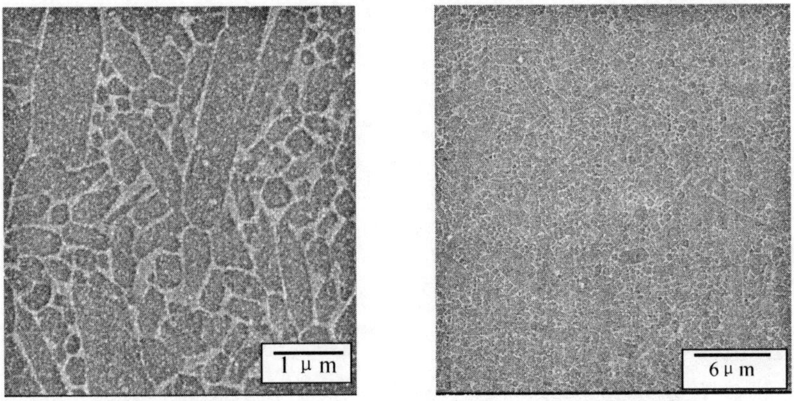

(a) (b)

Figure 1: Microstructure of Ceralloy 147-31N SRBSN after plasma etching.

In order to examine the relationship between rolling contact performance of Ceralloy 147-31N Si_3N_4 and its surface integrity, the as-sintered ingot was machined into three batches (with ten specimens in each batch) of 12.7 mm diameter balls each having different machining parameters. Table 1 shows the applied machining procedures. These three conditions are referred as coarse, fine and conventional. Roughing, the first step of three step machining process are the same for all three conditions.

Table 1: Machining procedures applied to finish 12.7 mm diameter balls.

Specimen Type	Finishing Step	Diamond Size	Removal
Coarse	1 (roughing)	accepted practice	
	2 (induce damage)	100 grit	0.1000 mm
	3 (finishing)	600 grit	0.0127 mm
Fine	1 (roughing)	accepted practice	
	2 (induce damage)	180 grit	0.1000 mm
	3 (finishing)	600 grit	0.0127 mm
RCF-Conventional	Use the "accepted" practice for RCF finishing (using 180, 220, 320 grit diamond, and 600 and 1200 grit diamond paste in sequence)		

The second step of the grinding process was intended to produce different depths of machining damage in the ball's near-surface-volume. The variation on the second step produces Si_3N_4 ball specimens with different subsurface qualities, such as the depth of subsurface damage, population of the micro-cracks etc. The final step of grinding guarantees that coarse and fine specimens have the same surface roughness. The conventional specimens have a better surface finish by introducing 1200 grit diamond paste on the final finish. Five balls in each batch were tested on a modified four ball tester to compare their rolling contact fatigue performance, and the other five balls in each batch were machined into C-sphere test specimen to examine the variation in hoop tensile strength caused by machining. The design consideration, the geometry of the C-sphere and the loading scenario are explained in the next section.

3 C-sphere test

A C-sphere flexure strength test specimen was developed to enable the study of and measurement of surface strength and linked flaw size [5]. The design of the C-sphere specimen was inspired by the C-ring specimen, which is used to evaluate the strength of ceramic tubes [6]. Enabling the identification of a flaw type (usually surface- or near-surface located) and measurement of its size in finished ceramic balls is obviously important for the study of strength, but perhaps more importantly, for the study and predictability of RCF performance (a response limited by surface- or near-surface-located flaws or in changes thereof).

Three batches of Ceralloy 147-31N Si_3N_4 balls with a diameter of 12.7 mm, as described in Table 1 were machined into C-sphere flexure strength specimens, which are shown in Figure 2(a). Grinding of the slot was performed in a two step process using a Type 1F1 diamond plated grinding wheel (127 mm diameter * 6.35 mm thick * 3.175 mm R) for the final grinding. The geometry and tolerance of the C-sphere specimen is shown in Figure 3.

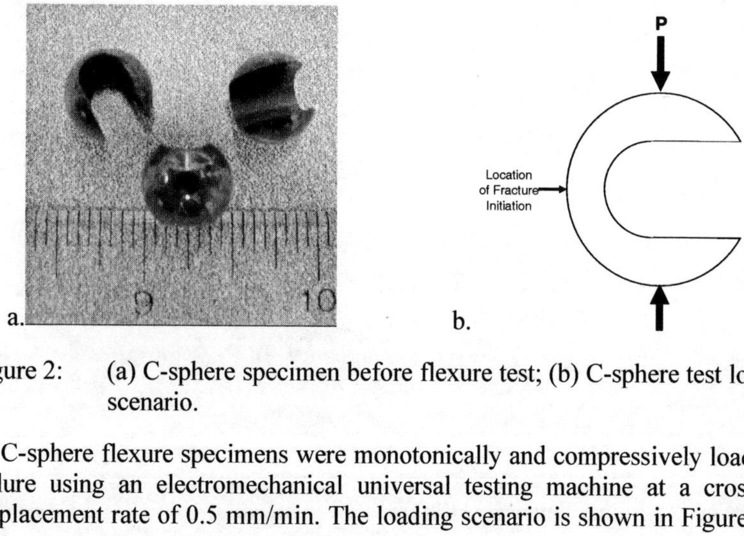

Figure 2: (a) C-sphere specimen before flexure test; (b) C-sphere test loading scenario.

C-sphere flexure specimens were monotonically and compressively loaded to failure using an electromechanical universal testing machine at a crosshead displacement rate of 0.5 mm/min. The loading scenario is shown in Figure 2(b).

A special jig was used to horizontally align the C-sphere slot prior to loading. Load to fracture was recorded and combined with the failure load-failure stress relationship from FEA to determine C-sphere flexure strength. Weibull strength distributions were determined using commercially available software. Optical fractography was also conducted on all specimens to identify failure location and the fracture surfaces of a select few specimens were examined with SEM.

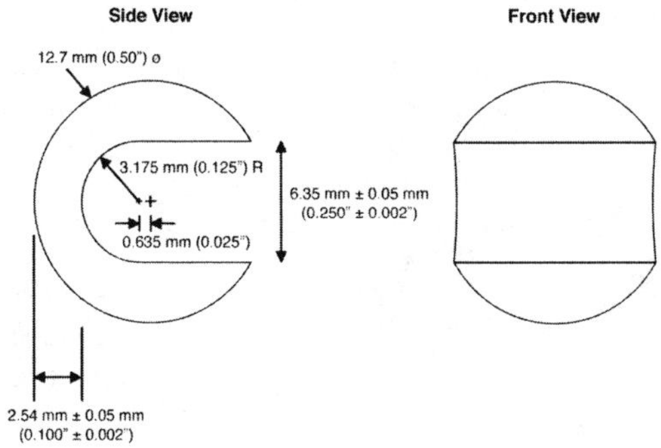

Figure 3: C-sphere specimen geometry.

4 Rolling contact fatigue test of silicon nitride balls with artificial cracks

Bearing grade HIPed Si_3N_4 excellent rolling contact performance, however, the RCF performance of SRBSN is less understood. In order to compare the rolling contact performance of coarse, fine, conventional Ceralloy 147-31N Si_3N_4 balls, artificial cracks are created and positioned into the contact path to accelerate the failure of the test specimens. A modified Plint TE 15 impact tester is used to generate the surface cracks. A test specimen is attached to a pendulum, and released at 60 degree angle to collide with a fixed contact ball. The energy absorbed during the contact can be calculated by the potential energy loss of the pendulum. The contact ball used for the artificial crack generation is TSN-03H HIPed Si_3N_4. Figure 4(a) shows the artificial crack under ultra-violet (UV) light after dye penetration. Figure 4(b) shows SEM image of the centre of the crack where the two crack edges have an approximate gap of 2.5 micron.

A modified four ball machine is used to perform the test [7]. Figure 5(a) shows the schematic of modified four ball contact with Si_3N_4 fixed in the collet as upper driving ball and three steel balls in the cup as lower contact balls. The centre of contact path on the upper Si_3N_4 ball situates at 1.17 mm above the bottom of the ball. In order to classify the position of the crack within the contact path, the positioning parameters of the crack, and some typical positions of the crack are shown in Figure 5(b) [8]. As shown in the schematic, β measures the angle between ring crack and the contact path centre line. 2a is the width of the contact path, and R represents the radius of the ring crack. δ measures the distance between the centre of the hypothetic ring crack circle and the centre line of the contact path. For all the tests, Castrol 75w90 transmission oil is selected as the lubricant due to the chemical compatibility, potential automotive application of the Si_3N_4 bearing and the lubricant's relatively high viscosity to maintain a lubrication film to reduce the contact between silicon nitride surface asperity and the steel ball surface. The applied Hertzian contact stress between the upper ball and lower balls is 5.6 GPa, and the spindle speed is 5000 rpm, which results in 11250 stress cycle per minute.

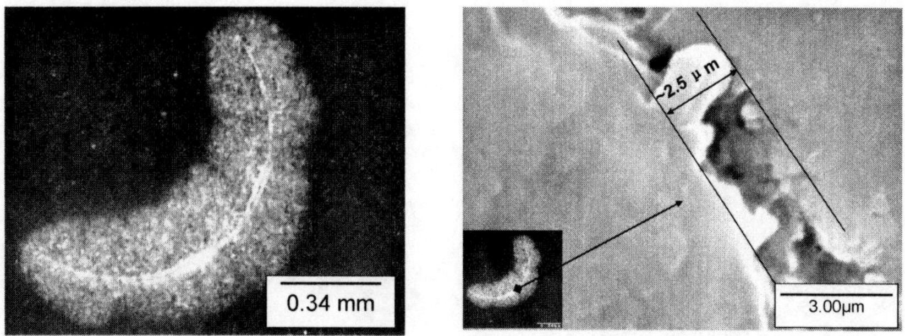

(a) (b)

Figure 4: (a) Visualizing crack under UV light after dye treatment; (b) approximate crack gap size measured in SEM examination.

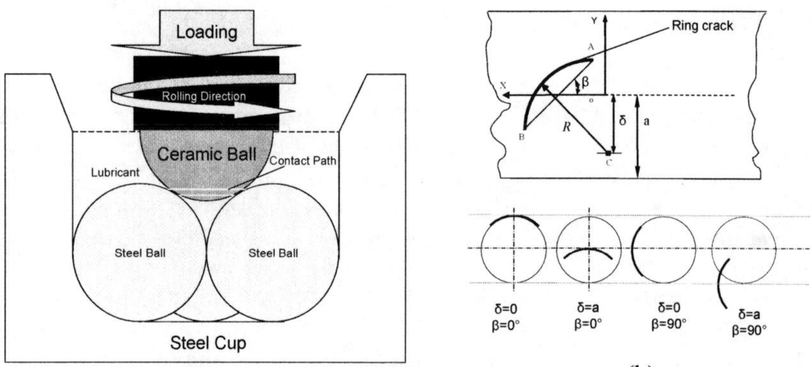

(a) (b)

Figure 5: (a) Schematic of modified four ball test – contact geometry; (b) crack positioning parameter and typical locations of the crack.

5 Results and discussion

5.1 C-sphere strength

Five C-Sphere specimens machined from each of three batches of coarse, fine and conventional balls were tested. The maximum stress is calculated in ANSYS from the mechanical load used to break the specimens. The two-parameter

Weibull analysis of the results is show in Figure 6. There is a trend of increasing characteristic strength from coarse, fine to conventional conditions. The coarse specimen has a characteristic strength of 796 MPa, while fine and conventional specimens have strengths of 802 MPa and 897 MPa respectively. The fracture surface of a tested C-sphere specimen is shown in Figures 7. Figure 7(a) shows a typical fractography of a ceramic fracture surface, where the fracture origin is identified. The magnified fracture origin is shown in Figure 7(b). According to the C-sphere fractography, the fracture origin is mostly on the surface, sometimes in the near surface area, although it is not necessary to appear on the maximum stress area on the centre of outer fibre of C-sphere specimen. According to Griffith's Criterion, for a fracture toughness of 6 MPa√m and crack geometry factor of 1.5, the estimate strength limiting flaw sizes are around 27.9, 27.6 and 22.1 microns for coarse, fine and conventional conditions respectively. This result reflects the machining effect on the surface/subsurface integrity of the specimens. Aggressive machining is perceived to generate deeper subsurface damage and higher population of micro-cracks, which are considered to act as one type of the strength limiting flaw. This perception is preliminarily verified by C-sphere flexure test, however, more tests should be done to make such a statistically confident conclusion.

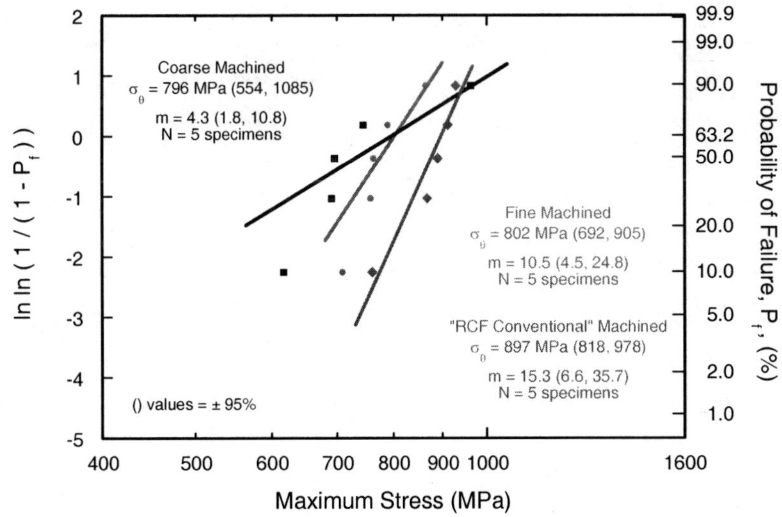

Figure 6: Weibull analysis of C-sphere strength. Values in parentheses represent ±95% confidence interval.

5.2 RCF test

Five balls in each batch of coarse, fine and conventional conditions are tested, and their fatigue lifetimes relative to the positioning parameters are summarised in Table 2. There are two specimens, Fine-05 and Conventional-05, which are not positioned accurately in the contact path, and these two specimens are excluded from the analysis. From the Weibull analysis results shown in Figure 8, we can see that conventional specimens show an extended lifetime compared with fine and coarse specimen with identical artificial crack on the surface and cracks positioned in the contact path with the same positioning parameter.

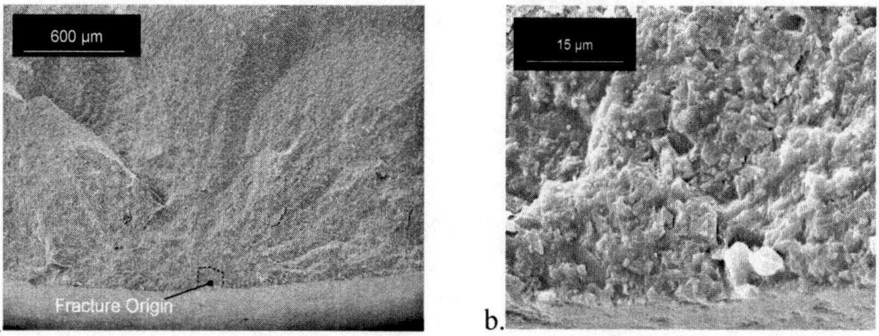

Figure 7: (a) Fractography of Si_3N_4 fracture surface; (b) fracture origin.

Table 2: RCF results of Ceralloy 147-31N SRBSN.

Specimen ID	Crack Position Parameter β	Crack Position Parameter δ	Fatigue Life (No. of Stress cycles)
Coarse-01	90°	0	$2*10^5$
Coarse-02	90°	0	$3.7*10^5$
Coarse-03	90°	0	$2.3*10^5$
Coarse-04	90°	0	$4*10^5$
Coarse-05	90°	0	$3.3*10^5$
Fine-01	90°	0	$4.1*10^5$
Fine-02	90°	0	$3.0*10^5$
Fine-03	90°	0	$4.2*10^5$
Fine-04	90°	0	$3.4*10^5$
Fine-05	90°	0.5a	$1.2*10^7$
Conventional-01	90°	0	$4.8*10^5$
Conventional-02	90°	0	$8*10^5$
Conventional-03	90°	0	$1.5*10^6$
Conventional-04	90°	0	$4.5*10^5$
Conventional-05	45°	0.5a	suspended

The coarse specimen has the worst RCF performance, with a characteristic stress cycles to failure $C_f = 3.4*10^5$. However, in order to improve the reliability of RCF results, more tests should be run to give a statistically confident conclusion. Figures 9 shows the spall profile of specimens Fine-02 and Conventional-02. Although there are variations on the geometry of the spalls from specimen to specimen, there is evidence that secondary cracks are developed in the spalling process.

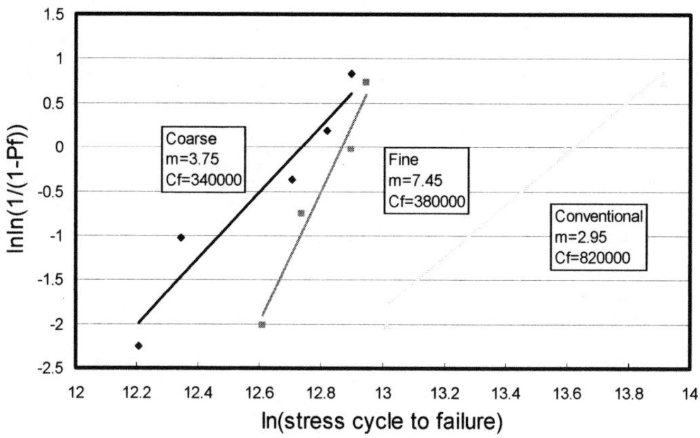

Figure 8: Weibull analysis of RCF test results of coarse, fine and conventional conditions.

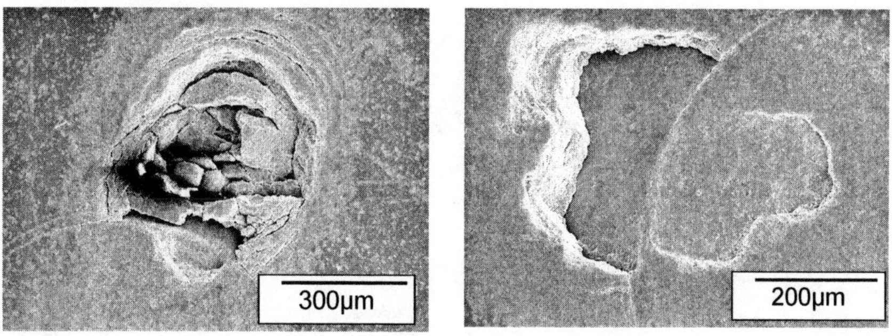

Figure 9: (a) Spall profile of specimen Fine-02; (b) spall profile of specimen Conventional-02.

5.3 The relationship between RCF and C-sphere

The C-sphere and RCF tests reveal the same trend of increasing strength and fatigue life of coarse, fine and conventional specimens. This correlation can be explained by the analysis of stress field of C-sphere specimen and four-ball contact. The failure of the C-sphere specimen is perceived to be micro-crack propagation when the outer fibre is subject to tensile stress. During the specimen machining process, coarse specimens are the most aggressively machined, which result in a higher density and greater depth of induced micro-cracks. As described earlier, the size of the flaw where fracture initiates determine the strength of C-sphere. For the coarse condition, there is a higher probability that deeper strength-limiting flaws (micro-cracks) are located at the maximum tensile stress area, and the size of the flaw cause a weaker strength of coarse C-sphere specimen.

In modified four ball test, the upper Si_3N_4 ball is subject to Hertzian contact against the contact ball. For a perfect Si_3N_4 ball without any surface cracks (natural or artificial) positioned in the contact path, the maximum tensile stress generated on the surface due to Hertzian contact is not high enough to initiate a crack. As a result, the RCF life for a perfect Si_3N_4 ball is very long, and it doesn't normally fail within a reasonable testing time (100 million stress cycles). However, for a pre-cracked Si_3N_4 ball, due to the existing of an artificial crack, the stress field on the surface is changed, which is explained in Figure 10(b). There is a gap existing between the two crack edges. For the type of artificial cracks created in the RCF tests in this study, as described earlier, the width of the gap is measured at 2.5 micron. As a result of the gap, when the section of the ball surface to the left of the crack is subject to contact stress, it creates a bending force which generates a tensile stress field on the specimen surface. Together with the stress field created due to Hertzian contact, when the overall tensile stress on the surface reaches a threshold, the micro-cracks will propagate to form a secondary crack on the surface. Due to the existence of a secondary crack, it created a tertiary crack under a similar scenario, and so on so forth, as illustrated in Figure 10(a).

a.

b.

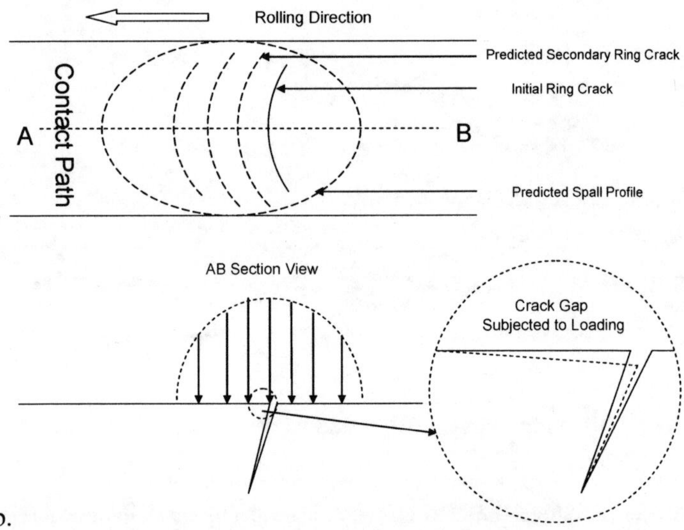

Figure 10: (a) Location of original crack and secondary cracks in the contact path; (b) Mechanism of secondary crack creation.

The secondary and tertiary crack propagates to meet the original crack and meet each other, which form a spall type of failure. Figure 11 shows the secondary cracks of Fine-02 specimen and associated spall failure. The mechanism of forming a secondary crack on the surface in modified four ball test is quite similar to micro-crack propagation in C-sphere test when it is interpreted by tensile stress on the surface, which explains the correlation between RCF results and C-sphere results. The boundary element modelling of the artificial crack under Hertzian pressure was carried out by Wang and Hadfield, and the simulation results show the same conclusion [9, 10].

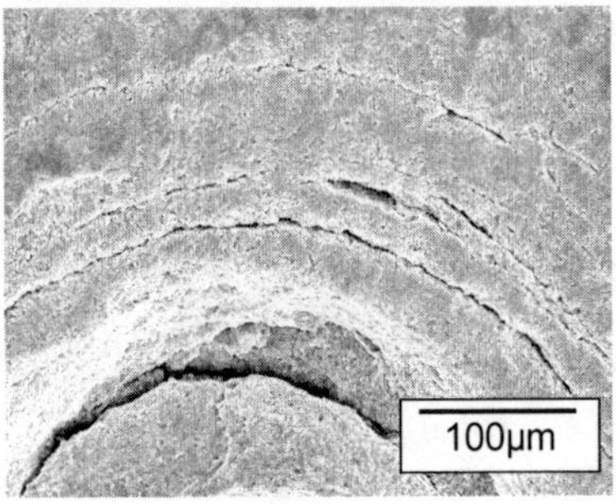

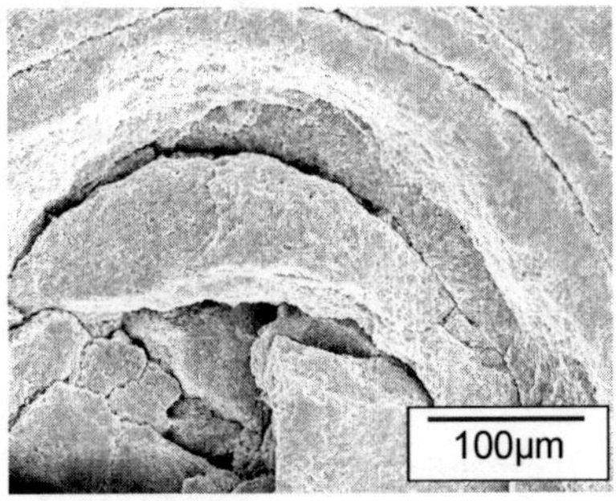

Figure 11: Secondary cracks created on the surface of Fine-02 specimen.

6 Conclusion

I. The C-sphere flexure strength results of Ceralloy 147-31N SRBSN show an increased strength comparing from coarsely, fine to conventionally machined conditions.

II. The RCF result reveals an increasing fatigue lifetime among three batches of specimens with the subsurface finished to coarse, fine and conventional condition. The conventional condition shows the longest rolling contact fatigue life, however, the coarse condition has the shortest lifetime under the same test condition.

III. The trend of C-sphere and rolling contact fatigue results are explained by the similarity of the tensile stress field created in C-sphere and rolling contact fatigue test which result in the eventual failure. Due to the positive relationship between C-sphere and rolling contact fatigue results, C-sphere strength can be used as a predictor of fatigue lifetime of Si3N4 balls. It can help designers to screen out "weak" candidate silicon nitride material prior to rolling contact fatigue test, which can make reasonable savings on time and cost. Additionally, because flaw types can be identified, C-sphere strength testing can be used by Si_3N_4 manufactures and ball finishers to assess the quality of their product.

References

[1] Riley, F. L., Silicon nitride and related materials, *Journal of the American Ceramic Society*, 83, pp. 245-265, 2000.

[2] Effner, U. and Woydt, M., "Slip-rolling and Machining of Engineering Ceramics," BAM, Berlin 259, 2002.

[3] Quinn, G. D., Ives, L. K., and Jahanmir, S., "On the Fractographic Analysis of Machining Cracks in Ground Ceramics: A Case Study on Silicon Nitride," NIST 996, 2003.

[4] Strakna, T. J., Jahanmir, S., Allor, R. L., and Kumar, K. V., Influence of grinding direction on fracture strength of silicon nitride, *Journal of Engineering Materials and Technology-Transactions of the Asme*, 118, pp. 335-342, 1996.

[5] Wereszczak, A. A., Wang, W., Jadaan, O. M., and Kirkland, T. M., Strength of A C-Sphere Flexure Specimen, *Ceramic Science and Engineering Proceedings*, 27, pp. 281-293, 2006.

[6] C1323, *Annual Book of ASTM Standards*, vol. 15.01: ASTM International, 2001.

[7] Tourret, R. and Wright, E. P., *Rolling Contact Fatigue Performance Testing of Lubricants: Papers Presented at The International Symposium Organized By the Institute of Petroleum*. London: Heydon, 1976.

[8] Wang, Y. and Hadfield, M., Influence of ring crack location on the rolling contact fatigue failure of lubricated silicon nitride: Experimental studies, *Wear*, 243, pp. 157-166, 2000.

[9] Wang, Y. and Hadfield, M., Life prediction for surface crack initiated contact fatigue of silicon nitride bearing balls, in *Tribological Research and Design for Engineering Systems*. Amsterdam: ELSEVIER SCIENCE BV, pp. 349-358, 2003.

[10] Wang, Y. and Hadfield, M., A mechanism for nucleating secondary fractures near a pre-existing flaw subjected to contact loading, *Wear*, 254, pp. 597-605, 2003.

Section 4
Contact mechanics

Numerical analysis of the physical phenomena in the working zone in the rolling process of the round thread

L. Kukielka & K. Kukielka
Department of Mechanical Engineering,
Koszalin University of Technology, Poland

Abstract

Thread rolling is a very complicated technological process. To improve the quality of the product and reduce production cost of the round thread, we should know the physical phenomena existing in the contact zone between rolls and deform work pieces. Therefore, this paper presents the physical and mathematical models of deformations (displacements and strains) and stress in the cold process of round thread rolling. The process is initially considered in a geometrically and physically non-linear regime, as well as a boundary value problem. The physical phenomena on a typical incremental step were described using a step-by-step incremental procedure, with an updated Lagrangian formulation. The state of strains was described by Green–Lagrange's tensor, while the state of stress by the second symmetrical Pioli–Kirchhoff's tensor. The object was treated as an elastic (in the reversible zone) and visco-plastic body (in the non-reversible zone) with mixed hardening. The variational equation of motion in three dimensions for this case was proposed. Then, the finite elements methods (FEM) and dynamic explicit method (DEM) were used to obtain the solution. The application is developed for the method of finite elements in the ANSYS programme, which provides a complex time analysis for displacement, strains and stresses occurring in the object. The effective discrete computable model which counts minimum degrees of freedom and a guide to convergence of solutions for the maximum value of stresses and strains, is proposed. Examples of simulation of the influence on various process conditions on the states of strain and stress are presented.
Keywords: round thread, rolling process, model investigation, equation of motion, FEM, ANSYS, numerical analysis, DEM, state of strain, state of stress.

WIT Transactions on Engineering Sciences, Vol 55, © 2007 WIT Press
www.witpress.com, ISSN 1743-3533 (on-line)
doi:10.2495/SECM070121

1 Introduction

The development of industry, especially in the areas of motorization and construction engineering, results in the fact that there are new requirements for products on a higher level. They require an improvement of the technological quality and an enlarged corrosion resistance. One of the methods of mechanical working is thread rolling, which gives a surface layer with profitable and usable properties. Using this modern technology gives basic advantages, such as an increase in the physical and mechanical properties of surface layer, higher wear resistance, enlarged dimension and shape accuracy of the screws and the increased efficiency of the thread rolling process. The properties beneficial for exploitation purposes are the result of plastic deformations in the rolled surface layer (fig. 1(b)).

Round threads with quick pitch make up the specific group. The screw joint folding from the nut and screw (fig. 1a) is used mainly in the construction of communication tunnels and construction engineering to support ceilings and in earth works for the protection of excavations. The screw threads are used with a nominal diameter of $d = 31$ and $d = 38$ mm, the pitch of $P = 12{,}56$ and $P = 12{,}78$ mm respectively and typical lengths of $L = 6000$ mm.

Figure 1: The screw joint (a) and the surface layer of the round thread after rolling (b).

The basic problems in designing the rolling process of the round thread on a pipe are elaborate proper construction of the tool (threading head) and selection of the processing condition for providing technical requirements and property of the surface layer of the screw, simultaneously increasing the tool life and process productivity. Presently, this technology is not used in Polish industry, with no base of scientific knowledge about this process or guidelines to the selection of process conditions. Therefore, at the Koszalin University of Technology, scientific research is working out the round thread rolling method on pipes and also the realization with high velocities and high temperatures.

2 Introducing investigations

The aim of the introductory investigations were to check the possibility of making the round thread on a pipe with a rolling method on a typical rolling mill and to determinate the important influential factors on the quality of the thread.

The research has shown that the thread rolling process on pipes is very complicated technological process. The influencing factors on the rolling process and the quality of screw can be divided into three principal groups: materials property, geometrical property of the thread and tool, and technological parameters of the rolling. Process characterizing is with the high instability, load of the pipe with low texturing stiffness, high force of working, causes that pose difficult problems to produce the screw with the expected quality.

Also, there is a big problem for each kind of material, dimension of pipe (outline diameter, partition gauge and accuracy dimensional – shape) and state of the surface layer (roughness, state of internal stresses), also selection of optimal working outline surface of the rolls, the kind of lubricate factor and the rolling factor (velocity of moving rolls, velocity of rolling, material and setting of support, force of pressure) to aim for a desirable quality of the thread and productivity of the process. Not satisfy these conditions can cause such defects as: pitting, irregular outline, crack, overlapping, incomplete outline, curving and dimensional deviation.

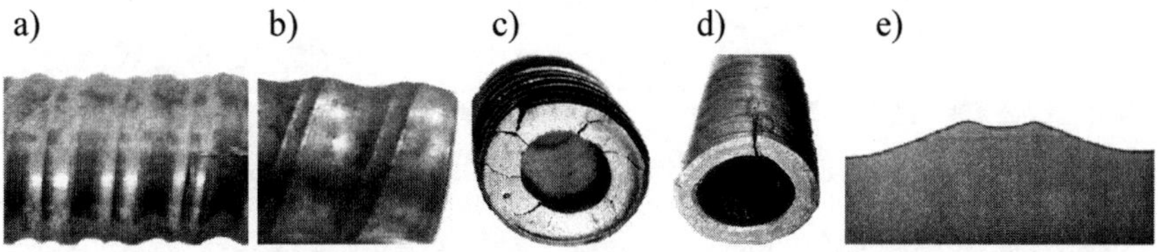

Figure 2: Defects formed by the thread rolling process: irregular outline (a), pitting (b), crack (c), overlapping (d), incomplete outline (e).

Achieving a high quality of thread requires a very precise manufacturing process together with the right design of the tool and other important factors, all of which have an influence on the final product.

Elaborate experimental knowledge of the rolling process and optimization with a high numbers of factors is a task that is time-consuming and makes expensive experimental research, where the theory of experiment planning must be applied. Therefore, the problem was approached using a theoretical model to elaborate the pipe rolling process, where the variational and the finite element method, were applied. Elaborating this model is necessary because of the complexity and the analysis of the physical phenomena for the real condition of the rolling process (geometry of the tools, values of the technological factors, the friction conditions, and so on). For this condition, known occurrences of the displacement, strain and stress is used to obtain the pressures and forces of the rolling, also to describe the properties of the thread surface.

The computational calculations were made with an elaborated application in ANSYS system. Exemplary results of numerical simulation concern the influence of friction conditions in the contact zone tool - work pieces (pipe – steel *C55*) on the outline of thread and states of displacement, strains and stresses in the surface layer, for the thread rolling process on cold (nominal diameter $d = 31$ mm).

3 Mathematical model of process

A mathematical model of the process is formulated in increments and contains the following: a material model, an equation of motion and deformation, with initial and boundary conditions.

3.1 Material model

3.1.1 Incremental model of yield stress

Yield stress σ_y is the most important parameter characterizing the resistance of a visco-plastic deformation. The incremental model of the yield stress for a typical step time $t \rightarrow \tau = t + \Delta t$ was defined as [2]:

$$\Delta\sigma_y = F_2(y)\Delta y + \frac{\partial F_1[\bullet]}{\partial \dot{\varepsilon}_{eq}^{(VP)}}\Delta\dot{\varepsilon}_{eq}^{(VP)} + \frac{\partial F_1[\bullet]}{\partial \sigma_{st}} F_3(\varepsilon_{eq}^{(VP)})\Delta\varepsilon_{eq}^{(VP)}, \tag{1}$$

where $\Delta\varepsilon_{eq}^{(VP)}$, $\Delta\dot{\varepsilon}_{eq}^{(VP)}$ are the incremental of effective visco-plastic strain and strain rate, $F_2(y)\Delta y$ is the component of change in the initial yield stress with a change of chemical composition, $[\partial F_1[\bullet]/\partial\sigma_{st}]F_3(\varepsilon_{eq}^{(VP)})\Delta\varepsilon_{eq}^{(VP)}$ is the component of change in the temporary yield stress σ_y with change of the visco-plastic strain, $[\partial F_1[\bullet]/\partial\dot{\varepsilon}_{eq}^{(VP)}]\Delta\dot{\varepsilon}_{eq}^{(VP)}$ is the component of change in the temporary yield stress with change of the visco-plastic strain rate, σ_{st} is the state stress depending on the accumulated effective visco-plastic strain and time.

3.1.2 Elastic/visco-plastic material model

A new model of mixed hardening for isotropic material which includes the combined effects of elasticity (a reversible domain), visco-plasticity (a non-reversible domain) (*E/VP*) is used. The model takes into account the history of the material.

The constitutive equation of increment components of a total strain tensor takes form:

$$\Delta\varepsilon_{ij} = \frac{1}{1-\widetilde{S}^{**}}[D_{ijkl}^{(E)}\Delta\sigma_{kl} - A] \tag{2}$$

and of increment components of the total stress tensor:

$$\Delta\sigma_{ij} = C_{ijkl}^{(E)}\Delta\varepsilon_{kl} - \psi\widetilde{S}_{ij}^{*}[\widetilde{S}_{ij}C_{ijkl}^{(E)}\Delta\varepsilon_{kl} - A], \tag{3}$$

where:

$$\widetilde{S}^{**} = \widetilde{S}_{ij}^{*}C_{ijmn}^{(E)}\widetilde{S}_{mn}, \tag{4}$$

is a positive scalar variable,

$$\widetilde{S}_{ij}^{*} = \frac{\widetilde{S}_{ij}}{\widetilde{S}_{ij} C_{ijkl}^{(E)} \widetilde{S}_{kl} + \frac{2}{3}\sigma_y^2 (\widetilde{C} + E_T)}, \tag{5}$$

is a component of a stress tensor,

$$A = \frac{2}{3}\sigma_y \frac{\partial \sigma_y}{\partial \dot{\varepsilon}_{eq}^{(VP)}} \Delta \dot{\varepsilon}_{eq}^{(VP)} , \tag{6}$$

is a positive scalar variable, $\Delta\sigma_{ij}$ is the increment component of the second Piola-Kirchhoff stress tensor, $D_{ijkl}^{(E)}$ are the components of tensor $\boldsymbol{D}^{(E)} = [\boldsymbol{C}^{(E)}]^{-1}$ in time t, $\Delta\varepsilon_{ij}$ is the increment component of Green-Lagrange strain tensor, $C_{ijkl}^{(E)}$ are the components of elastic constitutive tensor $\boldsymbol{C}^{(E)}$.

3.2 Incremental model of motion and deformation

In this section we develop the equation of a motion and deformation of the object in the updated Lagrangian formulation. Assuming that numerical solutions are obtained at discrete time *t*, the solution for $t+\Delta t$ is to be obtained. Using the conditions of stationary of functional $\Delta \boldsymbol{J}(\Delta\ddot{u}_i, \Delta\dot{u}_i, \Delta u_i) = \Delta \boldsymbol{J}(\cdot)$ where $\Delta u_i, \Delta\dot{u}_i, \Delta\ddot{u}_i$ are the *ith* increment components of the displacement, velocity and acceleration vectors, respectively and a finite element method, we can write an equation of motion and deformation in the form:

$$[\boldsymbol{M}]\{\Delta\ddot{\boldsymbol{r}}\} + [\boldsymbol{C}_T]\{\Delta\dot{\boldsymbol{r}}\} + ([\boldsymbol{K}_T] + [\Delta\boldsymbol{K}_T])\{\Delta\boldsymbol{r}\} = \{\Delta\boldsymbol{R}\} + \{\Delta\boldsymbol{F}\} + \{\boldsymbol{F}\} + \{\boldsymbol{R}\}, \tag{7}$$

where mass matrix $[\boldsymbol{M}]$, damping matrix $[\boldsymbol{C}_T]$, stiffness matrix $[\boldsymbol{K}_T]$ and force vector $\{\boldsymbol{F}_T\}$ are known at time t. However, increment stiffness matrix $[\Delta\boldsymbol{K}_T]$, external incremental load vector $\{\Delta\boldsymbol{R}\}$, internal incremental forces vector $\{\Delta\boldsymbol{F}\}$, incremental vectors of displacement $\{\Delta\boldsymbol{r}\}$, velocity $\{\Delta\dot{\boldsymbol{r}}\}$ and acceleration $\{\Delta\ddot{\boldsymbol{r}}\}$ of finite element assembly at a typical step time are not known. In order to solve this problem we apply the integration methods - central difference method (DEM), which is one of the methods of direct integration of the equation (7).

4 DEM solution

Assuming that an increase of temporary step Δt is very small, it is possible to execute a linearization of equation (7) and using the incremental decomposition we obtain:

$$[\boldsymbol{M}]\{{}^t\ddot{\boldsymbol{r}}\} + [\boldsymbol{C}_T]\{{}^t\dot{\boldsymbol{r}}\} + [\boldsymbol{K}_T]\{{}^t\boldsymbol{r}\} = \{{}^t\boldsymbol{F}_T\} + \{{}^t\boldsymbol{Q}\}. \tag{8}$$

Then using the central difference method (DEM), in which it is assumed that:

$$\{{}^{t}\dot{\boldsymbol{r}}\} = 0{,}5(\{{}^{t+\Delta t}\boldsymbol{r}\} - \{{}^{t-\Delta t}\boldsymbol{r}\}) / \Delta t\,,\ \{{}^{t}\ddot{\boldsymbol{r}}\} = (\{{}^{t+\Delta t}\boldsymbol{r}\} - 2\{{}^{t}\boldsymbol{r}\} + \{{}^{t-\Delta t}\boldsymbol{r}\}) / \Delta t^{2} \tag{9}$$

and substituting the relations in (9) into (8) we obtain:

$$[\widetilde{\boldsymbol{M}}]\{{}^{\tau}\boldsymbol{r}\} = \{{}^{t}\widetilde{\boldsymbol{Q}}_{T}\}, \tag{10}$$

where:

$$[\widetilde{\boldsymbol{M}}] = [\boldsymbol{M}] / \Delta t^{2} + 0{,}5[\boldsymbol{C}_{T}] / \Delta t,$$

$$\{{}^{t}\widetilde{\boldsymbol{Q}}_{T}\} = \{{}^{t}\boldsymbol{F}_{T}\} + \{{}^{t}\boldsymbol{Q}\} - [\boldsymbol{K}_{T}]\{{}^{t}\boldsymbol{r}\} + \frac{2\{{}^{t}\boldsymbol{r}\} - \{{}^{t-\Delta t}\boldsymbol{r}\}}{\Delta t^{2}}[\boldsymbol{M}] + \frac{\{{}^{t-\Delta t}\boldsymbol{r}\}}{2\Delta t}[\boldsymbol{C}_{T}]. \tag{11}$$

The integration method requires that the time step Δt is smaller than critical value Δt_{kr}, which can be calculated from the mass and stiffness properties of the complete element assemblage: $\Delta t \le \Delta t_{kr} = T_N / \pi$, where T_N is the smallest period of the finite element assemblage with N degrees of freedom.

5 Model investigation

The model investigation was conducted in order to settle the course deformation layer top sample executed from the plastic material, as well as with the aim to qualify boundary conditions for displacements indispensable to numeric analysis of the round tread rolling process.

To improve accuracy in elaboration of the displacement vector component of the node, the real part of the thread was substituted by a rectangular model with proper magnification. The model dimension and dimension of the rectangular grid on the model satisfy the criterions of the geometrical similarity, however the material model (plasticine) and real material (steel *C55*) were criterions of the physical similarity.

Two samples were joined by sides with a plot mesh, and were closed in a metal form. Then the samples were subjected to the deformation by a perpendicular shift of rectilinear motion in the model stamp of an outline of round thread (fig. 3). The exemplary view of deformed samples with finite element grid for three causes of friction coefficient is presented in fig. 4.

On figure 4 observe that increasing the value of the friction coefficient on the contact surface between tool - work pieces causes an increase in the adhesion zone and decrease of material sliding on the contact surface. That has an influence on curving vertical lines of grid to the bottom of the thread. This curving is improved together with an increase in the friction coefficient.

6 Numerical simulation of state of displacement, strain and stress of material during round thread rolling

The application developed with regard to the method of finite elements in ANSYS program provides a complex time analysis of physical phenomena during the rolling process. Digital computing for the process was carried out with the use of two methods. The first method requires introducing the boundary conditions for displacements in the contact zone determined by the model investigation, whereas the second one requires the adequate determination of the contact zone without an introduction of boundary conditions. The main aim of the simulation was to define the influence of friction coefficient on the state of deformation (displacements and strain) and stress in the surface layer of the object. The numerical analysis for 2-D states of deformation and 3-D states of stress was applied on the example of steel *C55* (DIN) (Fig. 5). The stand is a rigid body $E\to\infty$, however the material model as an elsto/visco-plastic body with non-linear hardening. The model has discretized by finite element *PLANE183* with nonlinear function of the shape. In the second calculating method the contact tool with work pieces was modeling by element *TARGE169* and *CONTA171*. Computational model contain 213266 degrees of freedom.

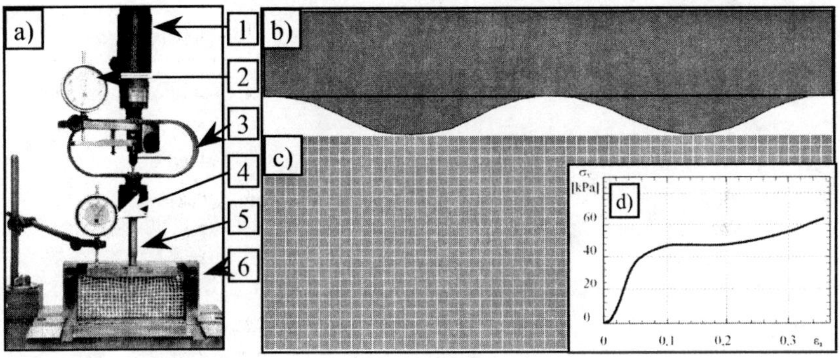

Figure 3: The view of the stand for model investigation (a), the outline of used stamp (b) and the meshed sample before deformation (c), the stress-strain curve for the material model (d): 1 - remote control hydraulic cylinder, 2 - dial gauge for force pressure on stamp, 3 - dynamometer, 4 - dial gauge for measure displacement, 5 - stamp, 6 - container.

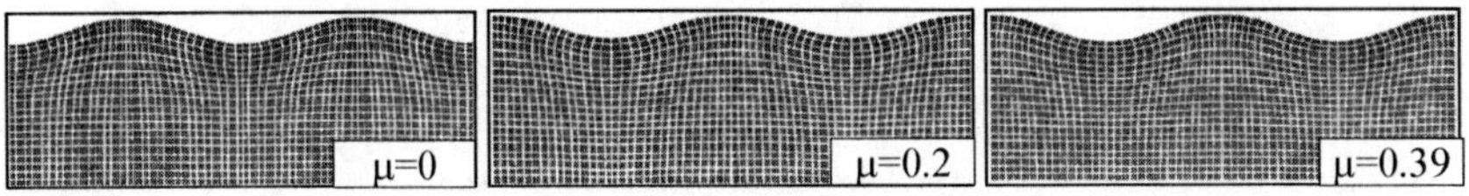

Figure 4: The mesh after deformation for μ=0 (a), μ=0,2 (b) and μ=0,39 (c).

Exemplary results of the numerical simulation are present on figures 6 and 7. Analyzing the intensity distribution of strain, stress and deformation of the finite element grid, where the influence of the lubrication condition is observed.

For $\mu = 0$ in the contact zone tool – workpieces (fig. 6(a)), during the forming of the outline of the thread, material is not broken by the tool and slide through the contact surface. The curving of the vertical line of the finite element grid is invisible. On the other side, an increase in the friction coefficient causes an increase in the braking of the material. For high value of the friction coefficient (fig. 6(c)) there occurs a strong braking of material in the contact zone, also from the adhesion zone of the material. That causes higher displacements of material in the zone placed father from the contact zone. Then the line of the finite element grid are stronger curved. The results of the numerical analysis are comparable with results from model investigation presented in section 4.

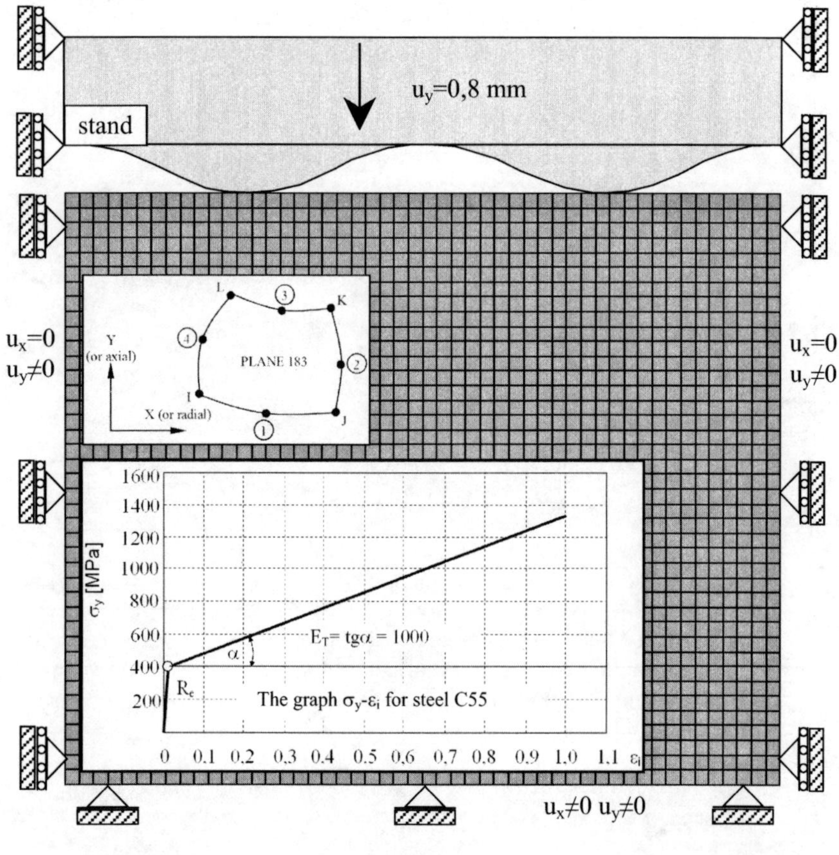

Figure 5: The discretized model with boundary conditions for displacements.

The friction coefficient also has influence on value and distribution of stress. For $\mu = 0$ the maximum value of stress intensity count $\sigma_i = 932$ MPa and is placed on the bottom of thread, for $\mu = 0{,}39$ is less and count $\sigma_i = 729$ MPa (MX1, Fig. 6(c)). For $\mu > 0$ appear local maximum of stress on the sides of the thread (MX2), where the value increase with increase of friction coefficient from value $\sigma_i = 829$ MPa, for $\mu = 0{,}2$ (Fig. 6(b)) to $\sigma_i = 945$ MPa for $\mu = 0{,}39$ (Fig. 6(c)).

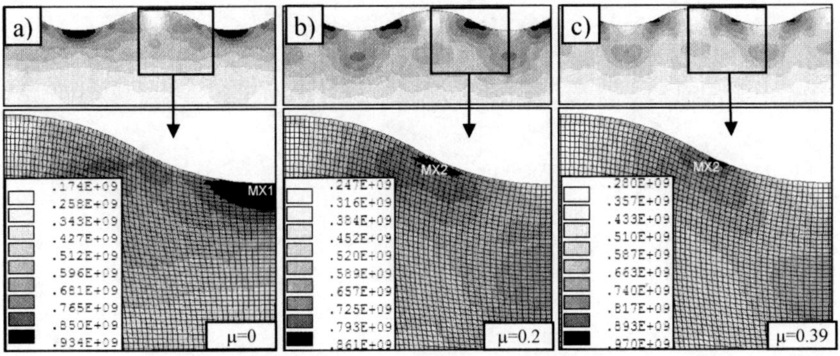

Figure 6: The deformation of grid and the stress intensity maps for various value of frictions coefficient.

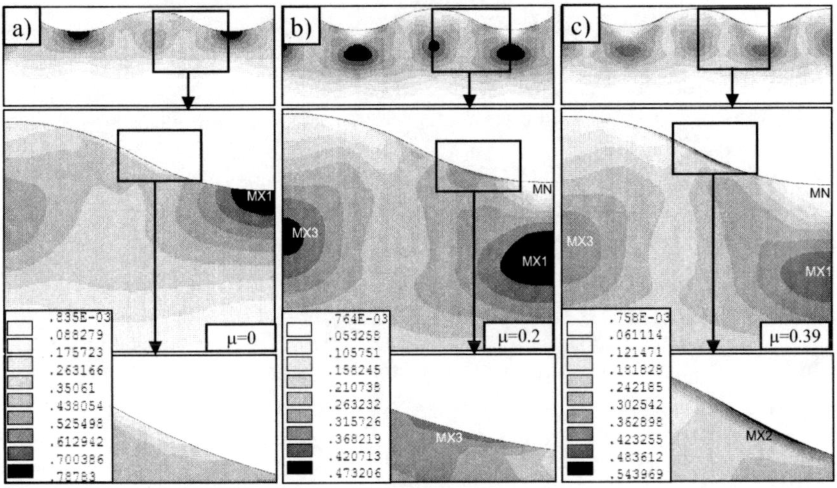

Figure 7: The strain intensity maps for various value of friction coefficient.

The friction coefficient has a high influence on value and distribution of strain. For $\mu = 0$ the maximum strain intensity $\varepsilon_i = 0{,}78$ is located on the bottom of the thread, close to the contact surface (MX1, Fig. 7(a)). For $\mu > 0$ appear an adhesion zone of material in the bottom of the thread, which take the characteristic shape of a wedge. In this zone the value of strain is very small. For $\mu = 0{,}2$ - $\varepsilon_i = 0{,}1$ (MN, Fig. 7(b)) and for $\mu = 0{,}39$ strains are closer to the contact surface and getting smaller to value $\varepsilon_i = 0{,}0016$ (elastic strains) (MN, Fig. 7(c)). However, the local maximum strains (MN, Fig. 7(c)) are more and more moving down from the contact surface. Two local maximums of the strains appear. The first one (MX2) is placed close to the contact zone of the side of the thread, where a higher value of friction coefficient increase strains value from ε_i=0,4 for $\mu = 0{,}2$ (Fig. 7(b)) to value $\varepsilon_i = 0{,}54$ for $\mu = 0{,}39$. The next one, local maximum (MX3), is located in the depth of material on a symmetry axis that passes through the top of the thread. Here, strains are getting smaller together with increasing of the friction coefficient from value $\varepsilon_i = 0{,}44$ for $\mu = 0{,}2$ (fig. 7(b)) to $\varepsilon_i = 0{,}42$ for $\mu = 0{,}39$ (fig. 7(c)).

7 Conclusions

The round thread rolling process on the pipes is a geometrical, physical and thermal non-linear and boundary problem. Measurement of the process parameters that influence the technological quality, such as: a displacement zone, a temperature, stress, structural change etc. during the thread rolling process is impossible without today's modern techniques of a measurement.

An application of modern numerical methods and computing systems allows an analysis of complex physical phenomena occurring in the process under investigation. The application developed in the ANSYS system enables a time analysis of the rolling process with the consideration of the changeability of the lubrications conditions. On the course of physical phenomena in the working zone we can forecast a technological quality of the round thread.

The results obtained of the computer simulation of the thread rolling process show that the friction coefficient influence the states of displacements, strains and stresses in the surface layer of the thread, also that it is one of the factors affecting the technological and the exploitation quality. The best operational quality of the thread is received during the rolling process with high lubrication conditions ($\mu = 0$).

The simulation results for the condition of lubrication can be of use while designing the round thread rolling process: making a selection of the process condition and the kind of lubrication factor in the aspect of the technological quality of the screw.

References

[1] Łyczko K.: T*he technology of tool and the female thread rolling.* Częstochowa University of Technology, 1999 (in polish).

[2] Kukielka L., Kukielka K.: Numerical analysis of the process of trapezoidal thread rolling, *III International Conference on High Performance Structures and Materials*, 3-5 May 2006, Ostand, Belgium. WIT*PRESS* Southampton, Boston, 2006, pp. 663-672.

[3] Kukielka K., Kukielka L.: Modeling And Numerical Analysis Of The Thread Rolling Process, *77th Annual Meeting of the Gesellschaft für Angewandte Mathematik und Mechanik*, March 27th - 31st, Technische Universität Berlin, 2006.

Optimal shape of fibers in composite structure using Inverse variational principles

P. Procházka
Czech Technical University in Prague, Czech Republic

Abstract

Inverse variational principles proved their importance in shape optimization of structures. In this paper they are applied to searching for the optimal shape of fibers in a composite structure. As the boundary element method seems to be more promising than other modern numerical methods applied to the search for optimal shape, in the submitting paper the boundary element method is redefined to enable one to use such an approach, which leads to possibility for the optimal interfacial energies and, hence, to the optimal bearing capacity of the composite structure. Necessary discretization of the domain, which occurs in the finite elements, is suppressed in our case. Standard procedure in the finite elements leads to dependence of the stiffness matrix on the shape of the fibers. In this case, following a basic idea for homogenization and localization, concentration factors have to be calculated in terms of the boundary element method instead. These terms are dependent on the shape of the fibers. It appears that the procedure is still not convergent (we solve a strongly nonlinear problem) and additional constraint has to be involved in the formulation. In order to formulate and solve this problem, the idea of Inverse variational principles is applied here for expressing necessary quantities. The paper concentrates on the calculation of quantities, which are necessary to formulate the optimization problem. The main attention is focused on calculation of concentration factors, which play the most important role in the approach proposed.
Keywords: discrete element method, boundary element method, dynamical equilibrium.

WIT Transactions on Engineering Sciences, Vol 55, © 2007 WIT Press
www.witpress.com, ISSN 1743-3533 (on-line)
doi:10.2495/SECM070131

1 Introduction

Conventionally, the optimal shape design problem consists of minimization of an appropriate cost functional with certain constraints, such as equilibrium and compatibility conditions and design requirements. The formulation of the cost functional depends of the concrete intention of a designer. One of a reasonable and practical form of the cost functional respects the minimization of the strain energy of the body subjected to a specific load. Such a problem can easily be formulated in terms of inverse variational principles, which assure that the surface energy attains its minimum.

The inverse variational principles are naturally connected with finite element method, which starts with energetic formulation. But, the FEM is less suitable for the problems involving the problem like moving boundary, optimal shape, etc. On the other hand, the direct connection of the BEM with the variational principles is not seen at first sight (see [1]). In the latter paper optimization problems based on the inverse variational principles solved by boundary elements is formulated. This approach is extraordinarily advantageous, as no internal mesh has to be generated.

In [10] the optimal composite is obtained by using a two-step procedure: (i) first an ideal structure of the matrix material is found by weakening the polymer by an optimal arrangement of pores, and (ii) the rods are embedded in the matrix. The design parameters are the shape, volume fraction, and spatial arrangement of the piezoceramic rods, and the structure of the matrix material. It turns out that the optimal matrix is highly anisotropic and is characterized by negative Poisson's ratios in certain directions.

Since we are concentrated on optimization of composite structures using homogenization, the theory for periodic media given by Suquet [2] is used in this paper. Similarly to Suquet´s examples symmetric problems are considered.

The way on how to formulate the problem starts with the idea of Hashin-Shtrikman variational principles according to [3], which were worked out into integral form.

First, homogenization and localization, having the principal meaning in the approach introduced in this paper, is discussed using the boundary element method. Then the optimization of shape of fibers is formulated and solved using the information from the previous sections. Some examples are discussed at the end of this paper.

2 Localization and homogenization of symmetric periodic structures

Localization and homogenization is concisely described in Suquet [2]. Recall some basic consumption which we use later in the integral formulations. First, we denote quantities used in this text. Two different scales will naturally be introduced. The macroscopic scale, the homogeneous law in which is sought, will be described in coordinate system $\boldsymbol{x} \equiv \{x_1, x_2, x_3\}^{\mathrm{T}}$ and the microscopic scale - heterogeneous - is characterized in the system of coordinates $\boldsymbol{y} \equiv \{y_1, y_2, y_3\}^{\mathrm{T}}$. The medium is generally heterogeneous, but locally - in the microscopic scale – is assumed to be periodic, thus a representative volume element may be cut out from the structure and the periodicity conditions can be introduced on the boundary of this element.

Let us distinguish the quantities under study in dependence of the macroscopic or microscopic scale in the following manner: The displacements in the macroscopic level will be denoted as $\boldsymbol{U} \equiv \{U_1, U_2, U_3\}^{\mathrm{T}}$ while in the microscopic level as $\boldsymbol{u} \equiv \{u_1, u_2, u_3\}^{\mathrm{T}}$. Moreover, in macroscopic level, let us denote strains as $\boldsymbol{E} \equiv \{E_{ij}\}$, i, j = 1,2,3 and stresses as $\boldsymbol{S} \equiv \{S_{ij}\}$, i, j = 1,2,3. In the microscopic level let us denote strains as $\boldsymbol{\varepsilon} \equiv \{\varepsilon_{ij}\}$, i, j = 1,2,3 and stress as $\boldsymbol{\sigma} \equiv \{\sigma_{ij}\}$, i, j = 1,2,3. Define also the microscopic-macroscopic relation of the averaged stresses and strains by

$$S_{ij} = \frac{1}{\mathrm{meas}\Omega} \int_{\Omega} \sigma_{ij} \, \mathrm{d}\Omega = < \sigma_{ij} >, \qquad E_{ij} = \frac{1}{\mathrm{meas}\Omega} \int_{\Omega} \varepsilon_{ij} \, \mathrm{d}\Omega = < \varepsilon_{ij} > \tag{1}$$

where < . > stands for the average, Ω is the representative volume element, and meas Ω is its volume, $\Omega = \Omega^{\mathrm{f}} \cup \Omega^{\mathrm{m}}$, $\Omega^{\mathrm{f}} \cap \Omega^{\mathrm{m}} = 0$, Ω^{f} denotes the domain of fiber and Ω^{m} is the domain of matrix. As usual, meas Ω is set to unity. Note that average usually means homogenization, but one should use that term with care: there are many kinds of averaging.

The elasticity system (equilibrium equations, kinematical conditions and Hooke's law) is defined as (small deformation theory is imposed):

$$\operatorname{div} \boldsymbol{\sigma}(\boldsymbol{y}) = 0, \quad \boldsymbol{\sigma}(\boldsymbol{y}) = \boldsymbol{L}(\boldsymbol{y}) : \boldsymbol{\varepsilon}(\boldsymbol{y}), \quad \boldsymbol{\varepsilon} = \frac{1}{2}(\nabla \boldsymbol{u} + \nabla^{\mathrm{T}} \boldsymbol{u}) \qquad \text{in } \Omega \tag{2}$$

and periodic boundary conditions along the boundary of the unit cell $\partial\Omega$ are given.

Localization consists of the solution of system of elasticity system (equilibrium equations, kinematical conditions and Hooke's law) on the representative volume element (or unit cell) for concentration factors $\boldsymbol{A}^{\mathrm{f}}$ of fibers and $\boldsymbol{A}^{\mathrm{m}}$ for matrix:

$$\begin{aligned} \varepsilon_{ij}^{\mathrm{f}}(\boldsymbol{u}(\boldsymbol{y})) &= A_{ijkl}^{\mathrm{f}}(\boldsymbol{y}) E_{kl}, \quad \boldsymbol{y} \in \Omega^{\mathrm{f}} \\ \varepsilon_{ij}^{\mathrm{m}}(\boldsymbol{u}(\boldsymbol{y})) &= A_{ijkl}^{\mathrm{m}}(\boldsymbol{y}) E_{kl}, \quad \boldsymbol{y} \in \Omega^{\mathrm{m}} \end{aligned} \tag{3}$$

Periodic boundary conditions will be employed on $\partial\Omega$. If $\boldsymbol{n} \equiv \{n_1, n_2, n_3\}$ is outward unit normal to $\partial\Omega$, it holds:

stress: components of tractions $p_i = \sigma_{ij} n_j$ are opposite on the opposite sides,
strains: the local strain tensor $\boldsymbol{\varepsilon}(\boldsymbol{u})$ is split into its average $\boldsymbol{E}$ and a fluctuating term $\bar{\boldsymbol{\varepsilon}}(\boldsymbol{u})$ as:

$$\boldsymbol{\varepsilon}(\boldsymbol{u}) = \boldsymbol{E} + \boldsymbol{\varepsilon}^*(\boldsymbol{u}), \quad \boldsymbol{\varepsilon}^*(\boldsymbol{u}) = \boldsymbol{\varepsilon}(\boldsymbol{u}^*), \quad < \boldsymbol{\varepsilon}(\boldsymbol{u}^*) >= 0 \tag{4}$$

displacements: u_i^* are the same at opposite sides. Hence, the fluctuating displacement $\boldsymbol{u}^*$ may be considered a *periodic field*, up to a rigid displacement that will be disregarded. The geometry and denotation is obvious from Fig. 1 for 2D case. Interfacial surface between fiber and matrix is denoted by Γ.

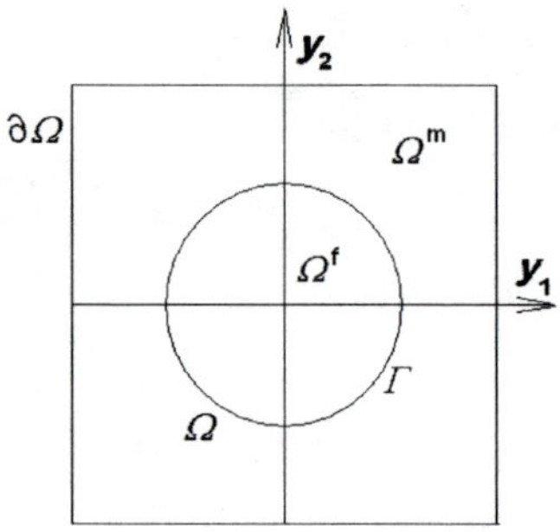

Figure 1: Unit cell used in the study.

As we concentrate on symmetric problems, and linear elasticity is considered (hence superposition is admitted), the periodicity conditions can be substituted according to Figs. 3, 4, 5, where only first quarter is considered with different boundary conditions, describing symmetry or antisymmetry of particular problems.

In Fig. 2 the geometry, supports and loading for the response of E_{11} is depicted, in Fig. 3 the same for E_{22} and in Fig. 4 for E_{12}. For the sake of simplicity two dimensional case is drawn. The triangles denoting supports are rollers.

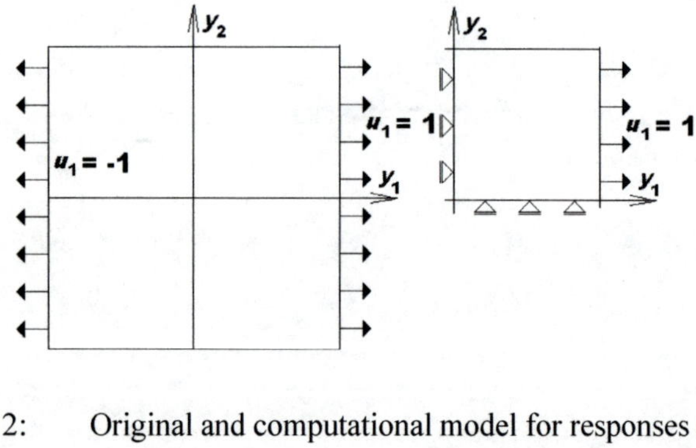

Figure 2: Original and computational model for responses of E_{11}.

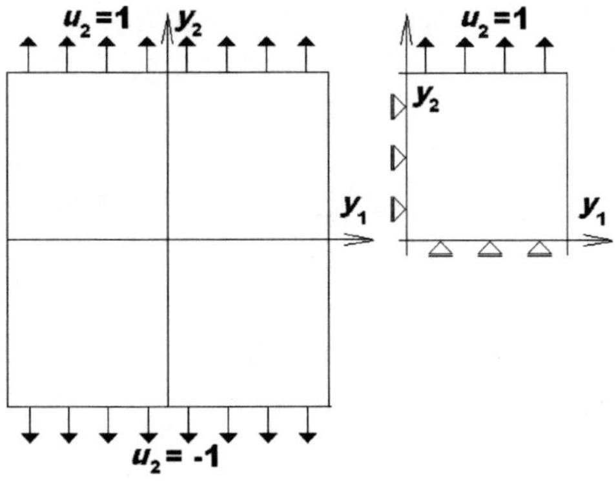

Figure 3: Original and computational model for responses of E_{22} .

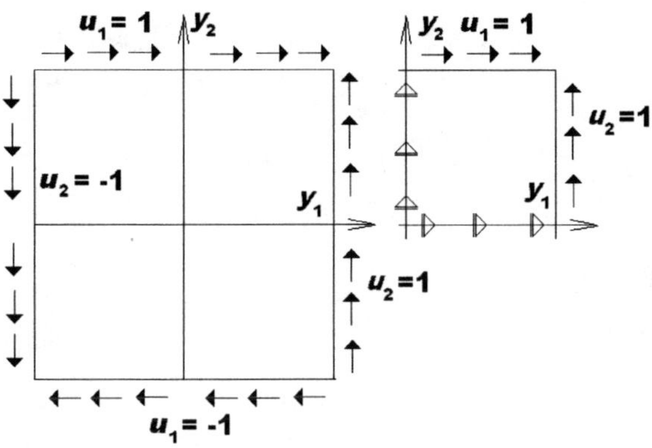

Figure 4: Original and computational model for responses of E_{12} .

Under the above described circumstances Hill's energy condition holds valid, as proved, e.g., by Suquet, [2]:

$$< \sigma_{ij}(y)\varepsilon_{ij}(y) >= S_{ij}E_{ij} \tag{5}$$

Using (1), (2_2) and (3) the components of the overall stresses are written in the following way:

$$S_{ij} =< \sigma_{ij}(y) >=< L_{ijkl}(y)\varepsilon_{kl}(y) >= (< L^{\mathrm{f}}_{ijkl}A^{\mathrm{f}}_{kl\alpha\beta}(y) >_{\mathrm{f}} + < L^{\mathrm{m}}_{ijkl}A^{\mathrm{m}}_{kl\alpha\beta}(y) >_{\mathrm{m}})E_{\alpha\beta} \tag{6}$$

where $<.>_{\mathrm{f}}$ stands for average on fiber and $<.>_{\mathrm{m}}$ is the average on matrix. This averaging process is made in such a way that the integrals are taken over fiber and matrix, respectively, but the denominator generally remains meas Ω , see (2).

By definition, the homogenized stiffness matrix $\boldsymbol{L}^{*}$ is written as:

$$S_{ij} = L^{*}_{ijkl} E_{kl} \tag{7}$$

Comparing (6) and (7) the overall stiffness matrix follows as

$$L^{*}_{ijkl} = < L^{\mathrm{f}}_{ijkl} A^{\mathrm{f}}_{kl\alpha\beta}(y) >_{\mathrm{f}} + < L^{\mathrm{m}}_{ijkl} A^{\mathrm{m}}_{kl\alpha\beta}(y) >_{\mathrm{m}} \tag{8}$$

It is worth noting that the homogenized stiffness matrix is symmetric with similar properties as that of the classical stiffness matrix in the problem defined in the microscale.

3 Localization using BEM

Without lack of generality, let us consider a symmetric unit cell depicted in Fig. 1, for example. The overall strain E_{ij} is assumed to be given independently of the shape of the unit cell and of the shape of the fiber. The loading of this unit cell will be given by unit impulses of E_{ij}, i.e. we successively select $E_{i_0 j_0} = E_{j_0 i_0} = 1$; $E_{ij} = 0$ for either $i_0 \neq i$ or $j_0 \neq j$. How to select the unit impulses of the overall strain components will be discussed later on.

Now we concentrate our attention on the approach of computing the concentration factors, which play the most important role in our solution of the optimal problem.

First let us specify the boundary conditions, being equivalent to the unit impulses of the overall strain components. In elasticity it is possible to prescribe the overall strain field all over the domain Ω of the unit cell. Then the solution of responses to the unit impulses is given by the periodicity conditions and tractions along the interfacial surface Γ between fibers and matrix. This approach is little bit cumbersome in case of debonding is admitted. The latter case is not considered here, but we apply more general form of introducing the unit impulses. It is well known that because of identity, see (2_3), and Green's theorem it holds:

$$\begin{aligned} E_{ij} &= \int_{\Omega} \varepsilon_{ij}(y)\,\mathrm{d}\Omega = \frac{1}{2}\int_{\Omega}\left[\frac{\partial u_i}{\partial y_j}(y) + \frac{\partial u_j}{\partial y_i}(y)\right]\mathrm{d}\Omega = \\ &= -\frac{1}{2}\int_{\partial\Omega}[u_i(y)n_j(y) + u_j(y)n_i(y)]\,\mathrm{d}\gamma(y) \end{aligned} \tag{9}$$

From (9) it immediately follows that the unit responses are given by prescribed displacements along the boundary of the unit cell. Moreover, using symmetry assumed in the beginning of this paper, we can solve the problem only on one quarter of the unit cell and get for the normal components of strains boundary conditions of the first quarter by Fig. 2 and Fig. 3 and for the shear strains boundary conditions according to Fig. 4.

The procedure is split into two steps. Assume the above described surface displacements to be prescribed along the entire boundary $\partial\Omega$ and there are no body forces here. In the first step, the unit cell obeys static equilibrium equations and linear homogeneous Hooke's law (homogeneous and isotropic medium):

$$\sigma_{ij}^0 = L_{ijkl}^0 \varepsilon_{kl}^0, \text{ in } \Omega, \quad \text{boundary conditions are fulfilled on } \partial\Omega \text{ in the sense of individual states} \tag{10}$$

L_{ijkl}^0 are components of not yet determined material stiffness matrix (stiffness tensor). These components will be stated later. Such a medium is called comparative one.

The solution of (10) is easy, as the comparative medium is homogeneous and isotropic:

$$u_i^0 = E_{ij} y_j, \quad \varepsilon_{ij}^0 = E_{ij} \quad \text{in } \Omega, \quad p_i^0 = \sigma_{ij}^0 n_j = L_{ijkl}^0 E_{kl} n_j \quad \text{on } \partial\Omega$$

In the second step a geometrically identical unit cell is considered. Also the loading and boundary conditions on $\partial\Omega$ remain valid. Define

$$\bar{u}_i = u_i - u_i^0 = u_i = E_{ij} y_j, \ \bar{\varepsilon}_{ij} = \varepsilon_{ij} - \varepsilon_{ij}^0 = \varepsilon_{ij} - E_{ij}, \ \bar{\sigma}_{ij} = \sigma_{ij} - \sigma_{ij}^0 = \sigma_{ij} - L_{ijkl}^0 E_{kl} \tag{11}$$

Our next aim is to determine primed quantities, components of displacement vector $\bar{u}_i$ and components of strain and stress tensors $\bar{\varepsilon}_{ij}$ and $\bar{\sigma}_{ij}$. In order to do so, system of fifteen equations of elasticity (2) has to be formulated for the primed set. We start with Hooke's law, which is valid for heterogeneous medium:

$$\sigma_{ij}(y) = L_{ijkl}(y) \varepsilon_{kl}(y) \quad \text{in } \Omega \tag{12}$$

Since the material stiffness tensor appears to be nonhomogeneous and unisotropic, idea used in [3], among others, will be adapted also here:

$$\sigma_{ij}(y) = L_{ijkl}^0 \varepsilon_{kl}(y) + \tau_{ij}(y) \quad \text{in } \Omega \tag{13}$$

where τ_{ij} are components of polarization tensor and the direct relation between stresses and strains becomes homogeneous and isotropic, so that integral formulation of elastic problem may be formulated. Subtracting (13) and (12) yields:

$$\tau_{ij} = [L_{ijkl}]\varepsilon_{kl}, \qquad [L_{ijkl}] = L_{ijkl} - L^0_{ijkl} \tag{14}$$

which can be considered a definition of polarization tensor. Moreover, transformation to the primed system will not disturb the direct relation stresses – strains, as after substituting (13) to (11_3) gives:

$$\bar{\sigma}_{ij} = \sigma_{ij} - \sigma^0_{ij} = L^0_{ijkl}\varepsilon_{kl} + \tau_{ij} - \sigma^0_{ij} = L^0_{ijkl}\varepsilon_{kl} + \tau_{ij} - L^0_{ijkl}E_{kl} = L^0_{ijkl}\bar{\varepsilon}_{kl} + \tau_{ij} \tag{15}$$

Since both σ_{ij} and σ^0_{ij} are statically admissible, it holds (the following equations must be defined in the sense of distributions):

$$\frac{\partial(L^0_{ijkl}\bar{\varepsilon}_{kl} + \tau_{ij})}{\partial y_j} = 0 \quad \text{in } \Omega, \quad \bar{u}_i = u_i - u^0_i, \quad \bar{p}_i = p_i - p^0_i \quad \text{on } \partial\Omega \tag{16}$$

Following the assumption of the same prescribed boundary conditions, some of the terms in (16) disappear.

Owing to constant distribution of L^0_{ijkl} in Ω, the equivalent integral formulation can be written as:

$$\begin{aligned}\bar{u}_m(\xi) = & \int_{\partial\Omega} p^*_{mi}(y,\xi)\bar{u}_i(y)\,\mathrm{d}\gamma(y) - \int_{\partial\Omega} u^*_{mi}(y,\xi)\bar{p}_i(y)\,\mathrm{d}\gamma(y) + \\ & + \left([L^{\mathrm{f}}_{ijkl} - L^0_{ijkl}] \int_{\Omega^{\mathrm{f}}} + [L^{\mathrm{m}}_{ijkl} - L^0_{ijkl}] \int_{\Omega^{\mathrm{m}}} \right) \varepsilon^*_{mij}(y,\xi)\bar{\varepsilon}_k(y)\,\mathrm{d}\Omega \end{aligned} \qquad \xi \in \partial\Omega \tag{17}$$

$$\begin{aligned}c_{mn}(\xi)\bar{u}_n(\xi) = & \int_{\partial\Omega} p^*_{mi}(y,\xi)\bar{u}_i(y)\,\mathrm{d}\gamma(y) - \int_{\partial\Omega} u^*_{mi}(y,\xi)\bar{p}_i(y)\,\mathrm{d}\gamma(y) + \\ & + \left([L^{\mathrm{f}}_{ijkl} - L^0_{ijkl}] \int_{\Omega^{\mathrm{f}}} + [L^{\mathrm{m}}_{ijkl} - L^0_{ijkl}] \int_{\Omega^{\mathrm{m}}} \right) \sigma^*_{mij}(y,\xi)\bar{\varepsilon}_k(y)\,\mathrm{d}\Omega \end{aligned} \qquad \xi \in \partial\Omega \tag{18}$$

where c_{mn} are components of a tensor depending on position $\xi \in \partial\Omega$ and the quantities with asterisks are given kernels.

Differentiating (17) by ξ_n, applying Hooke's law, $L^{\mathrm{f}}_{mnkl} = L^0_{nmkl}$ and putting $\xi \in \partial\Omega$ provides

$$\begin{aligned}\bar{\varepsilon}_{mn}(\xi) = & \int_{\partial\Omega} P^*_{nmi}(y,\xi)\bar{u}_i(y)\,\mathrm{d}\gamma(y) - \int_{\partial\Omega} U^*_{nmi}(y,\xi)\bar{p}_i(y)\,\mathrm{d}\gamma(y) + \\ & + \int_{\Omega^{\mathrm{m}}} \left([L^{\mathrm{m}}_{mnkl} - L^0_{mnkl}]\right)\Sigma^*_{nmij}(y,\xi)\bar{\varepsilon}_{kl}(y)\,\mathrm{d}\Omega \end{aligned} \tag{19}$$

and after discretization of the above equation relation (3) can easily be obtained. As sum of concentration factors on domain Ω is unity tensor, we can write:

$$L^*_{ijkl} = < L^{\mathrm{f}}_{ijkl} A^{\mathrm{f}}_{kl\alpha\beta}(y) >_{\mathrm{f}} + < L^{\mathrm{m}}_{ijkl} A^{\mathrm{m}}_{kl\alpha\beta}(y) >_{\mathrm{m}} \tag{20}$$

and the energy functional is formulated as, see, e.g., [1]:

$$\Pi(\boldsymbol{u}, \Omega^{\mathrm{f}}) = \frac{1}{2}[L^{\mathrm{f}}_{ijkl} < A^{\mathrm{f}}_{kl\alpha\beta}(p_s) >_{\mathrm{f}} + L^{\mathrm{m}}_{ijkl} < A^{\mathrm{m}}_{kl\alpha\beta}(p_s) >_{\mathrm{m}}] E_{ij} E_{\alpha\beta} - \lambda(\int_{\Omega^{\mathrm{f}}} d\Omega - C)$$

In the sense of the Inverse variational principle the lagrangian multiplier remains the same along the interfacial boundary, which provides algorithm described in [4].

4 Example

Unit cell is considered with fiber volume ratio equal to 0.21 ($\pi/16$). Since we compare energy densities at nodal points of the interfacial boundary, the relative energy density λ may be regarded as the comparative quantity influencing the movement of the interfacial boundary. One phase possesses the following material properties: Young's modulus of the first one is $E^1 = 210$ MPa, Poisson's ratio $\nu^1 = 0.16$; and the second $E^2 = 17$ MPa, and $\nu^2 = 0.3$.

In Fig. 5 the starting shape and the final, optimal shape are depicted for stiff fiber (phase 1) and weaker matrix (phase 2). In Fig. 6 similar picture is presented for weak fiber (phase 2) and stiffer matrix (phase 1). In both cases, the stiffer phase "tries" to occupy larger area exposed to loading.

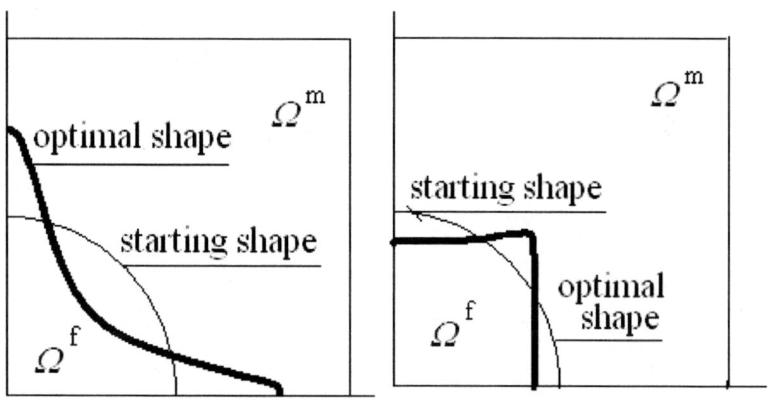

Figure 5: Optimal shapes for stiff and weak fibers.

5 Conclusions

In this paper inverse variational principle has been applied to the solution of optimal fiber shape design on a unit cell of periodic composite structure. When searching for optimal shape design of fibers in composite structures, many formulations have been used in the past. They very often start with minimum strain energy function. This assumption is in Inverse variational principles fulfilled implicitly. A natural requirement is the restriction to the constant volume or area in 2D or volume in 3D. Periodic distribution of fibers is considered in this paper.

The requirement of the constant volume or area seams to be restrictive, particularly when expecting application of Inverse variational principles to larger range of problems. Actually, it is not so. The constant C may change, too. Thus the formulation has to be extended in such a way that C is involved into the problem as a new variable and may be variated (differentiated) in some reasonable way.

Acknowledgement

The financial support of Grant agency of the Czech Republic, project No. 103/07/0304 is greatly appreciated.

References

[1] Prochazka, P., Shape optimal design using Inverse Variational Principles, submitted to EABE

[2] Suquet P.M., Elements of homogenization for inelastic solid mechanics, Lecture Notes in Physics, 272 - Homogenization Technique for Composite Media, 1987

[3] Prochazka P., Sejnoha, J., Behavior of composites on bounded domain. BE Communications, 7, 1, 6-8, 1996

[4] Prochazka, P., Shape optimization of composites based on minimum potential energy, OPTI 2007, New Forest, UK

Analytical solution of adhesion contact for a rigid sinusoidal surface on a semi-infinite elastic body

R. R. A. Sriwijaya, K. Takahashi & K. Jatmiko
Department of International Development Engineering, Tokyo Institute of Technology, Japan

Abstract

An analytical solution of adhesion contact for a rigid sinusoidal surface on a semi-infinite elastic body is presented. The solution for an equilibrium condition of the system for a combination of the work of Johnson [*International Journal of Solids and Structures,* 32(3–4), pp. 423–430, 1995] and Zilberman and Persson [*Solid State Communications,* 123(3–4), pp. 173–177, 2002; *Journal of Chemical Physics,* 118(14), pp. 6473–6480, 2003] under zero external pressure is obtained. The interfacial term of the total energy is calculated by considering the curvature of the contact area following the approach of Zilberman and Persson rather than the straight line of the contact area as Johnson. Our results agree with both the analytical result of Johnson for a slightly wavy surface and the numerical results of Zilberman and Persson for a largely wavy surface at the limitations of their assumptions. The equilibrium contact width is clearly expressed and the effect of the surface roughness is discussed.
Keywords: analytical solution, equilibrium condition, critical work of adhesion, sinusoidal surface, semi-infinite elastic body.

1 Introduction

The contact problems of a semi-infinite elastic body with a flat or a wavy surface have been investigated by some researchers. Johnson *et al.* [1] investigated a smooth contact problem of an elastic body with slightly wavy surface in contact with a rigid body with flat surface. They obtained a relation between the applied external pressure and the amplitude of roughness.

WIT Transactions on Engineering Sciences, Vol 55, © 2007 WIT Press
www.witpress.com, ISSN 1743-3533 (on-line)
doi:10.2495/SECM070141

Johnson [2] extended his work [1] by considering the adhesion effect, and solved it analytically. However, his solution can be applied only to wavy contact with adhesion with small amplitude roughness.

Zilberman and Persson [3, 4] investigated an adhesion contact of a largely wavy surface and solved it numerically. They considered the curvature rather than the straight line of the contact area in the calculation of interfacial term of the total energy. However, a local minimum as well as a local maximum of the system cannot be determined directly from their solution.

Considering the limitations of the work of Johnson [2] and Zilberman and Persson [3, 4], the present work is intended to obtain an analytical solution for an equilibrium condition of the system for combination of their works under zero external pressure. In addition, the effect of the thermodynamic work of adhesion as well as the effect of the surface roughness on the system is investigated.

2 Analytical method

2.1 Pressure distribution and displacement on the surface

A semi-infinite elastic body with initially flat surface subjected to a sinusoidal rigid surface is considered. It is assumed that the elastic body is homogeneous and isotropic, and the frictionless contact presents at the interface.

The surface pressure distribution and the surface displacement of the adhesion contact are the resultant of the surface pressure distribution and the surface displacement of two adhesionless contacts. The first is a semi-infinite elastic body subjected to a sinusoidal rigid surface while the second is a semi-infinite elastic body pulled by a flat rigid surface. In fact, the second adhesionless contact can be represented as a crack problem [5]. In the present work, the surface pressure distributions and surface displacements of Westergaard [6] and Koiter [7] are used.

The net surface pressure distribution, $p(x)$, upon the elastic body within the contact region is given by [5], i.e. $p(x)= p^s(x) + p^c(x)$, where $p^s(x)$ is the surface pressure distribution relates to the sinusoidal rigid surface, obtained by [6]

$$p^s(x) = \frac{2\bar{p}^s \cos\left(\frac{\pi x}{\lambda}\right)}{\sin^2\left(\frac{\pi a}{\lambda}\right)}\left[\sin^2\left(\frac{\pi a}{\lambda}\right) - \sin^2\left(\frac{\pi x}{\lambda}\right)\right]^{1/2}, \tag{1}$$

and $p^c(x)$ is the surface pressure distribution relates to the flat rigid surface, obtained by [7]

$$p^c(x) = \bar{p}^c\left[1-\left(\frac{\cos\left(\frac{\pi a}{\lambda}\right)}{\cos\left(\frac{\pi x}{\lambda}\right)}\right)^2\right]^{-1/2}, \tag{2}$$

where $\overline{p}^s$ is the mean pressure as in [6], $\overline{p}^c$ is the mean pressure as in [7], and a is the semi-contact width.

In the same manner as the net surface pressure distribution, $p(x)$, the net mean pressure is given by [2], i.e. $\overline{p} = \overline{p}^s + \overline{p}^c$.

Johnson *et al.* [1] obtained an expression for the mean pressure, $\overline{p}^s$, in one period, i.e. $\overline{p}^s = (\pi E^* h_o/\lambda)\sin^2(\pi a/\lambda)$, where h_o and λ are the amplitude of roughness and the wavelength of a sinusoidal rigid profile, respectively, and E^* is the plane strain modulus of the elastic semi-infinite body.

In the case of a rigid body in contact with an elastic body, the elastic modulus, E^* is given by $E^* = E/1-\upsilon^2$, where E and υ are Young's modulus and Poisson's ratio of the elastic body, respectively.

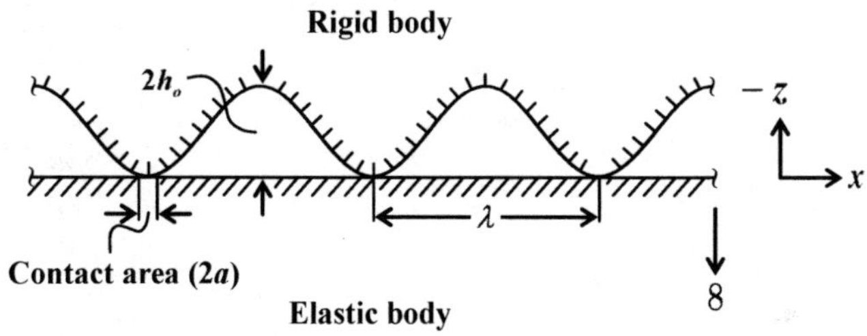

Figure 1: Geometry of the contact problem of a rigid body in contact with a semi-infinite elastic body.

The surface profile of the rigid body is expressed by $z(x) = h_o \cos(2\pi x/\lambda)$ (see Fig. 1).

The net surface displacement on the elastic body within the contact region is given by [3], [4], i.e. $u_z(x) = u_z^s(x) + u_z^c(x)$, where $u_z^s(x)$ is the surface displacement relates to the sinusoidal rigid surface, obtained by [6]

$$u_z^s = \frac{(1-\upsilon^2)\overline{p}^s\lambda}{\pi E \sin^2\left(\frac{\pi a}{\lambda}\right)} \cos\left(\frac{2\pi x}{\lambda}\right), \tag{3}$$

and $u_z^c(x)$ is the surface displacement relates to the flat rigid surface, obtained by [7]

$$u_z^c = \frac{2(1-\upsilon^2)\overline{p}^c\lambda}{\pi E} \left| \ln\left|\sin\frac{\pi a}{\lambda}\right|\right|. \tag{4}$$

Here, $u_z^c(x)$ within contact region is not zero, which is different from [3, 4].

2.2 Total energy of the present system

2.2.1 Elastic term in the total energy

The total free energy of the present system consists of the elastic term and the interfacial term. The elastic term is induced by the applied surface pressure distributions within the contact region. The total pressure distribution consists of Eqs. (1) and (2). The total elastic energy term, $U_{E\ total}$, over the whole semi-infinite elastic body in one period is obtained by

$$U_{E\ total} = \frac{1}{2}\int_{A_\lambda} p(x)\ u_z(x)\ dA, \tag{5}$$

where the parameters A_λ is the nominal contact area (i.e. λ^2). With Eqs. (1)-(4), Eq. (5) gives

$$U_{E\ total} = A_\lambda \left\{ \left[\frac{\overline{p}^{s\,2}\lambda}{4\pi E^*\sin^2\left(\frac{\pi a}{\lambda}\right)}\left(1+\cos^2\left(\frac{\pi a}{\lambda}\right)\right)\right] - \left[\frac{\overline{p}^s\overline{p}^c\lambda}{\pi E^*}\left|\ln\left|\sin\left(\frac{\pi a}{\lambda}\right)\right|\right|\right] \right. \tag{6}$$

$$\left. - \left[\frac{\overline{p}^s\overline{p}^c\lambda}{2\pi E^*\sin^2\left(\frac{\pi a}{\lambda}\right)}\cos^2\left(\frac{\pi a}{\lambda}\right)\right] + \left[\frac{\overline{p}^{c\,2}\lambda}{\pi E^*}\left|\ln\left|\sin\left(\frac{\pi a}{\lambda}\right)\right|\right|\right]\right\}.$$

Since we have no external pressure in the present system, the net mean pressure is equal to zero ($\overline{p}=0$), Eq. (6) can be represented as

$$U_{E\ total} = \frac{A_\lambda \pi E^* h_o^2}{4\lambda}\sin^4\left(\frac{\pi a}{\lambda}\right). \tag{7}$$

2.2.2 Interfacial term in the total energy

The interfacial term, U_I (i.e. energy change from the surface to the interface within the contact region [8]), of the system in one period is determined by considering the curvature of the rigid surface, given by $U_I = -A_\lambda \Delta\gamma s/\lambda$, where A_λ is the same parameter as in Eq. (5), and $\Delta\gamma$ is the thermodynamic work of adhesion, given by $\Delta\gamma = \gamma_1 + \gamma_2 - \gamma_{12}$, where γ_1 and γ_2 are the surface energies of the rigid body and the elastic body, respectively, and γ_{12} is their interfacial energy, and s is the surface length of the contact area. Considering the curvature of the surface roughness, s can be expressed by

$$s = 2\int_0^a \sqrt{1+\left(\frac{2\pi h_o}{\lambda}\right)^2 \sin^2\left(\frac{2\pi x}{\lambda}\right)}dx \cdot \tag{8}$$

The interfacial term, U_I, is

$$U_I = -\frac{2A_\lambda \Delta\gamma}{\lambda}\int_0^a \sqrt{1+\left(\frac{2\pi h_o}{\lambda}\right)^2 \sin^2\left(\frac{2\pi x}{\lambda}\right)}dx \cdot \tag{9}$$

Since Eq. (9) contains an elliptic integral of the second kind, consequently it is calculated by numerical methods.

2.2.3 Total energy of the system

The total energy of the system, U_{total}, (i.e. Gibbs free energy) in one period is given by

$$U_{total} = U_{E\,total} + U_I \cdot \tag{10}$$

Substituting Eqs. (7) and (9) into Eq. (10) gives the total energy of the system in one period, i.e.

$$U_{total} = \frac{A_\lambda \pi E^* h_o^2}{4\lambda}\sin^4\left(\frac{\pi a}{\lambda}\right) - \frac{2A_\lambda \Delta\gamma}{\lambda}\int_0^a \sqrt{1+\left(\frac{2\pi h_o}{\lambda}\right)^2 \sin^2\left(\frac{2\pi x}{\lambda}\right)}dx \cdot \tag{11}$$

Eq. (11) can be rearranged to

$$U_{total} = \frac{A_\lambda E^* h_o^2}{\lambda}\left\{\frac{\pi}{4}\sin^4\left(\frac{\pi a}{\lambda}\right) - \frac{\pi^2}{\lambda}\Delta\bar{\gamma}\int_0^a \sqrt{1+\left(\frac{2\pi h_o}{\lambda}\right)^2 \sin^2\left(\frac{2\pi x}{\lambda}\right)}dx\right\}, \tag{12}$$

where $\Delta\bar{\gamma}$ is the normalized thermodynamic work of adhesion, given by $\Delta\bar{\gamma} = \Delta\gamma/\left(E^* h_o^2 \pi^2/2\lambda\right)$.

2.3 Equilibrium of the system

The equilibrium of the system is given by minimizing the total energy, U_{total}, with respect to the semi-contact width, a. Therefore, the equilibrium contact width can be obtained by

$$\frac{\partial U_{total}}{\partial a} = \frac{A_\lambda E^* h_o^2}{\lambda}\left(\frac{\pi^2}{\lambda}\right)\left\{\sin^3\left(\frac{\pi a}{\lambda}\right)\cos\left(\frac{\pi a}{\lambda}\right) - \Delta\bar{\gamma}\sqrt{1+\left(\frac{2\pi h_o}{\lambda}\right)^2 \sin^2\left(\frac{2\pi a}{\lambda}\right)}\right\} = 0. \tag{13}$$

Eq. (13) can be represented by the normalized work of adhesion, $\Delta\bar{\gamma}$ i.e.

$$\Delta\bar{\gamma} = \frac{\sin^3\left(\frac{\pi a}{\lambda}\right)\cos\left(\frac{\pi a}{\lambda}\right)}{\sqrt{1+\left(\frac{2\pi h_o}{\lambda}\right)^2 \sin^2\left(\frac{2\pi a}{\lambda}\right)}} \quad . \tag{14}$$

This equation presents a necessary condition for equilibrium of the system.

3 Results and discussion

We have confirmed our results with the total energy calculated by Zilberman and Persson's equation [4] and the equilibrium condition calculated by Johnson's equation [2]. It is shown that our results conform to those of Zilberman and Persson and Johnson at the limitations of their assumptions.

Figs. 2(a) and 2(b) are plotted by Eq. (12) with the amplitude of the roughness, h_o/λ = 0.5 and the wavelength, λ = 50 Å, respectively. Fig. 2(a) shows the relation between the normalized total energy, $U_{total}/(A_\lambda E^* h_o^2/\lambda)$, and the normalized contact width, $2a/\lambda$ for a normalized work of adhesion, $\Delta\bar{\gamma} = 0.1$. Fig. 2(b) shows the relation between the normalized total energy, $U_{total}/(A_\lambda E^* h_o^2/\lambda)$, and the normalized contact width, $2a/\lambda$, for several normalized work of adhesion, $\Delta\bar{\gamma}_1$ = 0.06, $\Delta\bar{\gamma}_2$ = 0.8, $\Delta\bar{\gamma}_3$ = 0.1, $\Delta\bar{\gamma}_4$ = 0.12, $\Delta\bar{\gamma}_5 \approx 0.126\cdots$ and $\Delta\bar{\gamma}_6$ = 0.14. It is shown that the normalized total energy decreases as the normalized work of adhesion increases.

In Fig. 2(a), it shows that the normalized total energy curve has a local minimum and a local maximum. This suggests that when the elastic body contacts to the rough rigid body, the normalized contact width immediately snap into the local minimum, point A. And, when the local maximum, point B is reached, the normalized contact width immediately snap into complete contact.
In Fig. 2(b), each curve for $\Delta\bar{\gamma}$ = 0.06 - 0.12 has a local minimum (i.e. points A_1-A_4) and a local maximum (i.e. B_1-B_4), while the curve for $\Delta\bar{\gamma}_5 \approx 0.126\cdots$ has a horizontal inflection (i.e. point C). On the other hand, the curve for $\Delta\bar{\gamma}_6$ = 0.14 has no a horizontal inflection, neither local minimum nor local maximum. In the same manner with Fig. 2(a), this suggests that when the elastic body contacts to the rough rigid body, the normalized contact width immediately increases to the local minimum $\Delta\bar{\gamma}_5 < 0.126\cdots$, while for $\Delta\bar{\gamma} \geq 0.126\cdots$ the normalized contact width immediately increases to the complete contact. The same manner can also be explained for $\Delta\bar{\gamma}_6$ = 0.14. All of the local minima are stable equilibrium points, whereas, all of the local maxima and the horizontal inflection are unstable points.

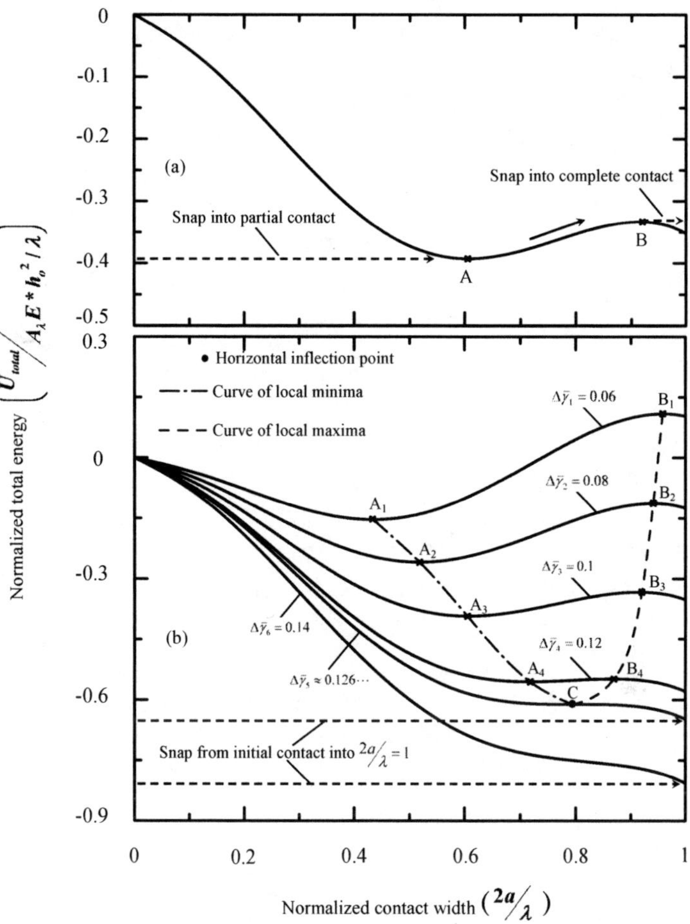

Figure 2: The relation between the normalized total energy and the normalized contact width. (a) curve is calculated for $\Delta\bar{\gamma} = 0.1$ and $h_o/\lambda = 0.5$, (b) curves are calculated for $\Delta\bar{\gamma} = 0.1$ and several h_o/λ.

Fig. 3 is plotted by Eq. (14) for h_o/λ = 0.5. It shows the relation between the normalized work of adhesion, $\Delta\gamma/(E^* h_o^2 \pi^2 / 2\lambda)$, and the normalized contact width, $2a/\lambda$. The curve of stable equilibrium points corresponds to the curve of local minima in the Fig. 2(b), while the curve of unstable points corresponds to the curve of local maxima. The critical normalized work of adhesion, $\Delta\bar{\gamma}_{crit}$, corresponds to the horizontal inflection point. In the same manner, points A_1-A_4, B_1-B_4 and C in Fig. 3 correspond to points A_1-A_4, B_1-B_4 and C in Fig. 2(b). The contact width of the stable equilibrium points and the unstable points can be obtained from the curves in Fig. 3 for a given normalized work of adhesion. If we could give such a normalized contact width larger than the curve of unstable points under zero external pressure condition, the normalized contact width immediately increases to snap into complete contact.

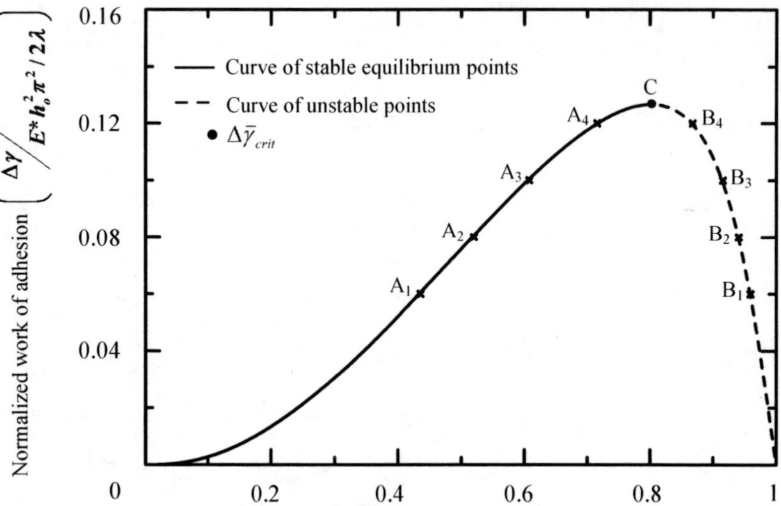

Figure 3: The relation between the normalized work of adhesion and the normalized contact width. The equilibrium curve is plotted for $h_o/\lambda = 0.5$ and several $\Delta\bar{\gamma}$.

Fig. 4 is plotted by Eq. (14) in the same manner as Fig. 3 for several h_o/λ. In the case of the normalized amplitude of roughness is close to zero, (i.e. $h_o/\lambda \approx 0$), the present solution agrees with the analytical solution of Johnson [2] for slightly wavy surface. On the other hand, if the normalized amplitude of roughness is large enough, the solution agrees with the numerical solution of Zilbermann and Persson [3,4] for largely wavy surface. The critical work of adhesion, $\Delta\bar{\gamma}_{crit}$, for each h_o/λ is given in Fig. 4. If a value of the normalized work of adhesion is larger than the $\Delta\bar{\gamma}_{crit}$, the normalized contact width immediately increases to snap into complete contact directly after initial contact because there is no equilibrium point within the system.

4 Conclusions

An analytical solution of adhesion contact for a rigid sinusoidal surface on a semi-infinite elastic body is presented. The solution for an equilibrium condition of the system for combination of Johnson's and Zilberman-Persson's works under zero external pressure is obtained. The interfacial term of the total energy is calculated by considering the curvature of the contact area in the same way as

Zilberman and Persson. Our results agree with both the analytical result of Johnson and the numerical results of Zilbermann and Persson at the limitations of their assumptions. The equilibrium contact width is clearly expressed and the effect of the surface roughness is discussed.

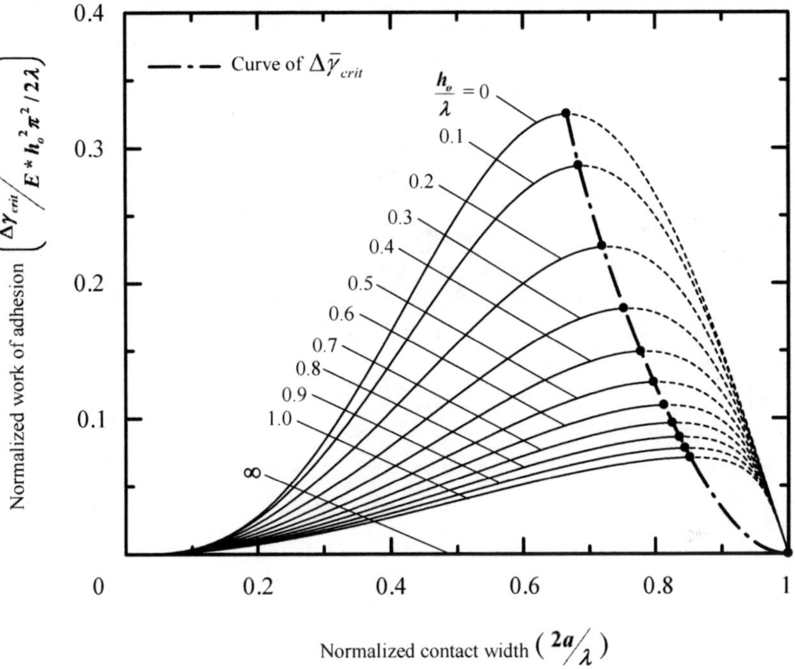

Figure 4: The relation between the normalized work of adhesion and the normalized contact width. The $\Delta\bar{\gamma}_{crit}$ are plotted for several h_o/λ.

Acknowledgements

First author is deeply grateful to AUNSeed-Net/JICA for their financial support to him during study at Tokyo Institute of Technology, Japan. The authors thank Prof. Shigeki Saito for his valuable comments, Dr. Silviu Zilberman for his numerical source code and Mr. Hemthavy Pasomphone for his help during preparing this manuscript.

References

[1] Johnson, K.L., Greenwood, J.A. & Higginson, J.G., The contact of elastic regular wavy surfaces, *International Journal of Mechanical Sciences*, 27(6), pp. 383-396, 1985.

[2] Johnson, K.L., The adhesion of two elastic bodies with slightly wavy surfaces, *International Journal of Solids and Structures*, 32(3-4), pp. 423-430, 1995.

[3] Zilberman, S. & Persson, B.N.J., Adhesion between elastic bodies with rough surfaces, *Solid State Communications*, 123(3-4), pp. 173-177, 2002.

[4] Zilberman, S. & Persson, B.N.J., Nano adhesion of elastic bodies: Roughness and temperature effects, *Journal of Chemical Physics*, 118(14), pp. 6473-6480, 2003.

[5] Kendal, K., An adhesion paradox, *Journal of Adhesion*, 5, pp. 77-79, 1973.

[6] Westergaard, H.M., Bearing pressure and cracks, *Journal of Applied Mechanics*, 6(2), pp. A49-A53, 1939.

[7] Koiter, W.T., An infinite row of collinear cracks in an infinite elastic sheet, *Ingenieur-Archiv*, 28(70), pp. 168-172, 1959.

[8] Takahashi, K., Mizuno, R. & Onzawa, T., Influence of the stiffness of the measurement system on the elastic adhesional contact, *Journal Adhesion Science and Technology*, 9(11), pp. 1451- 1464, 1995.

Investigation of the temperature behaviour of sliding rubber materials

O. Lahayne & J. Eberhardsteiner
Vienna University of Technology,
Institute for Mechanics of Materials and Structures (IMWS),
Vienna, Austria

Abstract

At the IMWS a testing device called Linear Friction Tester (LFT) was developed, which is mainly used to characterise and quantify the friction behaviour of tire materials under varying conditions. During the friction process a considerable amount of heat is generated at the contact surface. Starting from theoretical approaches, there is supposed to be a strong interrelation between the friction coefficient μ and the temperature T of the rubber. By means of temperature sensors, FE-calculations and a model for the heat conduction it was possible to measure and calculate the heating of the rubber with high accuracy. In this paper some examples are presented for results of such measurements. A model for the heat conduction is illustrated, and testing the correlation between measurement and calculations validates the practicability of the model.
Keywords: Rubber, Temperature, Friction, Tire.

1 The testing device

In 1997 the Linear Friction Tester has been constructed [3]. The test conditions at the LFT are well defined, since it is placed in an air-conditioned container, and in good agreement with realistic breaking situations. The core of the device is a linear drive that pulls a sledge with a rubber test specimen over the friction surface. A pneumatic system applies the vertical load (up to 2000 N or 10 bar). The test specimens have a size between 10 and 30 cm^2 and contain different arrangements of lamellae. The friction surfaces were cut out of real road surfaces. Also friction tests on snow and ice surfaces can be performed. For some experiments additionally a high-speed-camera recorded the deformation of the sample.

WIT Transactions on Engineering Sciences, Vol 55, © 2007 WIT Press
www.witpress.com, ISSN 1743-3533 (on-line)
doi:10.2495/SECM070151

Figure 1: Friction surface with pyrometer and rubber sample.

The temperature of the rubber samples was recorded in two ways: If the temperature at the friction surface was of interest, an infrared-Pyrometer was used. It has a resolution of 1,4 mm and a response time of 5 μs, which is adequate for the size and sliding velocities of the rubber samples.

The temperature within the samples was recorded with up to four thermocouples, which were implanted in the samples. In both cases the signals of the measurement device were synchronised with the position of the samples.

2 Theory

Basis for the analytical calculation of the heat conduction within the samples is Fourier's law of heat conduction:

$$\dot{T}(x,t) = \frac{\lambda_m}{c_m \rho_m} T''(x,t) = a_m T''(x,t) \tag{1}$$

In this one-dimensional version, T denotes the temperature, λ the heat conductivity, c the heat capacity, ρ the density, and a the temperature conductivity. x is the distance to the friction surface, as shown in Figure 2. If this one-dimensional version is oversimplifying the problem will be discussed later. The maximum temperature within the sample is called T_R; the environment temperature T_0. Based on the assumption that the carrier material will always have the temperature T_0, the stable end condition after a sufficient friction time is a linear decrease of the temperature within the sample between T_R and T_0, as plotted in Figure 2. v_{slide} is the sliding velocity, F_x the vertical load.

As a first step, the process of heat conduction was calculated by means of a heat-flow-simulation based on Fourier's equation. For a one-dimensional simulation, the distance between the friction surface and the carrier material was divided in intervals of the same length; for two dimensions the cross-section was divided into squares. For the example in Figure 3, the absorbed power was highest in the middle of the sample and zero at the corners.

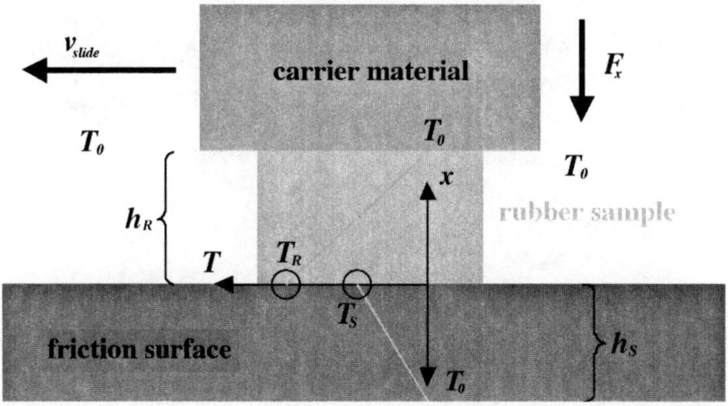

Figure 2: Schema of heat conduction during friction process.

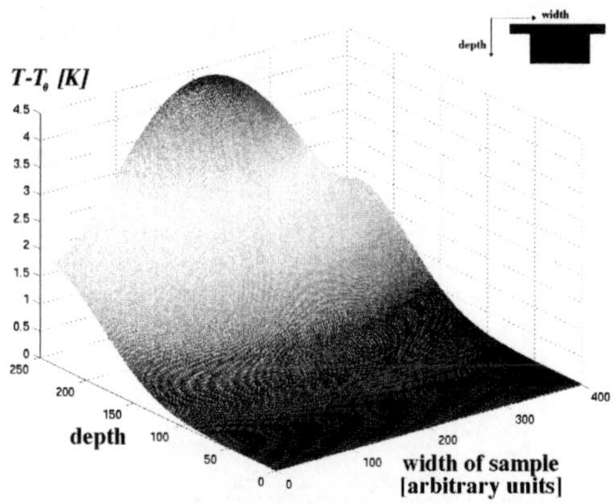

Figure 3: Result of simulation of heat conduction.

The results of the measurements and the heat-flow simulation were compared to each other and to an analytical solution of Fourier's equation. The boundary conditions for all methods were the same. For the analytical solution you have to distinguish between the process of heating and the process of cooling. Equation (2) shows the exact solution for Equation (1) for the proper boundary conditions.

$$T_W(x,t) = T_0 - U_m\left(1 - \tfrac{x}{h_G}\right) + \frac{8}{\pi^2}U_m \sum_{k=0}^{\infty} \frac{1}{(2k+1)^2}\cos(n_k x)\cdot e^{-n_k{}^2 at}$$
$$T_A(x,t) = T_0 + \frac{8}{\pi^2}U_m \sum_{k=0}^{\infty}\left[\frac{1}{(2k+1)^2} - \frac{1}{\pi}\sum_{l=0}^{\infty}\frac{1}{(2l+1)^2}\cdot e^{-n_l{}^2 at_r}\left(\frac{1}{l+k}+\frac{1}{l-k}\right)\right]\cos(n_k x)\cdot e^{-n_k{}^2 at} \quad (2)$$
$$n_k = \frac{2k+1}{h_G}\frac{\pi}{2},\ U_m = \frac{P_R h_R}{\lambda a}$$

P_R is the power generated by the friction process and absorbed by the rubber. The first line of Equation (2) is the solution for the process of heating; the second line the solution for the process of cooling.

Since these solutions are a bit unwieldy, for some tasks a crude approximation for the process of heating has been used:

$$T_W(x,t) = \frac{P_R}{\lambda A}\sqrt{2at} \cdot e^{-\frac{x}{\sqrt{2at}}} + T_0. \tag{3}$$

This approximation is usable only for $x << \sqrt{2at}$.

The power generated by the friction process is given by

$$P_{tot} = F_y \cdot \frac{s}{t} = F_y \cdot v_{Slide} = \mu \cdot F_x \cdot v_{Slide}. \tag{4}$$

Most of the generated power is absorbed by the friction surface; the rubber absorbs only about 1 % [4, 5]. Because the partition of P_{tot} between rubber and surface wasn't precisely known, it was for some tasks fitted to the results. Alternatively P_R can be calculated by means of FE- calculations, as Figure 5 will show.

3 Results

3.1 Temperature within the samples

It was helpful to test the correctness of the solution of Fourier's equation first not during a friction process, but by heating up a rubber sample at one side. In this way it is possible to reach higher temperatures without the vibrations of a friction process. For this, a small heating element was used, which affected only the surface of the sample, while the opposite surface was fixed on the carrier material. The temperature data were measured by four thermocouples. Figure 4 shows the sample and the position of the sensors.

Figure 4: Rubber sample with thermocouples.

Figure 5 shows the results of a test with this sample. The applied temperature was 55 K above T_0.

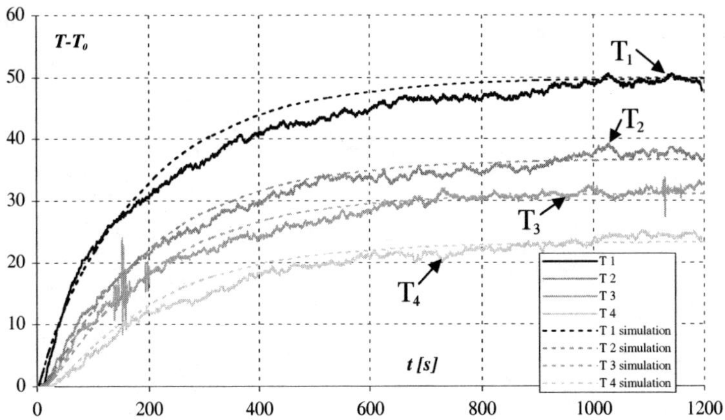

Figure 5: Temperature within the sample during heating.

The solid lines show the measured temperature, the dashes lines the results of the solution based on Equation (2) belonging to the position of the thermocouples T1-T4. The fit between measurement and calculation is satisfactory.

For this test the surface has been heated up homogeneous. For a friction process, though, an inhomogeneous distribution of the temperature at the surface has to be expected because of the deformation of the sample. Since the periods of heating up are much shorter during friction processes, the rise of the temperature is much smaller than in Figure 5, as Figure 6 demonstrates.

Figure 6 shows the results for two measuring points at distances of 0.3 mm (T3) and 0.7 mm (T1) from the surface. The data are compared with

- the results of a one-dimensional heat-flow-simulation (1D),
- the results of a two-dimensional simulation (2D), as shown in Figure 3,
- the analytical solution of Fourier's equation as shown in Equation (2).

As it can be seen, the results of the three methods are equivalent within the limits of the statistical spread of the measured data.

The length of the friction surface, as shown in Figure 1, was limited to 300 mm. Because of that only comparatively low temperatures were measured within the sample, especially if the distance of the sensor to the friction surface was higher than 1 mm. At the surface of the samples itself the temperature was always maximal. Because of that more precise tests were possible by the help of the pyrometer, which measured the temperature at the surface of the samples. This, though, was only possible after the end of the friction process because of the installation of the pyrometer.

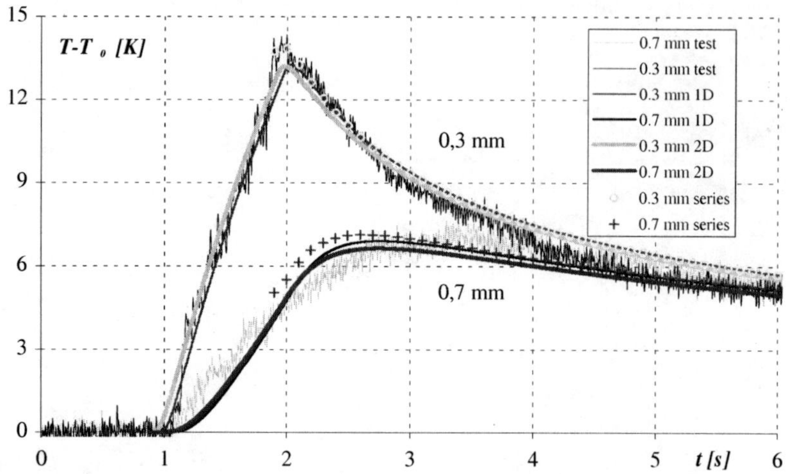

Figure 6: Temperature within the sample during friction process.

3.2 Temperature at the friction surface

The highest rises in temperature were measured for the roughest surfaces, such as asphalt. On this surface first a series of tests has been done to analyse the influence of the test parameters on the temperature. Not surprisingly, there is a strong influence of the vertical force, as can be seen in Figure 7. The vertical force was varied between 270 and 560 N, resulting in pressures between 1.7 and 3.5 bar. Only for 420 N the measuring points were marked with crosses. On the abscissa the sliding distance was plotted. The width of the sample was 20 mm.

In the same way the influence of other parameters was analysed, such as the influence of the sliding velocity, the sliding distance as well as the compound and the geometry of the sample. This demonstrates that the assumption for the heat distribution in Figure 3 is quite realistic.

These measurements were compared to results of FE-simulations of the temperature distribution, based on a material and friction model [5, 6]. Crucial for the characteristics of $T(s)$ is the deformation behaviour of the rubber sample. Figure 8 shows the temperature along the centre-line of the sample, based on FE-simulations and on measurements for the same parameters. Except for the maximum at the leading edge (left, -10 mm) which can't be resolved by the sensor, there is a good correlation between measurement and calculation.

This example alludes to tests on a very rough surface, which causes high deformations of the rubber sample, especially at the leading edge. For this reason there is usually a maximum for $T(s)$ at the leading edge (at about -12 mm) and a higher maximum near the middle of the sample.

For the result in Figure 7, the rubber sample has been moved over the pyrometer; in this way the spatial distribution of the temperature was measured. For measuring the temporal run of the temperature, the sample was usually stopped near the maximum of *T(s)*, i.e. at about 135 mm for the examples in Figure 7. Figure 9 shows two examples for *T(t)* for two sliding velocities, 300 and 700 mm/s.

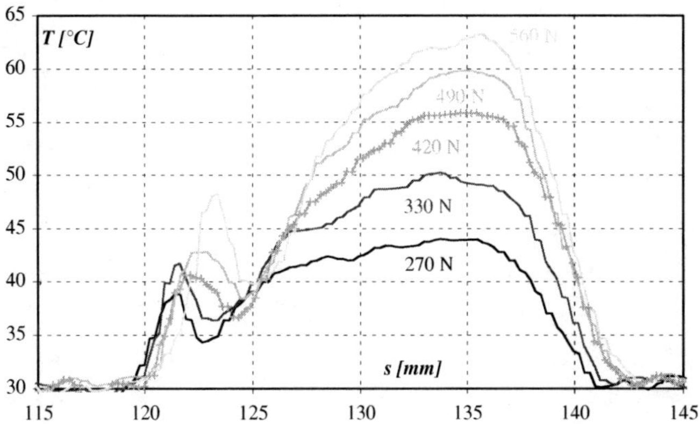

Figure 7: Influence of vertical force on the temperature.

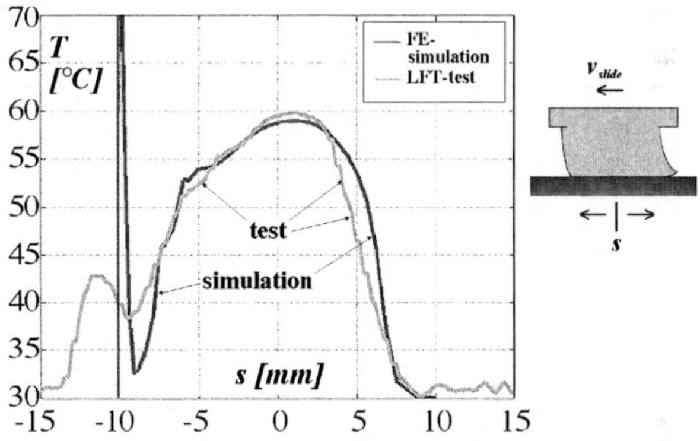

Figure 8: Comparison of FE- simulation and measurement.

The data of the measurements (solid lines) were compared to the results of the simulations, based on the parameters for the respective tests (circles). The results show that it is possible to describe and predict the temperature behaviour of a sliding rubber sample with satisfying accuracy, if the relevant parameters of the involved materials are given.

For some applications it is also possible to make good use of the approximation in Equation (3). In Figure 10, the mean value for three measurements for *T(t)* as in Figure 9 has been calculated (black circles), together with the standard deviation (grey bars). Using $t^{0,5}$ as x-axes, the diagram should show a linear run according to Equation (3).

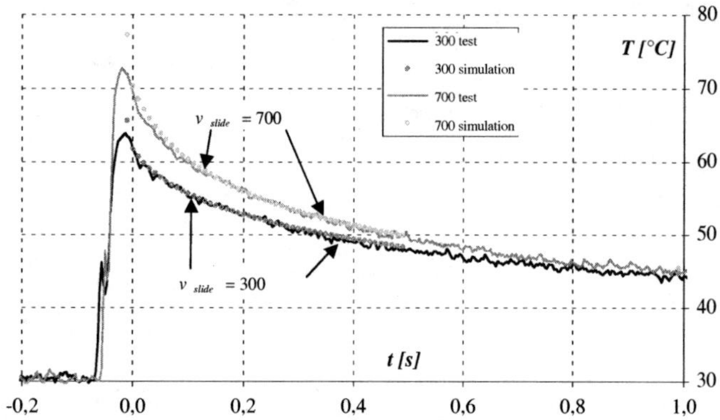

Figure 9: Examples for *T(t)* at the surface.

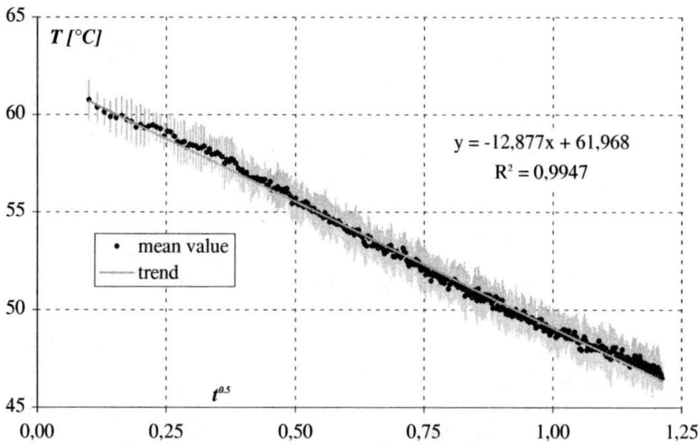

Figure 10: Example for approximation.

As can be seen, there is a good correlation between the straight line and the values for *T(t)*. Using this kind of analysis, the values of a and λ can be found.

These examples demonstrate the following:

- that the Equations (2) and (3) are correct solutions of Fourier's equation for heat conduction for these applications, and
- that the simulation of the heat conduction based on Equation (1) is equivalent to the exact solution in Equation (2).

Based on the results of the tests it is to be assumed that there is an inhomogeneous heat distribution over all cross-sections of the sample. In this case a higher number of thermocouples would be necessary to calculate the total flow of heat through the sample. Based on these results, it would be also possible to calculate the partition of the heat generated according to Equation (4) between surface and sample, which isn't known precisely so far. This affects only the total heat flow, and not the dependence of time and position within the sample.

4 Prospects

As mentioned before, the main focus of the tests at the LFT was the determination of the friction coefficient of tire materials. But tests at the LFT as well as theoretical studies have shown that the temperature of the sample has a great influence on the friction coefficient, and so also on the road performance of tires. For that reason it is also important to determine the heat generated during friction processes. In fact also for tires a high rise in temperature was measured, especially for breaking processes [7, 8]. Since it is easier to measure the temperature at the LFT these tests are helpful to improve the predictability of the road performance of tires.

The measurement of the temperature at the sample surface with a pyrometer is precise enough for our demands, but the tests with thermocouples can still be improved to gather more information about the three-dimensional heat flow within the samples during and after the friction processes. So far also a satisfactory description of the partition of the generated heat between sample and friction surface has to been found.

Acknowledgements

Most measurements for the mentioned experiments have been done with materials, which were produced and made available by the Continental AG, Hannover. This works wouldn't have been possible without the support of the colleagues of this company, especially Joachim Schramm, Bernhard Zernetsch and Reinhard Mundl. Therefore we want to express our gratitude for their support and cooperation and for their approval to make use of some of the results for this work.

References

[1] Olaf Lahayne: Experimentelle Reibungsuntersuchungen und Modellrechnungen zum Verhalten von Reifenmaterialien, PhD thesis, TU Wien 2007 (to be published)

[2] B.N.J. Persson: Role of the Flash Temperature; to be published 2007

[3] Harald Schwaiger: Entwicklung einer Prüfeinrichtung zur Untersuchung des Traktionsverhaltens von Gummiproben auf verschiedenen Oberflächen; Master thesis, TU Wien, 1996

[4] M. ten Bosch: Wärmeübertragung; Springer Verlag Berlin 1936

[5] K. Hofstetter: Thermo-mechanical Simulation of Rubber Tread Blocks during frictional Sliding, PhD thesis, TU Wien 2004

[6] K. Hofstetter, J. Eberhardsteiner, H.A. Mang, S. Del Linz: A Thermo-Mechanical Formulation Describing the Frictional Behavior of Rubber"; in: "Proceedings of the Fifth World Congress on Computational Mechanics (WCCM V) (Online-Proceedings http://wccm.tuwien.ac.at), H.A. Mang, F.G. Rammerstorfer, J. Eberhardsteiner (ed.); TU Wien, 2002

[7] Horst W. Stumpf: Reifenkonstruktion und Reifentechnik, lecture notes TU Wien, 2003
[8] N. Hagn: Messung der Temperaturverteilung im Latsch eines blockierten Reifens mittels Unterflurthermografie; TU Wien 1988/89

Progress on experimental and finite element studies of oblique elastic impact

P. P. Garland & R. J. Rogers
University of New Brunswick, Canada

Abstract

Oblique elastic impact of nonconforming bodies is a special case of stereomechanics that can occur in several engineering applications such as loose fitting joints, robotic tasks and granular assemblies. Of particular interest in this type of impact is the tangential force that develops between the colliding bodies due to friction and tangential compliance. The solution of this problem is made more difficult due to the constantly changing contact area over the duration of impact. These factors lead to highly nonlinear force-displacement relationship between the colliding bodies.

Analytical/numerical methods based on contact mechanics have shown the tangential force waveform experienced when spherical bodies collide at low velocities is dependent on the angle of incidence. At large angles of incidence, full sliding of the bodies will occur. In this case, the tangential force will be equal to the limiting Coulomb friction value. At smaller angles of incidence, the bodies may have varying periods of full sliding, partial slipping or complete sticking over the duration of impact. In these cases, the tangential force can be expected to oscillate during impact.

Experimental testing and finite element modeling are currently being used to verify the correctness of analytical/numerical solutions previously developed. The experimental study uses a simple pendulum with a spherical steel striker. Contact force data of the oblique impact event is collected using a single tri-axial piezoelectric force transducer. The finite element model of the impact event is coded using commercially available Abaqus/Explicit™ software. The finite element model also allows for exploration of the various stress distributions on the contact surface of the colliding bodies. Preliminary results from both methods indicate that the tangential force will oscillate for shallow angles of incidence. However, comparisons of these methods to one another and to the available solutions show significant characteristic differences.

Keywords: experimental impact, finite element analysis, contact forces, Coulomb friction, shear stress, partial-slip distribution.

WIT Transactions on Engineering Sciences, Vol 55, © 2007 WIT Press
www.witpress.com, ISSN 1743-3533 (on-line)
doi:10.2495/SECM070161

1 Introduction

Mechanical impact of bodies is a special case of transient contact problem which has been studied since the time of Newton. These problems are common in a number of areas of practical engineering interest including tube/support interaction, mechanical joints, fretting wear, robotic tasks and granular assemblies. Under the general classification of impact problems, there exist several cases which have been studied. These cases are characterised by some combination of material and geometric properties and initial conditions. In the discussion that follows, we limit ourselves to oblique elastic impact. For this case, the combined effects of tangential compliance and friction effects give rise to a resulting tangential force, which can oscillate for certain incidence angles within the impact duration (Maw et al. [1]). This oscillation is possible for incidence angles in which there is at least some relative sticking of coincidence points in the contact zone; for larger angles, relative sliding of coincidence points will occur.

Despite the significant body of literature available on the subject of oblique impact, the interaction of colliding bodies at the contact interface is still not well understood. The focus of the work discussed here is the development of methods that can be used to independently verify the results of a previously developed continuum model of oblique impact of identical spheres (Garland and Rogers [2]). The experimental work is aimed at verifying the overall characteristics of the contact forces, particularly the oscillation of tangential force possible at certain incidence angles, and the finite element model is aimed at verifying the stress distributions, particularly the surface shear stress component, which were assumed in developing the continuum model.

Previously developed elastic continuum models [1, 3] are not as prevalent in the literature as impulse-momentum techniques or discrete model approaches. Fewer still are experimental or finite element models for this specific case of impact. In 1981, Maw et al. [4] published experimental data of spherical disks impacting a rigid flat by using an air bed apparatus. The data consisted of incidence and rebound angles collected using high speed photography. The results of this study showed favourable agreement to the numerical work, the first to show tangential force oscillation, previously published by the same authors (Reference).

Osakue and Rogers [5] published experimentally obtained contact force waveforms using a simple pendulum apparatus. Their apparatus consisted of a steel sphere attached to an aluminum pipe which, when released, would strike a steel target block that housed a tri-axial piezoelectric force transducer. The force waveforms obtained showed reasonable agreement to expected results and verified the possibility for tangential force oscillation; however, the quality of the data obtained was not sufficient for detailed comparison to judge the validity of the subsequently developed model (Garland and Rogers [2]).

Existence of a three dimensional finite element model of oblique elastic impact of identical spheres appears to absent from the literature. Lim and Stronge [6] published results of a two dimensional finite element model of oblique impact of cylindrical bodies, which included both elastic and plastic loading cases. The elastic loading case included in this study showed tangential force oscillation that was characteristically different from what was expected. Unfortunately, results of the stress distributions were not included with this study.

The previously developed theoretical/numerical work of our oblique impact study (Garland and Rogers [2]) is an elastic continuum model that calculates the resulting tangential force waveform from the shear stress distribution at each time step throughout the impact duration. The shear stress distribution at each time step is calculated from analytical equations which relate the shear stress over the contact zone to elastic tangential displacement of points on the surface of the colliding sphere. This model is able to reproduce all of the shear stress distributions believed to be potentially present for oblique impact cases under a Coulomb friction assumption, which include full sliding or partial-slip distributions (Johnson [7]).

The following sections discuss the progress made in developing both the experimental and finite element results for use in verifying the above mentioned elastic continuum model. These portions of our study are not complete, and so only the preliminary results are available.

2 Experimental study

The following section discusses the experimental apparatus, data acquisition system and the preliminary results of contact force data.

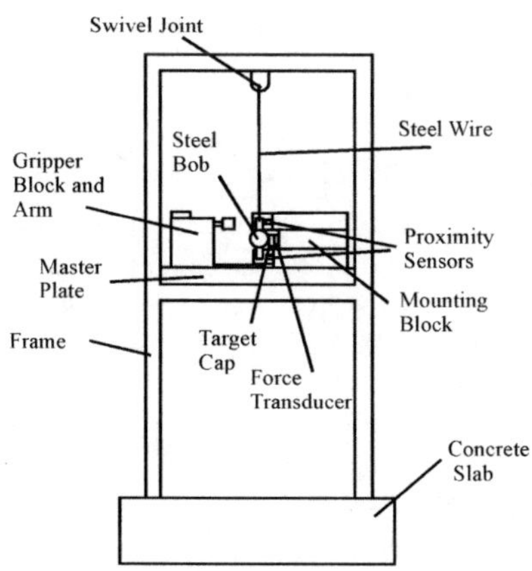

Figure 1: Schematic diagram of experimental apparatus.

2.1 Experimental apparatus

A schematic diagram of the experimental apparatus to be used in this study can be seen in Fig. 1. The apparatus is a simple pendulum setup with a spherical striker suspended on a steel wire. A magnetic gripper arm, which can be rotated to provide various horizontal incidence angles, is used to provide an initial offset of the pendulum. The mounting block is placed so that the sphere strikes the target cap at the lowest point of its trajectory. The vertical component of the force sensor is used to ensure this condition.

Upon release, the pendulum swings slightly downward and strikes a steel target cap with a spherical surface. A tri-axial piezoelectric force transducer is sandwiched between the target cap and a large steel mounting block. By varying the release distance and rotation angle of the gripper arm, various initial conditions can be set.

The contact forces during the impact will be measured using a single Kistler™ Type 9251 piezoelectric force transducer. This sensor is able to produce three charge signals proportional to the forces applied to the face of the transducer in three directions – normal and two tangential. These charge signals are converted to the appropriate voltage values using three Kistler™ 5010B charge amplifiers. A preload of about 10 times the maximum tangential component must be applied in the normal direction of the sensor to achieve the appropriate linear response.

For the preliminary results, the proximity sensors were not used. The exact pre-impact approach speed and incidence angle are not known. Instead, the maximum approach velocity ($\cong 100 \,{}^{mm}\!/_{s}$) and approximate angles judged from the graduated scale on the master plate are used.

The data from the force transducer is collected using a National Instruments™ PCI-6110 data acquisition board. This is a 4 channel simultaneous sampling board with a resolution of 12 bits. The experimental data was collected at 5 x 10^6 points per second.

2.2 Preliminary experimental results

Figure 2 shows a schematic diagram of the steel sphere striking the spherical target cap at incidence as seen from a top down view. Here, we can see that the incidence angle is measured between the normal (z-axis) and the horizontal tangential (x-axis). From the setup, we expect all tangential loads resulting from friction to take place in the x-axis direction.

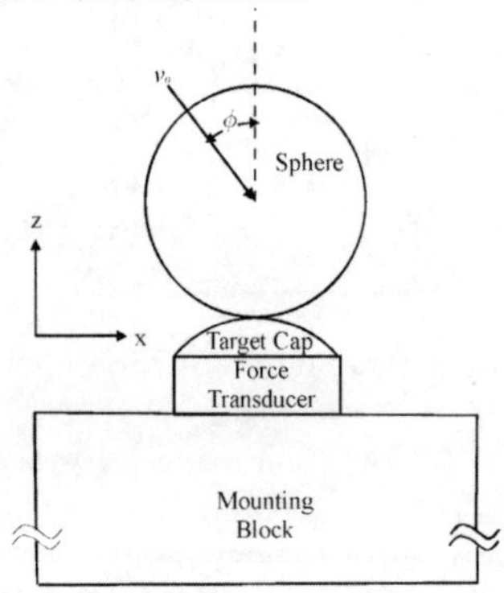

Figure 2: Configuration at incidence.

Figures 3 (a)–(l) show the contact force data collected during typical impact experiments. For these tests, the incidence angle was varied from 10° to 60° in steps of 10°. From these graphs, we can see that the tangential force does oscillate for the more shallow angles included in the study. For the largest angle of 60°, the tangential force has the same shape as the normal force vector, except that its amplitude appears to be scaled by the friction effect. If we assume a Coulomb friction model with a constant friction coefficient, this result is consistent with full sliding of the contact surfaces throughout the duration of impact.

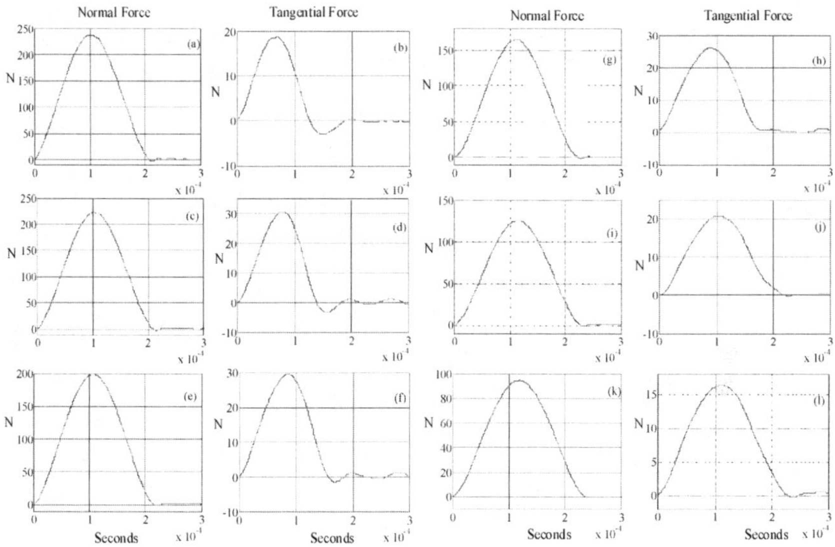

Figure 3: Experimental contact forces: (a) 10° normal, (b) 10° tangential, (c) 20° normal, (d) 20° tangential, (e) 30° normal, (f) 30° tangential (g) 40° normal, (h) 40° tangential, (i) 50° normal, (j) 50° tangential, (k) 60° normal, (l) 60° tangential.

During these tests, data collection continued past the end of impact, i.e., the zero crossing of the normal force vector. From results shown in Fig. 3, we can see that some post impact oscillation on both the normal and tangential force transducer signals. This oscillation is assumed to be caused by the natural system frequencies of the mounting block, and its effect appears to be more significant on the tangential force results. Since the apparently usable portion of the tangential waveform appears to extend slightly past the time of contact loss for many of the cases, it seems reasonable to assume that this ringing is affecting the tangential force results within the actual impact duration.

In order to compare these preliminary experimental results to those of the previously developed continuum model, we will normalize the tangential force waveforms and time scales by

$$Q = \frac{F_x}{\mu F_z}; \qquad \tau = \frac{t}{t_{\max}} \tag{1}$$

where Q is the normalized tangential force, τ is the normalized impact time, F_z is the measured normal force, F_x is the measured tangential force, t is the impact time, t_{max} is the time at the loss of contact and μ is the coefficient of friction. Using the assumption of Coulomb friction and the experimental data for the 60° incidence angle, the coefficient of friction is found to have a value of 0.178.

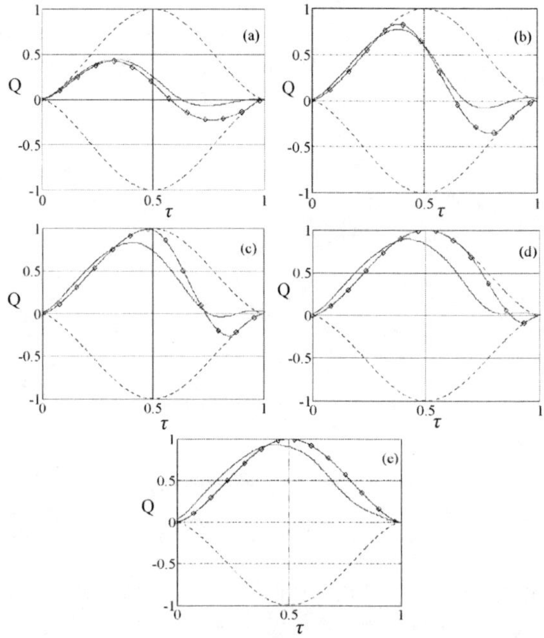

Figure 4: Normalized experimental tangential force waveforms: (a) 10°, (b) 20°, (c) 30°, (d) 40°, (e) 50° (friction envelope – dashed, experimental – solid, continuum – diamond).

Figures 4 (a) through (e) show the normalized experimental tangential force results along with the normalized tangential force obtained using the continuum model (Garland and Rogers [2]) for angles ranging from 10° to 50°, respectively. For these comparisons, the maximum experimental angle of 60° has not been included because the continuum simulation indicated that full sliding begins at 50°.

From these graphs, we can see that the experimental results match the previous numerical results reasonably well, but do show some differences in amplitudes and times of tangential force reversal (i.e. zero crossing). Also, not all of the experimental waveforms lie completely within the idealized friction envelope (μF_z). These differences are assumed to be related to natural system frequencies of the apparatus which act to contaminate the experimental contact force signal.

3 Finite element model

The following sections describe the finite element model used to simulate the spherical oblique impact problem and present preliminary results.

3.1 Modelling considerations

The 3-dimensional finite element model for this study has been developed in Abaqus/Explicit™ 6.6.1. We have simplified the identical sphere impact problem slightly by modelling a single deformable sphere impacting a perfectly rigid plate. From a Hertzian contact point of view, the problems are equivalent since the contact area remains circular with no distortion in the tangential plane.

The model uses an explicit dynamics approach with a coefficient of friction between the sphere and plate of 0.2. The contact formulation for both the normal and tangential behaviour employs the penalty contact method. The model employs 10 node tetrahedral elements.

In order to provide a sufficiently large contact zone for analysis of the stress distributions on the interface, an unrealistically large velocity is provided as the initial condition to the sphere; the ratio of maximum contact zone radius to sphere radius was 0.01, which does not violate the Hertzian contact condition. These large velocity values lead to forces which are not reasonable for the problem, and so only normalized tangential force (i.e. Eq. (1)) values are included in the results to follow.

3.2 Preliminary finite element results

Figures 5 (a)–(e) show the normalized finite element tangential force waveforms for incidence angles ranging from 10° to 50° along with those obtained from the numerical simulation (Garland and Rogers [2]). As we can see, tangential force oscillation does occur within the impact duration for certain angles of incidence. Comparison to the numerical results included in these figures shows that, once again, slight differences in the amplitudes and times of tangential force reversal are present between the two methods. The reason for these differences is unclear. One possible explanation may be the inability of the finite element model to reproduce the partial-slip shear stress distribution used in developing the continuum model.

A typical shear stress distribution along the centre line of the contact zone (i.e. line of symmetry) of the finite element model is shown in Fig. 6. This distribution corresponds to a non-dimensional time of 0.5 (i.e. half way through the impact). As can be seen, the stress distribution is plagued by significant contact noise. This raises a further mystery as to how Abaqus™ is able to produce a reasonable tangential force waveform from such noisy shear stress data. Comparison to the analytical shear stress model is not shown, but would certainly show large differences between the two methods.

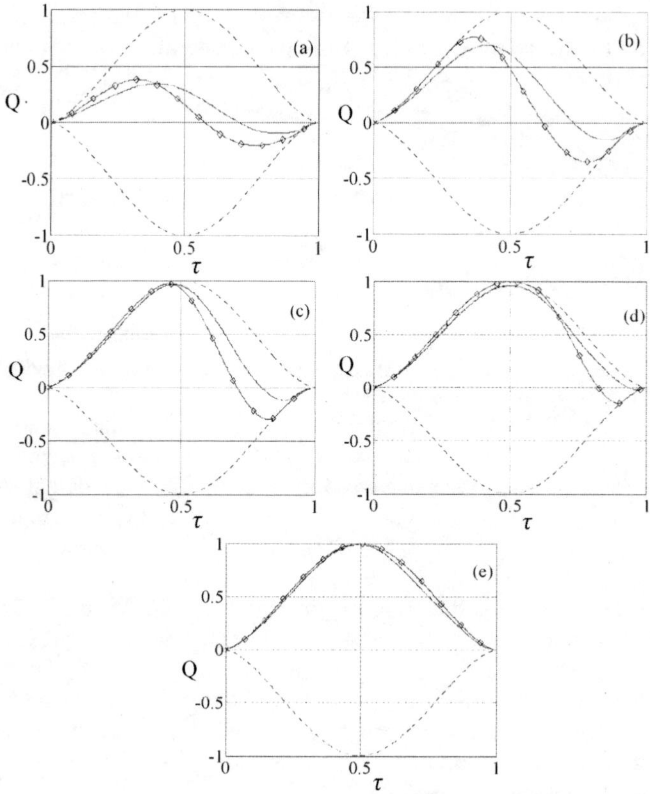

Figure 5: Normalized finite element tangential force waveforms: (a) 10°, (b) 20°, (c) 30°, (d) 40°, (e) 50° (friction envelope – dashed, finite element – solid, continuum – diamond).

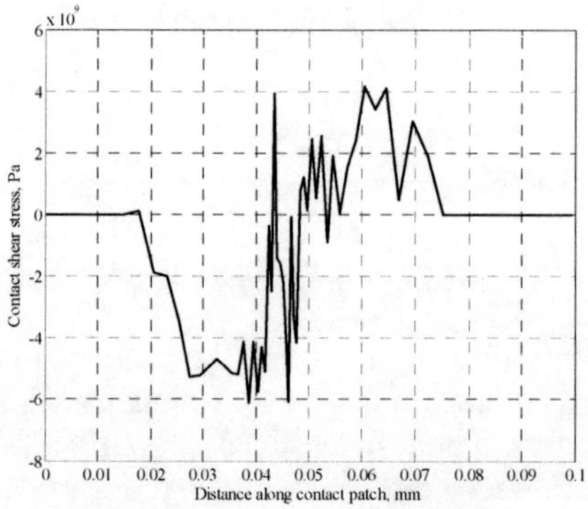

Figure 6: Typical finite element results of shear stress distribution along x-axis centre line of contact ($\phi = 7°$, $\tau = 0.5$).

The shear stress distributions obtained from the other simulations with different incidence angles or at other times through the impact duration show similar results to those of Fig. 6. As such, use of these stress distributions in their present state will not be useful in verifying the correctness of the assumptions used in the previously developed continuum model.

4 Conclusions and further work

The work presented in this paper is part of a larger study of the contact forces and stress distributions that are present in the oblique elastic impact of identical spheres. The goal of the experimental and finite element modelling methods is to verify the results of a previously developed numerical model (Garland and Rogers [2]). The preliminary results obtained from these methods have shown that this verification is not possible without some further refinement of the model.

Both the experimental and finite element results have shown that, for certain angles of incidence, the tangential contact force developed during oblique impact will oscillate. The differences between the experimental and the numerical model results are believed to be caused by interference of the experimental apparatus' natural system frequencies on the force transducer's signal. The source of the differences between the finite element model and the numerical model are not clear. The differences may be related to the contact noise, they could be inherent to the finite element contact formulation method employed by the software, or they could be due to approximations in the numerical model.

At present, the experimental apparatus is being redesigned in an attempt to distribute the mounting block's natural frequencies in such a way that the effect on the force transducer is minimized. These force signals may also need to be filtered to remove or reduce the effect of these interfering frequencies.

Several alternate methods of finite element modelling are currently being explored. These include development of an implicit impact model using Abaqus/Standard™ and exploration of other contact formulations such as Lagrange multiplier or augmented Lagrange methods. In an attempt to smooth the shear stress distribution results, the use of either time domain or spatial filtering is also being considered.

References

[1] Maw, N., Barber, J.R. & Fawcett, J.N., The oblique impact of elastic spheres, *Wear* 38, pp. 101-114, 1976

[2] Garland, P.P. & Rogers, R.J., An analytical solution for shear stress distributions during oblique impact of similar spheres, Transactions of the ASME, *Journal of Computational and Nonlinear Dynamics*, tentatively accepted, 2006

[3] Jaeger, J., *Elastic impact with friction*, PhD dissertation, Delft University, 1992

[4] Maw, N., Barber, J.R. & Fawcett, J.N., The role of elastic tangential compliance in oblique impact, Transactions of the ASME, *Journal of Lubrication Technology* 103, pp. 74-80, 1981

[5] Osakue, E.E. & Rogers, R.J., An experimental study of friction during planar elastic impact, Transactions of the ASME, *Journal of Pressure Vessel Technology* 123 (4), pp. 493- 500, 2001

[6] Lim, C.T. & Stronge, W.J., Oblique elastic-plastic impact between rough cylinders in plane strain, *International Journal of Impact Engineering Science* 37, pp. 97-122, 1998

[7] Johnson, K.L., *Contact Mechanics*, Cambridge University Press: Cambridge, 1985

Explosive pitting of 1018 steel witness plates

G. A. Walsh[1] & V. D. Romero[2]
[1]*Department of Materials Science and Engineering, New Mexico Institute of Mining and Technology, USA*
[2]*Department of Physics, New Mexico Institute of Mining and Technology, USA*

Abstract

1018 steel witness plates were put into contact with the explosives Ammonium Nitrate and Fuel Oil (ANFO), composition C4, dynamite, flake trinitrotoluene (TNT), smokeless powder and black powder. Explosive-induced pitting of the metal targets was observed in tests involving C4, dynamite, TNT and smokeless powder. The microstructure directly under pitting sites was examined using standard metallographic techniques. No sign of melting was observed under or around the pits, indicating that pitting is most likely caused by impingement of hard particles on the metal surface during combustion of the explosive material, or by jet formation from the collapse of voids at the explosive/target interface.
Keywords: explosives, explosive pitting, explosive-metal interactions.

1 Introduction

Pitting is a phenomenon that occurs on metal surfaces in contact with, or very close proximity to, detonating explosives [1]. This pitting of metals is often used in forensic investigations to indicate the occurrence of an explosive event [1], [2]. The cause of explosive pitting is, however, the subject of some controversy. Pitting has been said to be caused by the impact of high velocity particles (either unconsumed explosive or inert) [3]. Another theory is that small jets formed by the collapse of voids at the explosive / metal interface produces pitting [4]. Additionally, claims have been made indicating that the explosive event generates enough heat to melt small portions of the metal surface, creating small pits upon solidification [5].

www.witpress.com, ISSN 1743-3533 (on-line)
doi:10.2495/SECM070171

1.1 High velocity impact

The high velocity impact of a particle onto a metal surface causes pitting, or cratering, of the metal. An example microstructure of a copper target impacted by an aluminium projectile is shown in Figure 1 [6]. The target material's microstructure after such an impact is highly distorted; the grains have undergone dynamic recrystallization directly under the impact site, and appear flattened further into the sample.

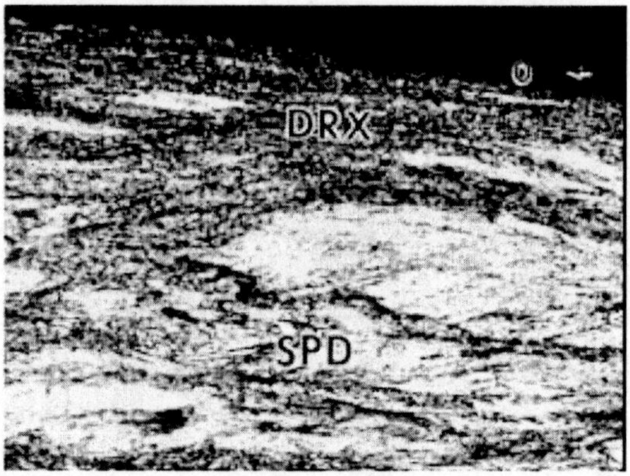

Figure 1: Dynamic recrystallization (DRX) and severe plastic deformation (SPD) under a crater in a copper target after impact by an aluminium projectile [6].

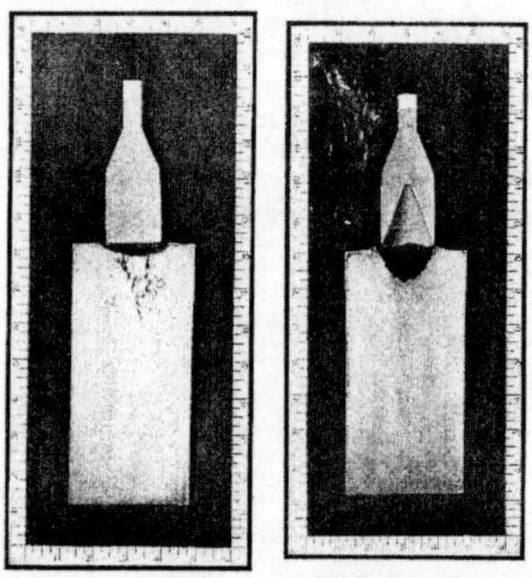

Figure 2: An illustration of the Monroe effect [7].

1.2 The Monroe effect

The Monroe effect, or the shaped charge effect, describes a focusing of the energy produced in an explosive detonation. This effect is illustrated in Figure 2 [7]. As the explosive material detonates, the cavity is collapsed, and a high energy jet is formed where the shock waves collide in the centre of the collapsing void. If a liner is inserted inside the cavity, the collapse of the liner generates a metallic jet, which is the basis for many modern anti-armour weapons [8].

The microstructure of a 1020 steel target impacted by a copper shaped charge jet is shown in Figure 3 [9]. In this figure, there is also a highly deformed grain structure and evidence of recrystallization.

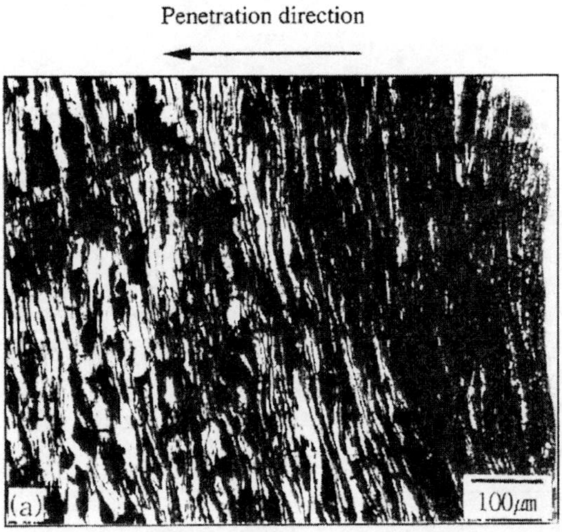

Figure 3: The effect of a copper shaped charge jet impact on the microstructure of a 1020 steel target [9].

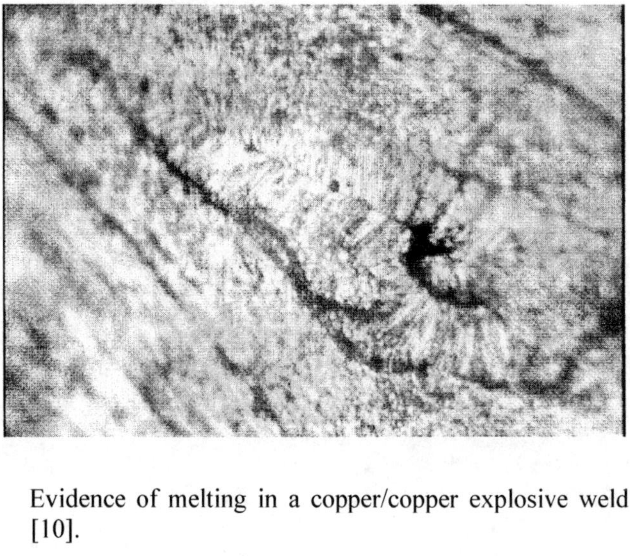

Figure 4: Evidence of melting in a copper/copper explosive weld interface [10].

1.3 Melting

The melting and re-solidification of metals creates a cast structure, such as that found in Figure 4, a micrograph indicative of melting at the interface of a copper/copper explosive weld [10]. This type of microstructure, or the presence of dendrites would be evidence of melting and re-solidification of a metal subjected to an explosive shock.

2 Experimental procedure

2.1 Explosive testing

In this study, six different types of explosives were used. The weights the explosives were varied in order to keep the same TNT equivalent weight of 0.82 lbs. By keeping the TNT equivalent weight the same for the various explosives, the air shock overpressure from all the blasts should be the same [11]. The explosives used and their weights are listed in Table 1. Each explosive was tested three times, for a total of 18 experiments.

Table 1: The types and weights of explosives used.

Explosive	Weight used (lbs)
Ammonium Nitrate and Fuel Oil (ANFO)	1.1
Composition C4	0.61
Dynamite	0.85
Flake Trinitrotoluene (TNT)	0.82
Black Powder	1.25
Smokeless Powder	0.64

The explosives were packed into 2” ID schedule 40 ASTM A53 steel pipe, the lengths of which were varied to accommodate the amount of explosive used. The pipes were threaded on one end, to accept a standard end cap. A non-electric detonator was inserted in the top centre of the explosive charge. The explosive was in direct contact with a 3” x 3” 1018 steel witness plate, which had been ground with 80 grit SiC paper to ensure a smooth, pit-free surface. The assemblies were taken to the Energetic Materials Research and Testing Center (EMRTC), at the New Mexico Institute of Mining and Technology, for testing. After testing, the plates were recovered and saved for future characterization. Two of the plates were not recovered after testing.

2.2 Metallography

After collection, the surface of each witness plate was photographed, and the number of pits on the surface was counted and recorded. The plates were then cross sectioned, and a portion of the sample was mounted, ground and polished using standard metallurgical techniques and etched with Nital. After etching, the samples were imaged with a Versamet 2 Metallograph for optical microscopy.

3 Results

3.1 Pitting frequency

The number of pits on the surface of each plate is listed in Table 2.

Table 2: Pitting frequency by explosive type (N/R = plate not recovered).

Explosive	Plate 1	Plate 2	Plate 3	Average
ANFO	0	0	N/R	0
Comp. C4	6	4	N/R	5
Dynamite	333	294	296	307.7
Flake TNT	157	235	392	261.3
Black Powder	0	0	0	0
Smokeless P.	431	471	392	431.3

3.2 Pit cross-sections

The microstructures of witness plates from the composition C4, dynamite, TNT and smokeless powder tests (Figure 5(a)–(d), respectively) show shallow pitting with severe deformation and recrystallization underneath the pit, whereas the reference sample (seen in Figure 5(e)) has a flat, uniform surface and equiaxed grain structure. One witness plate from an ANFO test revealed a microscopic pit upon metallographic examination, however, the damage to the grain structure is fairly limited, see Figure 6.

4 Discussion

4.1 Pitting frequency

Severe pitting was observed on witness plates subjected to the combustion of dynamite, flake TNT and smokeless powder. Minor pitting was noted on the plates after the detonation of C4. No visible pitting was seen on the surface of the ANFO witness plates; however, a small pit was found upon microscopic examination of the cross section of one plate. Black powder produced neither macroscopic nor microscopic pitting.

No clear correlation can be drawn between pitting frequency and explosive detonation velocity. This can be explained by differences in the consistency of the explosives used in this study. While C4 and ANFO are relatively homogeneous, soft materials, the other explosives studied consist of hard particles or, in the case of dynamite, have hard, particulate filler materials mixed into the explosive composition.

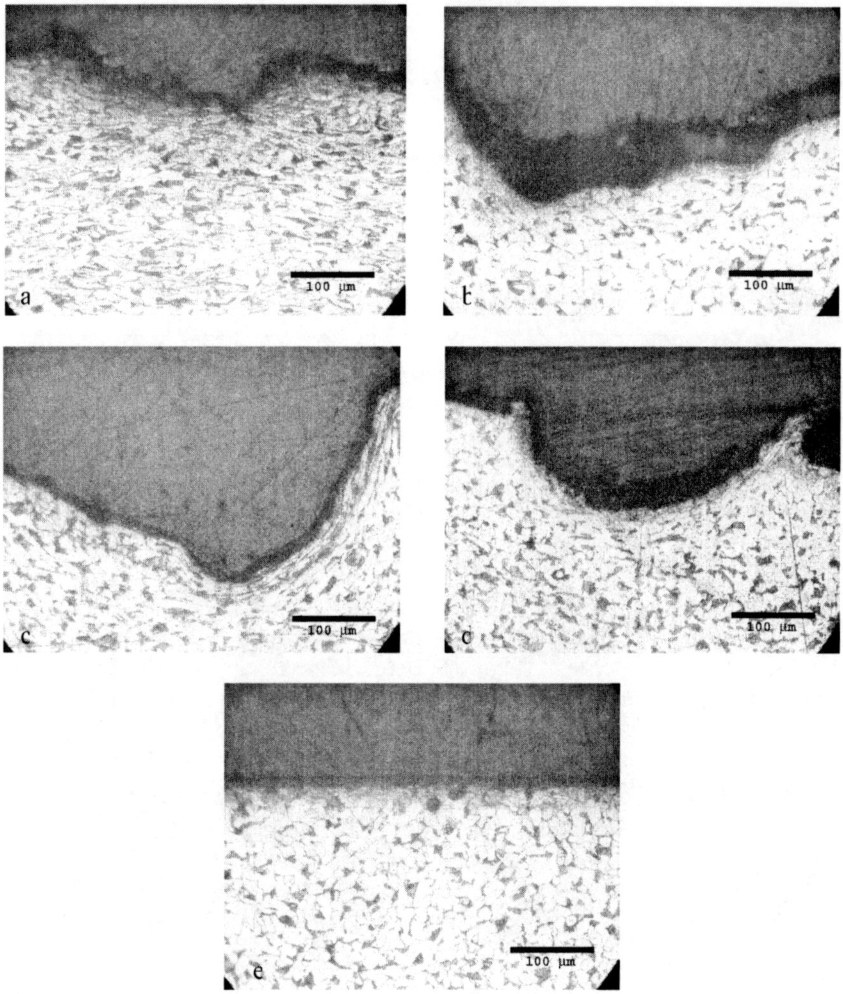

Figure 5: Recystallization and deformation of the witness plate grain structure from the (a) C4, (b) dynamite, (c) TNT and (d) smokeless powder tests; (e) is the reference sample.

4.2 Pit morphology

No evidence of melting and re-solidification was found in any of the plate cross sections. The microstructures obtained did appear similar to those found in previous work on both the impact of metals with hard objects and shaped charge jets.

Dynamite consistently supplied the largest pits, likely due to hard particles found within the dynamite. The dynamite material was removed from its casing and pressed into the metal pipe, flush against the witness plate prior to testing, making the probability of air pockets at the explosive/metal interface unlikely.

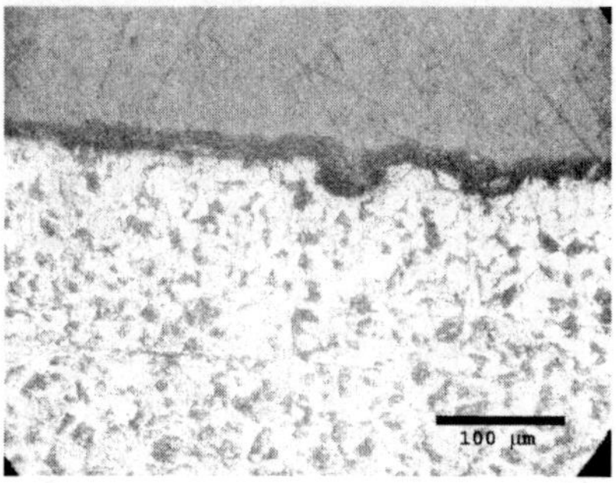

Figure 6: The microstructure under a small pit found in the ANFO witness plate.

Flake TNT and smokeless powder both generated a large number of pits. This pitting is likely due to the Monroe effect or a combination of the Monroe effect and denting. When preparing for testing, the flake TNT and smokeless powder were both poured into the pipe, and no effort was made to press them against the witness plate. This setup allows for the formation of many voids at the explosive/target interface, which could easily lead to Monroe effect-based jetting.

Composition C4 induced a small number of pits on the witness plates. The damage to the microstructure, however, was severe and penetrated fairly deep into the cross section of the material. The near-absence of pits is most likely due to the homogeneous and soft nature of C4. As in the dynamite test, this explosive was pressed against the witness plate to minimize the formation of air pockets at the target interface, but unlike dynamite, does not have hard particles in its makeup.

ANFO and black powder produced virtually no pitting, in these cases it is likely that low detonation velocity (in the case of ANFO) and burn rate (of black powder) minimized pitting. ANFO has a detonation velocity of 4.7 km/s, where the detonation velocities of the other high explosives used range from 6 to 8 km/s. The burn rate of black powder is on the order of tens of m/s. Additionally, ANFO is a soft, prill-type explosive, which is unlikely to dent a steel plate, even at high temperatures, pressures and velocities.

5 Conclusions

Pitting of 1018 steel witness plates was seen in dynamic tests involving composition C4, dynamite, flake TNT and smokeless powder. ANFO and black powder did not produce pitting. There is no clear correlation between pitting frequency and detonation velocity.

In examining the microstructure of metal targets after explosive experiments were carried out, no signs of melting were observed. The microstructures generated in this study were, however, consistent with those seen in the deformation of metal targets produced by the impact of hard projectiles and the impact of shaped charge jets.

Acknowledgements

The authors would like to thank the US Department of Homeland Security for funding this research under grant # RDJ9000. Also, thanks should be extended to the people of the Energetic Materials Research and Testing Center for their assistance in dynamic testing.

References

[1] United States of America vs. Terry Lynn Nichols. Oklahoma City Bombing Trial Transcript. CourtTV website. http://www.courttv.com/archive/casefiles/oklahoma/nichtranscripts/1125pm.html

[2] Loeb, B. Statement of Dr. Bernard S. Loeb, TWA Flight 800 Board Meeting, NTSB website. http://www.ntsb.gov/Speeches/s000822.htm

[3] United States of America vs. Timothy James McVeigh. Oklahoma City Bombing Trial – Transcripts. CNN website, Atlanta, GA. USA http://www.cnn.com/US/9703/okc.trial/transcripts/may/052097.eve.html?eref=sitesearch

[4] The FBI Laboratory: An Investigation into Laboratory Practices and Alleged Misconduct in Explosives-Related and Other Cases (April 1997). US Department of Justice website. http://www.usdoj.gov/oig/special/9704a/

[5] Smith, J.B. *Aircraft Accident Report – Air India Flight 182* pp. 40, 2001.

[6] Murr, L.E., Trillo, E.A., Pappu, S. & Kennedy, C. Adiabatic shear bands and examples of their role in severe plastic deformation. *Journal of Materials Science*, **37**, pp. 3337-3360, 2002.

[7] Birkhoff, G., MacDougall, P., Pugh, E. & Taylor, B. Explosives with cavities. *Journal of Applied Physics*, **19**, pp. 563-582, 1948.

[8] Petit, J., Jeanclaude, V. & Fressengeas, C. Breakup of copper shaped charge jets: experiment, numerical simulations and analytical modeling. *Journal of Applied Physics*, **98(12)**, 2005.

[9] Lee, S., Hong, M-H., Noh, J-W. & Baek, W.H. Microstructural evolution of a shaped-charge liner and target materials during ballistic tests. *Metallurgical and Materials Transactions A*, **33A**, pp. 1069-1074, 2003.

[10] Walsh, G.A.; Inal, O.T.; Lopez, D.H. & Gerity, P.F. Wave amplitude and frequency seen at explosively welded copper/copper interfaces. *Proceedings of Computer Methods and Experimental Measurements for Surface Effects and Contact Mechanics VII*, eds. deHosson, J.T.M.; Brebbia, C.A. and Nishida, S.I. WIT Press: Boston, MA., pp. 23-31, 2005.

[11] Cooper, P.W. *Explosives Engineering*. VCH Publishers, New York, Wienheim and Cambridge, pp. 405-406, 1996.

Numerical analysis of the influence of abrasive grain geometry and cutting angle on states of strain and stress in the surface layer of object

L. Kukielka & J. Chodor
Department of Mechanical Engineering,
Koszalin University of Technology, Poland

Abstract

Grinding is a very complicated technological process. To increase the quality of the product and minimize the cost of abrasive machining, we should know the physical phenomena which exist during the process. The first step to the solution of this problem is an analysis of a machining process with a single abrasive grain. In the papers [Kukielka and Kustra, *Surface Treatment VI Computation Methods and Experimental Measurements for Surface Treatment Effects*. WIT Press, 2003, pp.109–118; Kukielka et al, *Computer Methods and Experimental Measurements for Surface Effects and Contact Mechanics VII*. WIT Press, 2005, pp. 57–66.] the thermo-mechanical models of this process are presented, but in this work attention is drawn to the chip formation and its separation from the object. The influence of the tool geometry and the cutting angle on the states of strain and stress in the surface layer during machining is explained. The phenomena on a typical incremental step were described using a step-by-step incremental procedure, with an updated Lagrangian formulation. Then, the finite elements methods (FEM) and the dynamic explicit method (DEM) were used to obtain the solution. The application was developed in the ANSYS system, which makes possible a complex time analysis of the physical phenomena: states of displacements, strains and stresses. Numerical computations of the strain have been conducted with the use of two methodologies. The first one requires an introduction of boundary conditions for displacements in the contact area determined in the modeling investigation, while the second – a proper definition of the contact zone through the introduction of finite elements of TARGET and CONTACT types, without the necessity to introduce boundary conditions. Examples of calculations for the strain and stress field in the surface layer zones of object were presented.
Keywords: abrasive grain, single-grain machining, chip creation, yield stress, FEM, numerical analysis, state of strain, state of stress.

WIT Transactions on Engineering Sciences, Vol 55, © 2007 WIT Press
www.witpress.com, ISSN 1743-3533 (on-line)
doi:10.2495/SECM070181

1 Introduction

Grinding is considered to be a particularly complex and hard to execute way of machining, in which there occur many not yet fully investigated phenomena. This results from a large number of factors which have an influence on the course of the machining process. This requires the development of different varieties of the process and conducting of comprehensive cognitive research.

Grinding is characterised by many specific features, which make this process basically different from the ways of machining [4, 7]:

- an irregular arrangement of a very large number of abrasive grains on the working surface of the grinding wheel,
- diversified shapes of abrasive grains and negative working rakes of the cutting edges of grain apexes,
- different heights of the cutting edges of grain apexes on the active surface of the grinding wheel,
- an unspecified dependency between the thickness and the width of the chip removed with individual abrasive grains,
- non-isolation of the main and auxiliary machining edges,
- peculiar properties of abrasive grains: high hardness, resistance to the action of heat, sharpness, brittleness, and an ability to crack in the cleat plane, etc.,
- small penetration depths of abrasive grains into the material machined in comparison with their average sizes,
- large tangential velocities of micro-machining, which ensure the removal of a large quantity of chips in a unit of time.

The shape of a single abrasive grain shows a significant influence on the course of the machining process [9]. As it was found, the grains of abrasive materials, once their size has been reduced, are usually characterised by an irregular shape and a differentiated degree of the sharpness of machining corners and edges. While considering the work of such grains, their sizes, shape and geometry are subject to experiential investigation, and they are replaced with grains with a regular shape, which can be mathematically described. Most often, as a replacement model of the abrasive grain [1], the cone or the pyramid (with rounding or without rounding of its vertex) is accepted with the apex angle equal 2θ [2] and a sphere with radius ρ_k [10], while the spheroid with a constant semi-axis [8] is accepted less often.

A creation of abrasive wheels with abrasive grains geometrically correct and located properly in the binding material would substantially contribute to the change of the course of the work. The first stage is the recognition of the process of machining with a single abrasive grain. Learning about the topology of the abrasive grain and the mutual relations between its individual sizes should have a significant influence on the creation of its correct geometrical model. This model in combination with the suitable manner of giving specified shapes to grains in the process of their production would lead to the creation of a grinding wheel whose abrasive grains would perform e.g. the processes of the initial and finishing machining at the same time.

The process of the chip creation proves to have a substantial influence on the grinding process, together with the geometrical and kinematic dependencies in the contact zone of the grinding wheel-the object machined. It has a significant impact on the wear of the grinding wheel, the value of the components of the machining force, the temperature and the quality of the surface machined.

During the machining process, there occur large and fast plastic strains, which occur only in the part of the object machined. Under the influence of these strains in the material, its physical properties are subject to change: the hardness and strength increase. There occurs the so-called consolidation phenomenon of the material. The geometry of the zone of the chip creation during machining allows one to accept an assumption that in the area in question there is a plate state of strains.

For the correct modelling and analysis of the grinding process, the knowledge of the course of the physical phenomena occurring in the machining zone in real conditions proves to be necessary. For this purpose, an analysis of the process of machining with a single abrasive grain was conducted. This process was considered as a problem of a displacement of the model abrasive grain (Fig. 1) specified in paper [21] on the elastic/visco-plastic body.

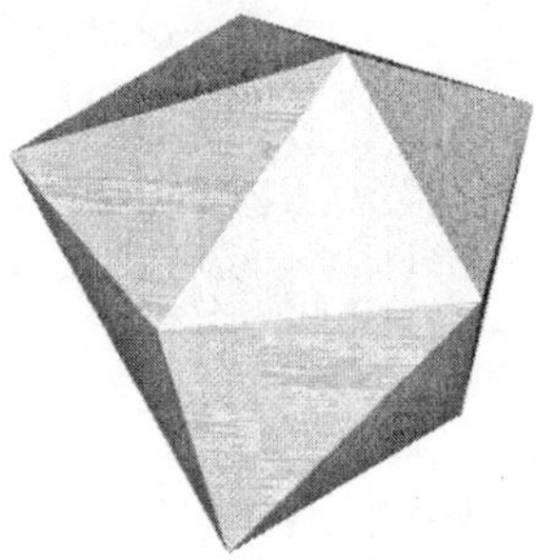

Figure 1: Geometry of the most probable abrasive grain [21].

An abrasive grain with the apex angle of $2\theta = 80 \div 120^0$ and the corner rounding $r = 0{,}001$ μm is tilted in relation to the foundation by angle $\alpha = 45 \div 65^0$ (Fig. 2). The allowance was $h = 0{,}01$ μm. The value of the real layer thickness of the material removed as a result of elastic strains was smaller and was ca. $h_r = 0{,}009$ μm.

It was assumed that the grain movement was kinematically forced and it slides horizontally on the surface of the elastic/adhesive-plastic body. The value of α angle determines whether there will occur a machining phenomenon (chip creation) or the strengthening process of the surface layer through burnishing (no chip).

In papers [21, 22], a thermal and mechanic model of the process of the grain displacement on the elastic/visco-plastic body was developed and the distributions of temperatures were determined together with the intensities of strain in the material machined in the initial chip creation phase.

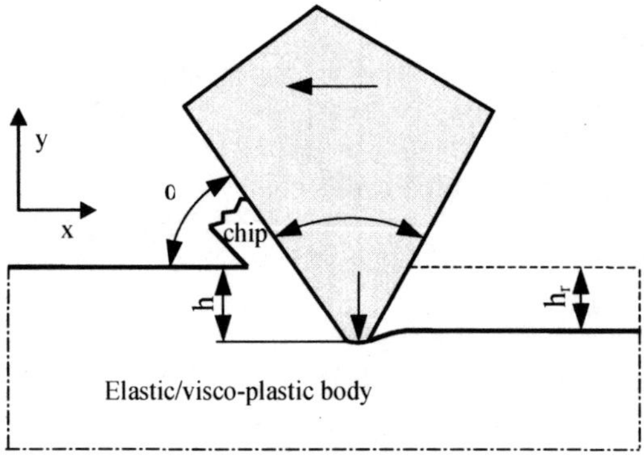

Figure 2: Diagram of the considered issue of the displacement of the abrasive grain on the elastic/visco-plastic body.

This study concerns the issues of the creation of the properties of the surface layer of a product, the chip formation and separation from the material of the part. The influence of the apex angle of the abrasive grain 2θ was determined together with the angle of its tool cutting edge angle α on the states of strains and tensions in the surface layer of the object at any time of the process. For the purpose of a description of phenomena at individual moments with the step-by-step method, an upgraded Lagrange's description was applied. In order to solve the problem, the finite elements method (FEM) and the dynamic explicit method (DEM) was used. Examples of solutions were presented in ANSYS programme for tensions and strains in the surface layer of the object.

2 Models of chip creation

Investigations concerning the chip creation process most often used the machining theory, which considers its creation as a result of the displacement of the material machined along the cutting plane. A modification of views in this area took into account machining with a grain with a spherical cutting blade (Table 1).

Models of chip creation during grinding with a conical cutting edge are an analogy to turning and milling – processes, which have been so far much better investigated than grinding. In many publications, these have been used in the description of the chip creation during grinding, while considering the features characteristic of this process [2, 6, 10, 20]. However, such an approach to the solution of the problem discussed was questioned as a result of numerous research studies. There are studies which treat the process of the chip creation during grinding as an extrusion of the material with a spherical cutting edge. These include the papers by Shaw [15, 23], which constitute an analogy to the hardness measurement with Brinell and Lortz's methods [3, 11], who based them on the theory described by Tomlenov [19] concerning the field of the path of shear for the perpendicular penetration into the material of a rounded punch with the consideration of friction. A detailed theoretical description supplied by Lortz of the chip creation by the extrusion of the material with a spherical cutting edge constituted the basis for further studies in this area [13, 14, 16-18]. Owing to the consideration of the so-called dead zone, the determination of the conditions in which the chip is created and the possibility to determine strains in the surface layer of the object machined, it can be stated that this description constitutes the most representative model of the chip creation process.

Table 1: Models of chip creation with single-grain machining [5].

Movement \ Shape		Grain with a spherical cutting blade		Grain with a conical cutting blade
		Chip creation by extrusion of material	Chip creation along cutting plane	
Grain movement in relation to the object	Rectilinear	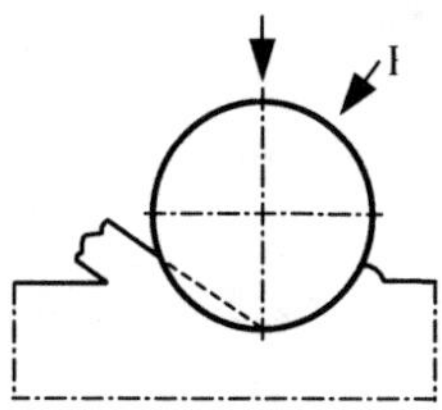	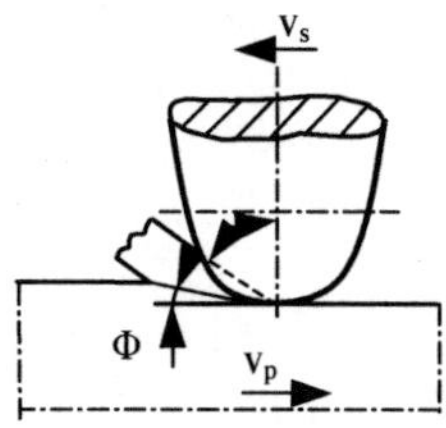	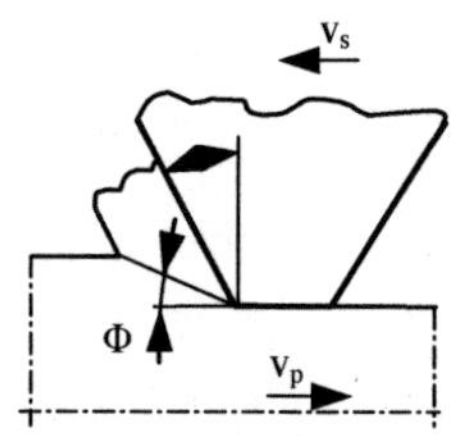
	Curvilinear	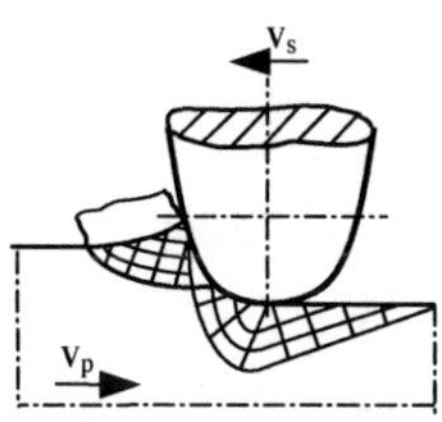	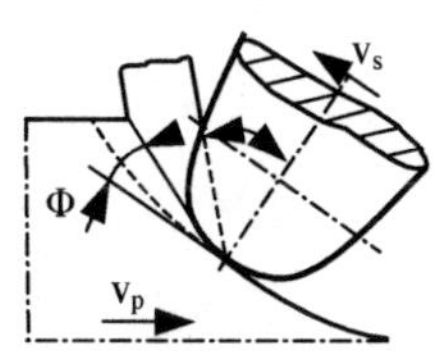	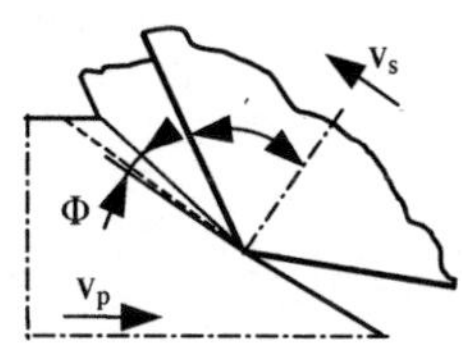

Legend: F – machining force, v_s – tangential speed of grinding wheel, v_p – tangential speed of the object machined, γ – tool rake angle, Φ – shear angle.

3 Data for computer simulation

It was accepted in the simulations that the abrasive grain is a non-deformable body, while the object is an elastic/visco-plastic body described with the aid of Cowper–Symonds' model. In the model, Huber–Mises–Hencky's plasticity model is used together with the associated flow right. The model takes into consideration the line-isotropic $(\beta = 1)$, kinematic $(\beta = 0)$ or mixed $(0 < \beta < 1)$ plastic strengthening as well as the influence of the intensity of the plastic strain speed, according to the involution dependence:

$$\sigma_Y = (R_e + \beta E_{tan} \varphi_i^{(p)})[1 + (\dot{\varphi}_i^{(p)}/C)]^m \text{, [MPa]} \tag{1}$$

where σ_Y – yield stress, R_e [MPa] – initial yield stress point, $\varphi_i^{(p)}$ $[-]$, $\dot{\varphi}_i^{(p)}$ $[s^{-1}]$ – intensity of strain and plastic strain rate respectively, $C\,[s^{-1}]$ – material parameter to determine the influence of the intensity of the plastic strain rate, $m = 1/P$ – material constant determining the sensitiveness of material on the plastic strain rate, $E_{tan} = E_T E/(E - E_T)$ – material parameter dependent of the module of plastic hardening $E_T = \partial\sigma_p/\partial\varphi_i^{(p)}$ and of Young's elasticity module E.

The following parameter values were accepted: density of body material $\rho = 7865$ kg/m^3, Poisson's number $\nu = 0{,}27$, limiting damaging strain $\varepsilon_f = 2{,}5$ and $E = 200$ GPa, $R_e = 310$ MPa, $E_{tan} = 763$ MPa, $C = 40$ s^{-1} and $P = 5$.

Also, constant values were accepted of static $\mu_s = 0{,}1$ and dynamic $\mu_d = 0{,}05$ friction coefficients. The apex angle of the grain changes in the range of $2\theta = 80 \div 120^0$, while the tool cutting edge angle $\alpha = 45 \div 65^0$. The rounding radius of the grain apex was $r = 0{,}001$ μm.

4 Method of solution

For the purpose of the solution of the problem, the dynamic explicit method, also known as the method of central differences, was used. In this method, the equation which describes the movement and deformation of the object investigated has the following form:

$$[\boldsymbol{M}]\{\ddot{\boldsymbol{r}}(\tau)\} + [\boldsymbol{C}]\{\dot{\boldsymbol{r}}(\tau)\} + [\boldsymbol{K}]\{\boldsymbol{r}(\tau)\} = \{\boldsymbol{R}(\tau)\}, \quad \tau \in [t_0, t_s] \tag{2}$$

where $[\boldsymbol{M}]$, $[\boldsymbol{C}]$ and $[\boldsymbol{K}]$ are matrices constant in time of: mass, damping and rigidity of the system respectively, $\{\boldsymbol{R}\}$ is the external load vector, and $\{\boldsymbol{r}\}$, $\{\dot{\boldsymbol{r}}\}$, $\{\ddot{\boldsymbol{r}}\}$ are vectors of the displacement, speed and acceleration of the nods of the system. This equation is integrated in relation to time with the step-by-step method and, additionally, is not rearranged during this operation. If it is assumed that the displacements, speeds and accelerations of the system are known at the beginning, at moment $\tau = t_0$ and equal $\{\boldsymbol{r}_0\}, \{\dot{\boldsymbol{r}}_0\}, \{\ddot{\boldsymbol{r}}_0\}$ respectively, than the whole interval is divided into parts with lengths Δt and on each step, a solution is sought for the abovementioned equation. This means that this equation is to be satisfied only in the selected times and not in the whole interval investigated. This means that for every moment, one can search the positions of the equilibrium of a system subject to external forces, force of inertia and forces of damping, while applying algorithms from a static analysis. The end of every moment of time is at the same time the beginning of another.

5 Results of numerical calculations

Numerical simulation in the ANSYS system was conducted for different apex angles 2θ and the tool cutting edge angle α of the grain. The object machined and the abrasive grain were digitised by elements of PLANE 162 type with a non-linear function of shape. The contact grain with body was modeling by element TARGE 169 and CONTA 171. The net of finished elements was concentrated in the contact area (Fig. 3). Sample simulation results are presented in Figs. 4 and 5.

While analysing the results obtained it was found that together with the change of the tool cutting edge angle α and the change of apex angle of the cutting edge 2θ, the values of strains and stresses are subject to change. Abrupt increases of stresses are the result of the chip creation phenomenon. Together with the increase of the tool cutting edge angle, the shear angle Φ of the material separated from the foundation increases, as well. It was found that both angles have a significant influence on the chip shape.

For the tool cutting edge angle $\alpha = 45^0$, we observe fast disturbances of the cohesion of the material between the neighbouring chip elements. This results in the fact that the chip drops off from the cutting edge in the form of separate elements – a segmental chip (Fig. 4(a), (b)).

(a) (b)

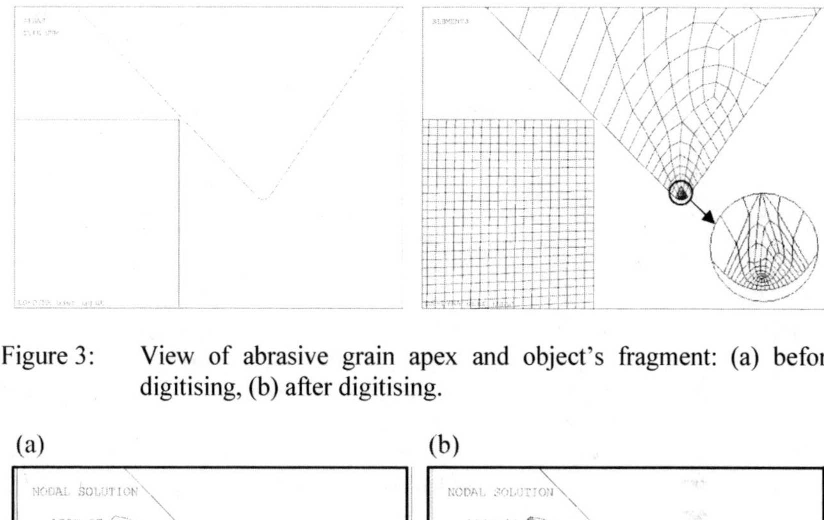

Figure 3: View of abrasive grain apex and object's fragment: (a) before digitising, (b) after digitising.

(a) (b)

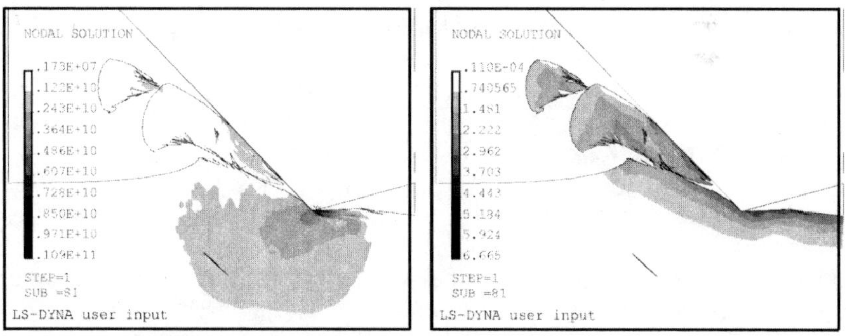

Figure 4: Maps of stress intensities (a) and strain intensities (b) in the chip creation phase for $2\theta = 120^0$, $\alpha = 45^0$, r = 0,001 μm.

For angle $\alpha = 65^0$, there occurs the phenomenon of chip curling (Fig. 5(a), (b)) in the direction of the foundation machined – a stepped chip. This is the result of the fact that the chip line from the side of the cutting edge action surface is longer than the chip line on its opposite side. For angle $\alpha = 55^0$, the chips created are segment chips. Fast cracking of the chip elements is observed.

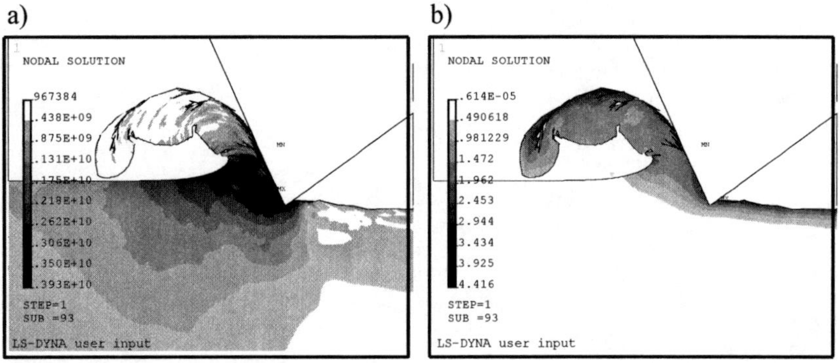

Figure 5: Maps of stress intensities (a) and strain intensities (b) in the chip creation phase for $2\theta = 80^0$, $\alpha = 65^0$, $r = 0{,}001$ μm.

During the grain movement on the surface, without a clear stage of the chip formation, the maximum intensity of stresses occurred in the contact place of the grain apex with the foundation of the material. It was ca. $\sigma_i = 9250$ MPa for $2\theta = 80^0$, $\alpha = 65^0$, $\sigma_i = 7100$ MPa for $2\theta = 80^0$, $\alpha = 55^0$, and $\sigma_i = 3700$ MPa for $2\theta = 80^0$, $\alpha = 45^0$ and it propagated in the direction of machining, and not deep into the material. The maximum intensity of stresses for the same values of angles and the same time steps of the simulation were: $\varepsilon_i = 17{,}87$, 4.85 and 2.22 respectively. Maximum stresses in the material at the moment of a distinct stage of chip creation for $2\theta = 80^0$, $\alpha = 65^0$ occurred in the cutting plane. At the moment of its formation, an increase of stresses was observed in this region from the value of ca. $\sigma_i = 6370$ MPa to the value of $\sigma_i = 8210$ MPa at the moment of the material cracking of separation from the foundation. For $2\theta = 80^0$, $\alpha = 55^0$, the values of maximum stresses $\sigma_i = 4190$ MPa were concentrated also in the cutting plane. For $2\theta = 80^0$, $\alpha = 45^0$, the maximum intensity of strains was ca. $\sigma_i = 4130$ MPa. At the moment of the chip being separated from the foundation, there occurred a distinct drop of stresses in the material. Stresses in the chip were also on a low level. Strains in the material after the grain had passed concentrated in the surface layer of the material of the foundation. Strains occurring in the material located before the cutting edge of the grain propagated not only in compliance with the machining direction but also inside the material. This was the result of the phenomenon of the creation of flashes in the initial phase and on further stages of the chip creation.

A characteristic zone in the material machined is the so-called dead zone. It plays the role of an additional cutting edge with a smaller angle of action than in the case of the proper cutting edge. The ratio of h_r/h decreased together with the increase of the apex angle, and in certain cases this resulted in the occurrence of a wave of the material under the grain without any creation of the chip (noticeable for $\alpha = 65^0$).

6 Conclusions

An application of modern numerical methods and computing systems allows an analysis of complex physical phenomena occurring in the process under investigation. The application developed in the ANSYS system enables a time analysis of the process of machining with a single abrasive grain, with the consideration of the changeability of the grain's apex angle and the angle of its action.

Investigations into the shape of the chips obtained after machining with a single abrasive grain lead to the conclusion that the temperature of the material machined, and strictly speaking, the thermal conditions in which plastic strains occur, have a significant influence on the creation process and the shape of the chip. Machining with a single abrasive grain thus gives one or two chips, shorter and thicker ones, and the groove cut is with large flashes. In a similar machining process with a single abrasive grain, but concerning material heated up to 700^{0}C, flashes are much smaller, while chips are in the number and shape similar to those obtained by machining with a grain tied up in the grinding wheel [21].

The obtained results of the computer simulation of the process of machining with a single abrasive grain with a geometry of $2\theta = 120^0$ and action angle $\alpha = 45^0$ coincide with the results obtained by Kita and Ido [12]. They made an investigation into the influence of the cutting edge apex angle 2θ on the course of the creation of a chip and its shape with dry machining and with the use of a cooling and lubricating liquid. They obtained various chip shapes. For example, for angles $2\theta = 100^0$, $\alpha = 40^0$, they obtained a stepped chip with shallow gaps, which is similar to the one obtained here with the aid of a numerical simulation (Fig. 4(a), (b)). The material flashes obtained before the grain cutting edge and its shapes similar to the results of experiential investigations confirm the justifiability of the use of computer simulations and their reliability.

The distributions of stresses and strains obtained for different grain geometries and action angles, on particular phases of the deformation process, can be made use of while designing machining: making a selection of the machining conditions and its optimising in the aspect of the technological quality of the product.

References

[1] Bajkalov A.K.: *Vvedenie v teoriju šlifovanija materialov*. Kiev, Naukova Dumka 1978.

[2] Korčak S.N.: *Proizvoditelnost' procesja šlifovanija stalnych detalej*. Moskva, Mašinostroenie 1974.

[3] Lortz W.: *Schleifscheibentopographie und Spanbildungmechanismus beim Schleifen*. Diss. T.H. Aachen 1975.

[4] Maslov E. N.: *Teorija šlifovanija materialov*, Moskva, Mašinostroenie 1974.

[5] Oczoś K., Porzycki J, *Szlifowanie. Podstawy i technika*. WNT, Warszawa 1986.

[6] Ostrovskij V.I.: *Teoretičeskie osnovy processa šlifovanija*. Leningrad. Izd. Lenin. Univ. 1981.

[7] Saljé E.: *Erkenntnesstand und Entwicklungstendenzen beim Schleifen und Honen. „Schleifen, Honen und Polieren. Verfahren und Maschinen"*, Essen, Vulkan-Verlag 1982.

[8] Steffens K., Lauer-Schmaltz H.: *Spanbildung und Trennpunktlage beim Schleifen*. Industrie-Anzeigen 1978.

[9] Thörmahlen K.H.: *Einfluss des Korundkorn-Art auf den Schleifprozess*. Diss. T.U. Braunschweig 1973.

[10] Werner G.: *Kinematik und Mechanik des Schleifprozesses*. Diss. T. H. Aachen 1971.

[11] Lortz W.: A model of the cutting mechanism in grinding. *Wear* 1979, v. 53, 115-128.

[12] Kita Y., Ido M.: The mechanism of metal removal by abrasive tool. *Wear* 1978, No 1.

[13] König W., Steffens K.: A numerical method to describe the kinematics of grinding. *Annals of the CIRP* 1982, No 1.

[14] König W., Steffens K.: Modellversuche zur Erfassung der Wechselwirkung zwischen Reibbedingungen und Stofffluss. *Industrie-Anzeiger* 1981, No 35.

[15] Shaw M.C., de Salvo G.J.: A new approach to plasticity and its application to blunt two dimensional indenters. *Trans. ASME* 1970. No 5.

[16] Steffens K.: Beschreibung eines Gleitlinienfeldes für die Deutung der Spanbildung beim Schleifen. *Industrie-Anzeiger* 1979, No 19.

[17] Steffens K., König W.: Closed loop simulation of grinding. *Annals of the CIRP* 1983, No 1.

[18] Steffens K., Lauer-Schmaltz H.: Spanbildungstheorie für das Schleifen, *Industrie-Anzeiger* 1978, No 64.

[19] Tomlenov A.D.: Vnedrenie zakruglennogo puansona v metall pri naličii trenija. *Vestnik Mašinostroenija* 1960, No 1.

[20] Hastings W.F., Oxley P.L.B.: Mechanics of chip formation for conditions appropriate to grinding. Proc. *17th Int. Mach. Tool Des. and Res. Conf.*, Birmingham 1976.

[21] Kukielka L., Kustra J.: Numerical analysis of thermal phenomena and deformations in processing zone in the centreless continuous grinding process. *Surface Treatment VI Computation Methods and Experimental Measurements for Surface Treatment Effects*. Ed. C.A. Brebbia, J.T.M. de Hosson, S-I.Nishida WIT*PRESS*, Southampton, Boston, 2003, pp.109-118.

[22] Kukielka L., Kustra J., Kukielka K.: Numerical analysis of states of strain and stress of material during machining with a single abrasive grain.

Computer Methods and Experimental Measurements for Surface Effects and Contact Mechanics VII. WIT*PRESS* Southampton, Boston 2005, pp. 57-66.

[23] Shaw C.: Fundamentals of grinding. New developments in grinding. *Proc. Int. Grinding Conf.*, Carnegie-Mellon Univ., Pittsburgh 1972.

Efficient modelling of contact interfaces of joints in built-up structures

L. Gaul[1] & M. Mayer[2]
[1] *Institute of Applied and Experimental Mechanics, University of Stuttgart, Germany*
[2] *Engineering and Test Center – Numerical Simulation, Gasoline Systems Division, Robert Bosch GmbH, Germany*

Abstract

This paper introduces an improved approach to model contact interfaces of fixed joints in finite element analysis (FEA) with regard to prediction of the vibration behaviour of built-up structures. The approach consists of two parts: (1) the development of a suitable, new contact model which considers the most important physical effects of wave propagation over the contact interface; and (2) the implementation of the contact model in commercial finite element software. The new model is based on the contact models of Hertz and Mindlin for the contact of a single spherical asperity with an elastic plane. The Hertzian microscopic model for normal contact is then generalized with a statistical approach for rough surfaces introduced by Greenwood to a macroscopic normal contact model for engineering surfaces. To model the macroscopic tangential contact, a new model based on Mindlins approach is introduced which accounts accurately for microslip effects and considers the dependence of tangential contact behaviour on the normal pressure. For implementing the contact model in FEA, a special isoparametric contact element, the so-called zero thickness element, is programmed. The use of this element compared with existing contact algorithms has some major advantages with regard to the application of modelling the contact in fixed joints considered in this paper. The introduced approach is verified by simulating the vibration behaviour of a built-up structure and proving the prediction quality by comparing simulation results with experimental data.
Keywords: contact mechanics, contact elements, joints, joint modelling, microslip, contact damping, hysteresis model, evolution equation.

WIT Transactions on Engineering Sciences, Vol 55, © 2007 WIT Press
www.witpress.com, ISSN 1743-3533 (on-line)
doi:10.2495/SECM070191

1 Introduction

For lightly damped, linear members of a structure, very good estimates of eigenfrequencies, modal damping values, and corresponding mode shapes can be achieved by Experimental Modal Analysis (EMA). Furthermore, by model updating of finite element models of members, very good predictions of the vibration behaviour up to high frequencies are possible [1].

If we now assemble single members into a built-up structure, prediction of the structural vibration behaviour can be quite involved, even though the behaviour of all single members is well-known [2]. This is due to the fact that the mechanical contact at joint interfaces is usually not modelled sufficiently. Effects like uneven contact pressure distributions over the contact area, microslip damping and gaping of contact regions remain unconsidered but these effects can have a major influence on the structural vibration behaviour.

To account for these effects, this paper shows that the roughness of contacting surfaces (figure 1) has to be considered at least integrally to predict the vibration behaviour of built-up structures meaning resonance frequencies, mode shapes and modal damping values.

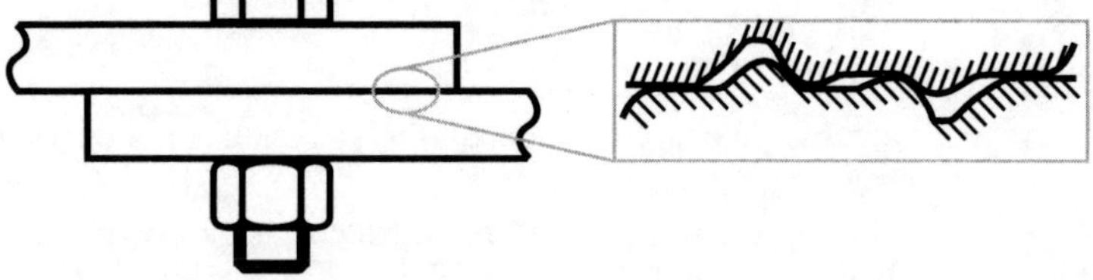

Figure 1: Although the contact area seems to be conforming on a macroscopic scale, the true contact consists of a multitude of non-conforming asperity contacts on a microscopic scale.

2 Contact mechanics

2.1 Normal contact of rough surfaces

For describing the normal contact of two rough surfaces, the contact model of Greenwood and Williamson is employed. This model is based on the Hertzian normal contact model for two elastic spheres [3] which is used to model the contact of single asperities. The Hertzian model leads to a circular contact region with radius r_A and a radial normal pressure distribution

$$p_\mathrm{N}(r) = p_\mathrm{max}\sqrt{1 - \frac{r^2}{{r_A}^2}} \qquad \text{with} \qquad p_\mathrm{max} = \frac{3F_\mathrm{N}}{2\pi {r_A}^2} \tag{1}$$

caused by normal force F_{N}. The approach α_{N} of facing points in the spheres which are far away from the contact region is given by

$$\alpha_{\mathrm{N}} = \frac{{r_A}^2}{r^*} = \left(\frac{9{F_{\mathrm{N}}}^2}{16 r^* {E^*}^2} \right)^{1/3} . \tag{2}$$

Based on this, Greenwood and Williamson [4] developed a model to describe the contact of a rough elastic with a planar rigid surface, see figure 2. The height distribution $\phi(z)$ of the rough surface is determined by the height distributions of the two real rough surfaces in contact. The height distribution $\phi(z)$ and also the

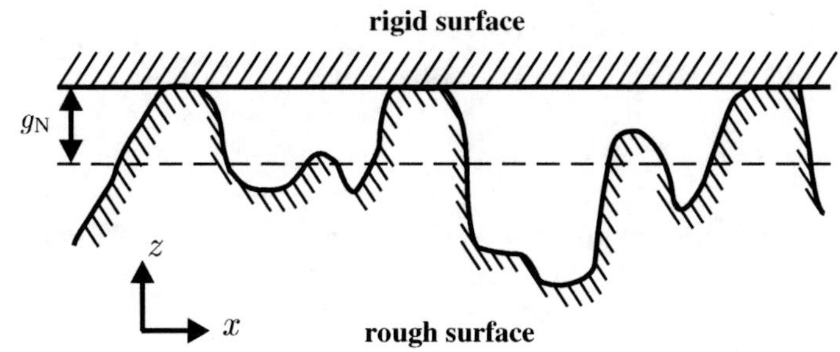

Figure 2: Contact of a planar rigid surface with a rough elastic surface. The dashed line represents the reference plane of the rough surface defined by the arithmetic mean value of the heights. g_{N} is the positive distance in normal direction between the rigid plane and the reference plane.

cumulative height distribution $\Phi(z)$ (Abbott-curve) can then be approximated by different distribution functions. For simplicity an exponential distribution function is used here:

$$\phi_{\mathrm{Exp}}(z) = \sigma^{-1} e^{-(z/\sigma)} \qquad \text{and} \qquad \Phi_{\mathrm{Exp}}(z) = \left[-e^{-(\tilde{z}/\sigma)} \right]_z^\infty = e^{-(z/\sigma)} . \tag{3}$$

This type of distribution is usually sufficient to describe the uppermost 25% of a measured cumulative height distribution [4] and leads to a relatively simple mathematical model. With this we get relations for the true area of contact A_{R} and the normal force F_{N} depending on the normal distance g_{N} of the two surfaces:

$$A_{\mathrm{R}} = \pi r_{\mathrm{A}} N \int_{g_{\mathrm{N}}}^{\infty} (z - g_{\mathrm{N}}) \sigma^{-1} e^{-z/\sigma} \mathrm{d}z = \pi r_{\mathrm{A}} N \sigma e^{-(g_{\mathrm{N}}/\sigma)} , \tag{4}$$

$$F_{\mathrm{N}} = \frac{4}{3} N E^* {r_{\mathrm{A}}}^{1/2} \int_{g_{\mathrm{N}}}^{\infty} (z - g_{\mathrm{N}})^{3/2} \sigma^{-1} e^{-z/\sigma} \mathrm{d}z = \pi^{1/2} N E^* {r_{\mathrm{A}}}^{1/2} \sigma^{3/2} e^{-g_{\mathrm{N}}/\sigma} , \tag{5}$$

where σ is the standard deviation of the height profile of the rough surface, N the overall number of peaks and E^* the average Young's modulus.

Normal pressure is obtained by dividing normal force by apparent area of contact A_0. By introducing new parameters the pressure can be expressed by [5]

$$p_\mathrm{N} = F_\mathrm{N}/A_0 = p_{\mathrm{N}0}\, e^{\,-\lambda(g_\mathrm{N} - g_{\mathrm{N}0})} \qquad \lambda = 1/\sigma > 0, \tag{6}$$

where $g_{\mathrm{N}0} > 0$ is the initial distance between the reference plane and the highest peak of the rough surface and $p_{\mathrm{N}0} > 0$ is the pressure value at initial contact $(g_\mathrm{N} = g_{\mathrm{N}0})$. The slope of this curve is given by $k_\mathrm{N} = -\lambda\, p_\mathrm{N}$ which corresponds to the normal contact stiffness. The asumption of an exponential height distribution leads to the interesting relation

$$p_\mathrm{N} \sim A_\mathrm{R} \qquad \text{with} \quad \frac{A_\mathrm{R}}{A_0} = \Phi(g_\mathrm{N} - g_{\mathrm{N}0}) = e^{\,-(g_\mathrm{N} - g_{\mathrm{N}0})/\sigma}\,. \tag{7}$$

2.2 A new hysteresis model for tangential contact

The introduced model is a new hysteresis model defined by an evolution equation which is based on Mindlin's approach for tangential contact of two spheres. Mindlin assumes that – by applying a tangential force F_T additionally to an acting normal force F_N on the spheres – contact area and normal pressure distribution remain the same as without a tangential force. Therefore, both variables can be described by Hertzian theory [3, chapter 7].

In the presliding or microslip regime $(0 \leq |F_\mathrm{T}| < \mu F_\mathrm{N})$ the relation between tangential force F_T and relative tangential displacement g_T is defined by

$$|F_\mathrm{T}| = \mu F_\mathrm{N} \left[1 - (1 - |g_\mathrm{T}|/g_\mathrm{T}^\mathrm{max})^{3/2}\right] \qquad \text{with} \qquad g_\mathrm{T}^\mathrm{max} = \frac{3}{2k_{\mathrm{T}0}}\mu F_\mathrm{N}, \tag{8}$$

where $k_{\mathrm{T}0}$ is the slope of the microslip curve at the origin and g_T is the relative tangential displacement of facing points in the spheres which are far away from the contact region. If $|F_\mathrm{T}|$ exceeds μF_N the whole contact area is slipping under the condition of Coulomb's law.

By differentiating equation (8) with respect to time where F_N is held constant we get the evolution equation

$$\dot{F}_\mathrm{T} = k_{\mathrm{T}0}\, \dot{g}_\mathrm{T} \left(1 - \frac{|g_\mathrm{T}|}{g_\mathrm{T}^\mathrm{max}}\right)^{1/2} = k_{\mathrm{T}0}\, \dot{g}_\mathrm{T} \left(1 - \frac{|F_\mathrm{T}|}{\mu F_\mathrm{N}}\right)^{1/3}. \tag{9}$$

This equation can be generalized by introducing an arbitrary exponent n which can be used to adapt the curve to measured data, see figure 3:

$$\dot{F}_\mathrm{T} = k_{\mathrm{T}0}\, \dot{g}_\mathrm{T} \left(1 - \frac{|F_\mathrm{T}|}{\mu F_\mathrm{N}}\right)^{n} \qquad \text{for} \quad 0 \leq |F_\mathrm{T}| < \mu F_\mathrm{N}, \qquad n \in (0,1). \tag{10}$$

The slope of this curve is given by $k_\mathrm{T} = \mathrm{d}F_\mathrm{T}/\mathrm{d}g_\mathrm{T} = \dot{F}_\mathrm{T}/\dot{g}_\mathrm{T}$ which corresponds to the tangential contact stiffness. This evolution equation is valid for increasing tangential loading. For decreasing $|F_\mathrm{T}|$ the relation between tangential force and tangential relative displacement is assumed to follow a linear elastic

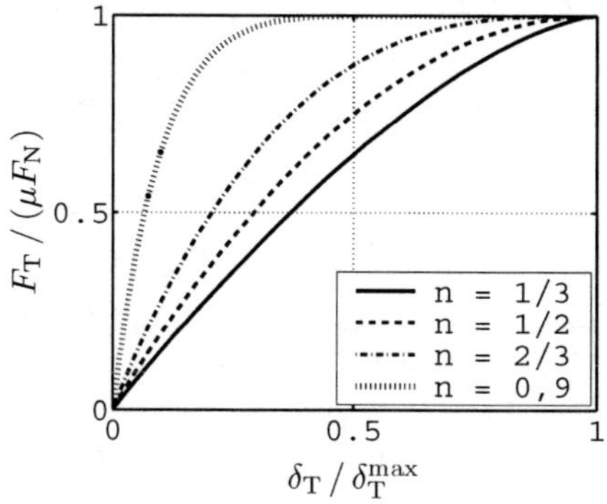

Figure 3: Hysteresis curve defined by equation (10) with different exponents n.

law, $\dot{F}_T = k_{T0}\ \dot{g}_T$. The dissipated energy per cycle for an oscillating tangential force of constant amplitude $\hat{F}_T$ and corresponding displacement amplitude $\hat{g}_T$ is given by

$$W_D = \frac{2\mu^2 F_N{}^2}{k_{T0}} \left\{ \frac{3}{2}\frac{\hat{g}_T}{g_T^{max}} + \frac{3}{2}\frac{1-n}{2-n}\left[\left(1-\frac{\hat{g}_T}{g_T^{max}}\right)^{(2-n)/(1-n)} - 1\right] \right.$$
$$\left. -\frac{1}{2}\left[1-\left(1-\frac{\hat{g}_T}{g_T^{max}}\right)^{3/2}\right]^2 \right\} \tag{11}$$

if the force amplitude is less than μF_N. Otherwise, equation (11) has to be evaluated for $\hat{g}_T = g_T^{max}$ and, additionally, the energy dissipated by Coulomb friction has to be considered:

$$W_D = \frac{2\mu^2 F_N{}^2}{k_{T0}} \left[\frac{n(3-n)}{2(2-3n+n^2)} + \frac{3}{2}\left(\frac{\hat{g}_T}{g_T^{max}} - 1\right)\right]. \tag{12}$$

3 Implementation of the contact model in FEA

The concept of zero thickness elements goes back to Goodman et al. [6] and is discussed in detail in Hohberg [7]. A zero thickness element is depicted in figure 4. The element consists of two four node quadrilateral elements which face each other. In each quadrilateral element, the three-dimensional displacement field $\{u\} = [u\ v\ w]^T$ is approximated by

$$\begin{Bmatrix} u \\ v \\ w \end{Bmatrix} = \begin{bmatrix} h_1\,[I] & h_2\,[I] & h_3\,[I] & h_4\,[I] \end{bmatrix} \begin{Bmatrix} \{u_1\} \\ \{u_2\} \\ \{u_3\} \\ \{u_4\} \end{Bmatrix} = [H]\,\{u\}_{nodal}, \tag{13}$$

where $\{u_i\} = [u_i\ v_i\ w_i]^T$ is the displacement vector of node i and h_i are the bilinear shape functions

$$\begin{aligned} h_1(\xi,\eta) &= \tfrac{1}{4}(1-\xi)(1-\eta) \qquad & h_3(\xi,\eta) &= \tfrac{1}{4}(1+\xi)(1-\eta) \\ h_2(\xi,\eta) &= \tfrac{1}{4}(1-\xi)(1+\eta) \qquad & h_4(\xi,\eta) &= \tfrac{1}{4}(1+\xi)(1+\eta) \end{aligned} \tag{14}$$

formulated in the natural coordinates ξ, η of the element [8]. The matrix $[H]$ contains the shape functions and $[I]$ is a 3×3 unit matrix. The choice of a natural coordinate system simplifies numerical integration of the element matrices.

Distinguishing between top and bottom quadrilateral and assuming that each quadrilateral is connected to the surface of the finite element mesh of one contacting body, we denote the corresponding displacement fields of the elements as

$$\{u\}^{\text{bottom}} = [H(\xi,\eta)]\,\{u\}^{\text{bottom}}_{\text{nodal}} \qquad \{u\}^{\text{top}} = [H(\xi,\eta)]\,\{u\}^{\text{top}}_{\text{nodal}}. \tag{15}$$

Since these elements are only two-dimensional, the traction vector in each element

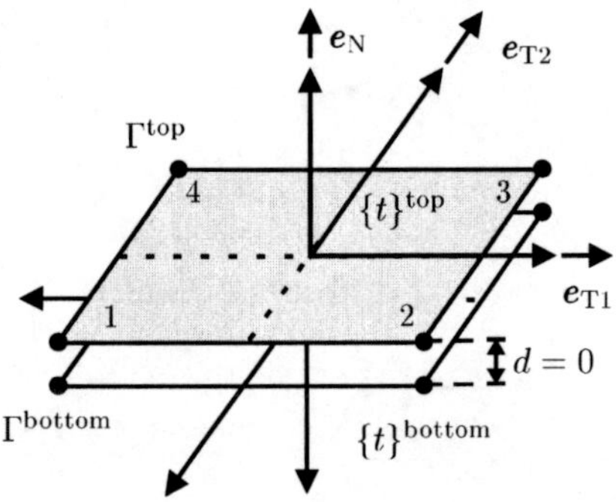

Figure 4: 8-node zero thickness element consisting of two 4-node quadrilateral elements which are connected to the surfaces of the contacting bodies.

describes the interface stresses

$$\{t\}^{\text{bottom}} = \begin{Bmatrix} t_{Tx} \\ t_{Ty} \\ t_N \end{Bmatrix}^{\text{bottom}} \qquad \{t\}^{\text{top}} = \begin{Bmatrix} t_{Tx} \\ t_{Ty} \\ t_N \end{Bmatrix}^{\text{top}} \tag{16}$$

and with this we can state the virtual internal work for each quadrilateral

$$\delta W_I^{\text{bottom}} = \int_0^{\ell_1}\int_0^{\ell_2} \delta\{u\}^{\text{bottom}\,T}\,\{t\}^{\text{bottom}}\,dx\,dy \tag{17}$$

$$\delta W_I^{\text{top}} = \int_0^{\ell_1}\int_0^{\ell_2} \delta\{u\}^{\text{top}\,T}\,\{t\}^{\text{top}}\,dx\,dy. \tag{18}$$

In contact mechanics, one is interested in the relative displacement field $\{g\}$ of the contacting surfaces. This is expressed for zero thickness elements through the relative displacement between the top and the bottom quadrilateral,

$$\{g\} = \{u\}^{\text{top}} - \{u\}^{\text{bottom}} = [H(\xi,\eta)]\left(\{u\}^{\text{top}}_{\text{nodal}} - \{u\}^{\text{bottom}}_{\text{nodal}}\right). \tag{19}$$

Furthermore, we know from Newton's third law that the traction vectors of the elements in contact must be equal in magnitude and opposite in direction,

$$\{t\} = \{t\}^{\text{top}} = -\{t\}^{\text{bottom}}. \tag{20}$$

We can now implement contact laws as incremented and linearized constitutive relations [9] between contact tractions and the relative displacement field,

$$\Delta\{t\} = \left[\frac{\partial\{t\}}{\partial\{g\}}\right]_{t+\Delta t\{g\}^{(j-1)}} \Delta\{g\} = {}^{t+\Delta t}[C]^{(j-1)}_{Tangential}\,\Delta\{g\}. \tag{21}$$

A description of different contact laws can be found in Gaul and Nitsche [10].

The virtual work of the contact tractions is given by the summation of the virtual internal work for each element, as given in equation (17) and equation (18):

$$\delta W_C = \delta W_I^{\text{top}} + \delta W_I^{\text{bottom}} = \int_0^{\ell_1}\int_0^{\ell_2} \delta\{g\}^T\,\{t\}\,dx\,dy. \tag{22}$$

Implementing the contact law in the virtual work expression yields the tangential contact stiffness matrix for a relative displacement quadrilateral element,

$$[K]_{Tangential} = \int_0^{\ell_1}\int_0^{\ell_2} [H(\xi,\eta)]^T\;{}^{t+\Delta t}[C]^{(j-1)}_{Tangential}\,[H(\xi,\eta)]\,dx\,dy\,. \tag{23}$$

These integrals are evaluated by applying the isoparametric concept and using Gaussian quadrature scheme [8]. The full stiffness matrix for the eight node zero thickness element is composed of the stiffness matrix of the quadrilateral element

$$[K] = \begin{bmatrix} [K]_{Tangential} & -[K]_{Tangential} \\ -[K]_{Tangential} & [K]_{Tangential} \end{bmatrix}. \tag{24}$$

The stiffness matrix $[K]$ is 12-times singular due to its composition, thereby causing 12 zero-energy modes. A zero-energy mode, or so-called hourglass mode, is a displacement mode that does not correspond to a rigid body motion, and it produces zero strain energy [8]. As zero thickness elements are always clamped between continuum elements, all 12 zero-energy modes are suppressed.

For applications of zero thickness elements in geomechanics, see, e.g., Beer [11] and for applications in model update procedures see Ahmadian et al. [12].

Figure 5: Airbag control unit generation 9 (Robert Bosch GmbH). Source: http://www.bosch-presse.de/TBWebDB/bosch-pbj/de-DE/start.cfm

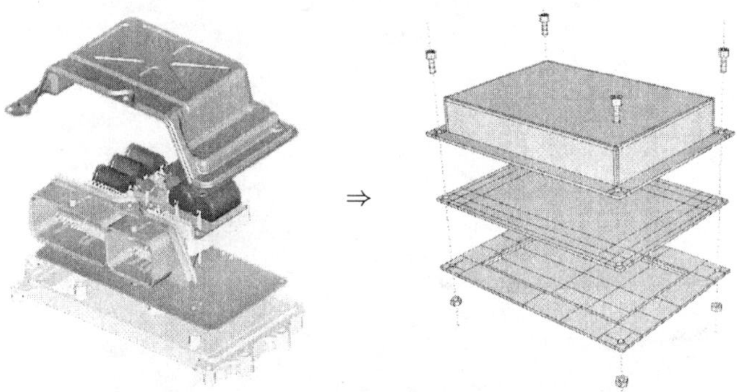

Figure 6: Exploded view of a control unit (left) and simplified test structure model (right) consisting of three simple aluminium parts connected by four bolts with nuts for investigating especially the influence of the contact region on the vibration behaviour.

4 Application to a structure with bolted joints

The contact model implemented in the presented zero thickness elements is applied to model the joint interfaces of a simplified control unit and thus predict the vibration behaviour. The simplified structure is depicted on the right of figure 6.

The linear elastic material parameters of the aluminium parts are experimentally determined to minimize errors in the vibration simulation. The simulation itself consists of two steps. In a first preloading step the bolts and nuts are tightened causing a non-homogeneous contact pressure distribution in the contact interface which is depicted in figures 7 and 8 and by it a non-homogeneous contact stiffness distribution. Next, following the experimental approach, the impulse response of the structure is simulated and resonance frequencies and modal damping values are determined by modal analysis of the simulated impulse responses at different locations of the finite element mesh.

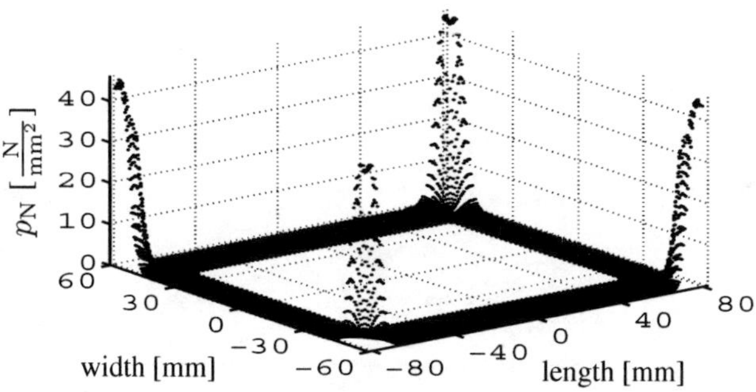

Figure 7: Normal pressure distribution p_N in the contact interface between upper cover and plate.

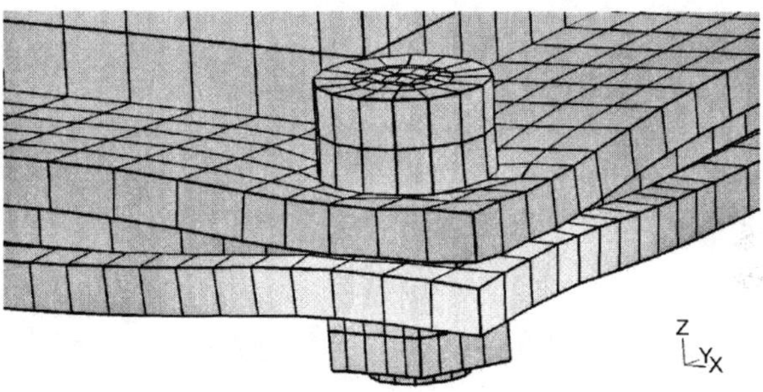

Figure 8: Gaping of the contact interfaces after applying the bolt load.

4.1 Conclusion

Table 1 compares simulated and measured results. Obviously, the new contact model accurately predicts the measured behaviour of this structure.

Table 1: Comparison of resonance frequencies and modal damping values up to 2 kHz determined by measuring the impulse response and following Experimental Modal Analysis (left) and by simulation of the impulse response, Fourier transform and following Modal Analysis (right).

	Measured		Simulated	
Mode 1	889 Hz	1.2 %	877 Hz	0.9 %
Mode 2	1101 Hz	0.8 %	1113 Hz	0.6 %
Mode 3	1349 Hz	1.1 %	1366 Hz	0.7 %
Mode 4	1424 Hz	0.9 %	1386 Hz	0.7 %
Mode 5	1521 Hz	0.7 %	1537 Hz	0.5 %
Mode 6	1645 Hz	0.6 %	1660 Hz	0.5 %
Mode 7	1766 Hz	0.7 %	1753 Hz	0.6 %
Mode 8	1960 Hz	0.6 %	1982 Hz	0.4 %

References

[1] Gaul, L., Albrecht, H. & Wirnitzer, J., Semi-active friction damping of large space truss structures. *Journal of Shock and Vibration*, **11**, pp. 173–186, 2004.

[2] Gaul, L. & Lenz, J., Nonlinear dynamics of structures assembled by bolted joints. *Acta Mechanica*, **125**, pp. 169–181, 1997.

[3] Johnson, K.L., *Contact Mechanics*. Cambridge University Press: Cambridge, 1985.

[4] Greenwood, J.A. & Williamson, J.B.P., Contact of nominally flat surfaces. *Proceedings of the Royal Society of London*, **Series A 295**, pp. 300–319, 1966.

[5] Willner, K. & Gaul, L., A penalty approach for contact description by fem based on interphase physics. *Proceedings of Contact Mechanics II*, eds. M.H. Aliabadi & C. Allesandri, Computational Mechanics Publications: Southampton, pp. 257–264, 1995.

[6] Goodman, R.E., Taylor, R.L. & Brekke, T.L., A model for the mechanics of jointed rock. *Journal of the Soil Mechanics and Foundations Division*, **94**, pp. 637–660, 1968.

[7] Hohberg, J.M., *A Joint Element for the Nonlinear Dynamic Analysis of Arch Dams*. Birkhaeuser: Basel, 1992.

[8] Bathe, K.J., *Finite Element Methods*. Springer-Verlag: Berlin and Heidelberg, 2002.

[9] Wriggers, P., *Computational Contact Mechanics*. John Wiley & Sons Ltd.: Chichester, 2002.

[10] Gaul, L. & Nitsche, R., The role of damping in mechanical joints. *Applied Mechanics Reviews*, **54**, pp. 93–106, 2001.

[11] Beer, G., An isoparametric joint/interface element for finite element analysis. *International Journal for Numerical Methods in Engineering*, **21**, pp. 585–600, 1985.

[12] Ahmadian, H., Jalali, H., Mottershead, J. & Friswell, M., Dynamic modeling of spot welds using thin layer interface theory. *Proceedings of the Tenth Int. Congress on Sound and Vibration ICSV10*, Stockholm, Sweden, pp. 3439–3446, 2003.

Domain decomposition based contact solver

J. Dobiáš[1], S. Pták[1], Z. Dostál[2] & V. Vondrák[2]
[1] *Institute of Thermomechanics, Prague, Czech Republic*
[2] *University of Ostrava, Ostrava, Czech Republic*

Abstract

The paper is concerned with the application of a new variant of the FETI domain decomposition method called the Total FETI to the solution of contact problems by the finite element method. The basic idea is that both the compatibility between adjacent sub-domains and Dirichlet boundary conditions are enforced by the Lagrange multipliers with physical meaning of forces, while the displacements are eliminated. We introduce the Total FETI technique to solve the equations and inequalities governing the equilibrium of system of bodies in contact. Moreover, we show implementation of the method into a code which treats the material and geometric non-linear effects. Numerical experiments were carried out with our in-house general purpose package PMD.
Keywords: contact, domain decomposition, non-linear, Lagrange multipliers, finite element method.

1 Introduction

Modelling contact phenomena is still a challenging problem of non-linear computational mechanics. The complexity of such problems arises from the fact that we do not know the regions in contact until we have run the problem. Their evaluations have to be part of the solution. In addition, the solution across the contact interface is non-smooth. In other words, a general contact problem is strongly non-linear and its reasonable solution in terms of a numerical technique, usually the finite element method, needs high quality software stemming from techniques exhibiting qualities like fast convergence rate, good parallel and numerical scalabilities, and so on.

In 1991 Farhat and Roux [1] came up with a novel domain decomposition method called FETI (Finite Element Tearing and Interconnecting method). This

WIT Transactions on Engineering Sciences, Vol 55, © 2007 WIT Press
www.witpress.com, ISSN 1743-3533 (on-line)
doi:10.2495/SECM070201

method belongs to the class of non-overlapping spatial decompositions. Its key concept is based on the idea that satisfaction of the compatibility between spatial sub-domains, into which a domain is partitioned, is enforced by the Lagrange multipliers with physical meaning of forces in this context. After eliminating the primal variables, which are displacements in the displacement based analysis, the original problem is reduced to a small, relatively well conditioned, typically equality constrained quadratic programming problem that is solved iteratively. The CPU time that is necessary for both the elimination and iterations can be reduced nearly proportionally to the number of processors, so that the algorithm exhibits the parallel scalability. This method has proved to be one of the most successful algorithms for parallel solution to problems governed by elliptic partial differential equations. Observing that the equality constraints may be used to define so called 'natural coarse grid', Farhat *et al.* [2] modified the original FETI algorithm in such a way that they were able to prove its numerical scalability.

The fact that sub-domains act on each other in terms of forces suggests that the FETI approach can also be naturally applied to solution to the contact problems. To this effect the FETI methodology is used to prescribe conditions of non-penetration between bodies. We shall obtain a new minimisation problem with additional non-negativity constraints which replace more complex general non-penetration conditions; see Dostál *et al.* [3]. It turned out that the scalability of the FETI methods may be preserved even for solution to the contact problems [3, 4].

A new variant of the FETI method, called the Total FETI (TFETI) method, was presented by Dostál *et al.* [5]. In this paper we are concerned with application of this method to solution to the contact problems while we in addition consider the material and geometric non-linear effects.

We briefly introduce theoretical foundations of the FETI and TFETI methods. Then we describe an algorithm, in which the TFETI based contact solver accounts for the inner loop, while the outer loop is concerned with the non-linear effects others than the contact. The numerical experiments were carried out with our in-house general purpose finite element package PMD (Package for Machine Design) [6].

2 Original FETI method

Consider the static case of a contact problem between two solid deformable bodies denoted as Ω_1 and Ω_2. We assume that their boundaries are subdivided into three disjoint parts Γ^u, Γ^f, and Γ^c with the Dirichlet, Neumann and contact conditions, respectively; see fig. 1(a). The governing equations are given by the equilibrium conditions along with the boundary conditions; see, e.g., Laursen [7] for comprehensive survey of formulations.

Fig. 1(b) shows a discretised version of the contact problem from fig. 1(a). Both bodies, i.e. sub-domains, are discretised in terms of the finite elements method. This figure also shows applied Dirichlet boundary conditions, some displacements, denoted as u, and the contact interface. The displacements are the primal variables in the context of the displacement based finite element analysis.

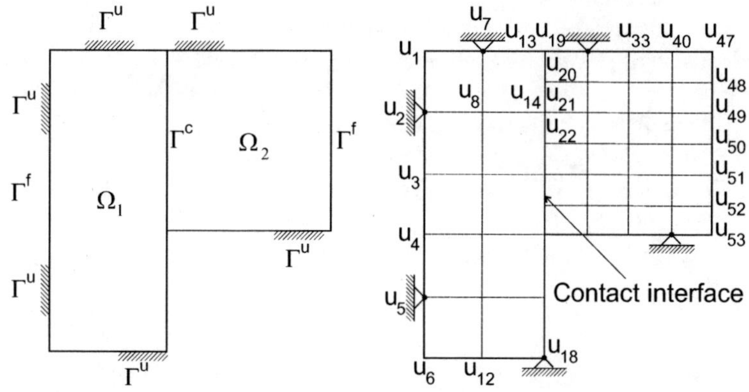

(a) Original problem. (b) Discretised problem.

Figure 1: Basic notation.

The result of application of the FETI method to the computational model from fig. 1 is depicted in fig. 2(a). The left sub-domain is decomposed into two sub-domains with fictitious interface between them. The fundamental idea of the FETI method is that the compatibility between sub-domains along both fictitious and contact interfaces is ensured by means of the Lagrange multipliers with the physical meaning of forces. They are also called the dual variables. λ^E denotes the forces along the fictitious interface and λ^I stands for the forces generated by contact.

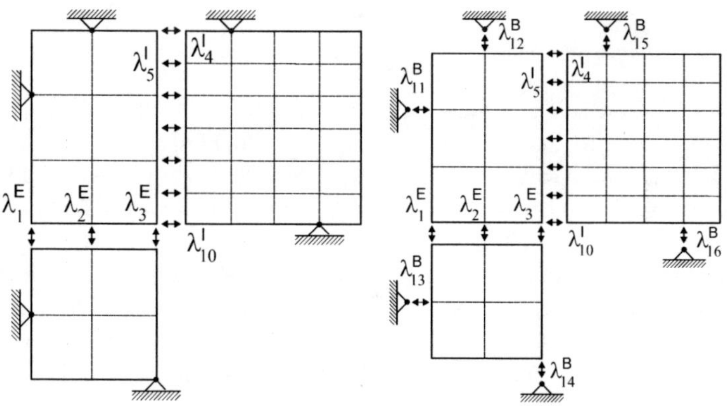

(a) FETI. (b) Total FETI.

Figure 2: Principles of FETI and TFETI.

Let N be a number of sub-domains and let us denote for $i = 1, \ldots, N$ by K_i, f_i, u_i and B_i the stiffness matrix, the vector of externally applied forces, the vector of displacements and the signed matrix with entries $-1, 0, 1$ defining the sub-domain interconnectivity for the i-th sub-domain, respectively. The matrix B is composed of matrices B^I and B^E, $B = \begin{bmatrix} B^I & B^E \end{bmatrix}$. B^E introduces connectivity conditions along the fictitious interfaces and B^I along the contact ones.

The discretised version of the problem is governed by the equation

$$\min \frac{1}{2} \mathrm{u}^\top K \, \mathrm{u} - f^\top \mathrm{u} \quad \text{subject to} \quad B^I \mathrm{u} \leq 0 \quad \text{and} \quad B^E \mathrm{u} = 0 \tag{1}$$

where

$$K = \begin{bmatrix} K_1 & & \\ & \ddots & \\ & & K_N \end{bmatrix}, \quad f = \begin{bmatrix} f_1 \\ \vdots \\ f_N \end{bmatrix}, \quad u = \begin{bmatrix} u_1 \\ \vdots \\ u_N \end{bmatrix}. \tag{2}$$

The FETI method assumes that Dirichlet boundary conditions are inherited from the original problem, which is shown in fig. 2(a). This fact implies that the magnitudes of defects of the stiffness matrices, K_i, may vary from zero, for the sub-domains with enough Dirichlet conditions, to the possible maximum (6 for 3D solid mechanics problems and 3 for 2D ones) in the case of the sub-domains exhibiting some rigid body modes. General solution to such systems requires computation of generalised inverses and bases of the null spaces, i.e. kernels, of the underlying singular matrices. The problem is that the magnitudes of the defects are difficult to evaluate because this computation is extremely disposed to the round off errors; see Farhat and Géradin [8].

3 Total FETI method

In this section we briefly review the main ideas the TFETI method stems from. To circumvent the problem of computing bases of the kernels of singular matrices, Dostál [5] came up with a novel solution. His idea was to remove all the prescribed Dirichlet boundary conditions and to enforce them by additional Lagrange multipliers denoted as λ^B in fig. 2(b). The effect of the procedure on the stiffness matrices of the sub-domains is that their defects are the same and their magnitude is known beforehand. From the computational point of view such approach is advantageous; see [8] for discussion of this topic.

The overall approach resembles the classic one by Farhat *et al.* [2] and others, e.g. [3]. The Lagrangian associated with the problem governed by eqn (1) is as reads

$$L(u, \lambda) = \frac{1}{2} u^\top K u - f^\top u + \lambda^\top B u. \tag{3}$$

This is equivalent to the saddle point problem

$$\text{Find} \quad (\bar{u}, \bar{\lambda}) \quad \text{so that} \quad L(\bar{u}, \bar{\lambda}) = \sup_\lambda \inf_u L(u, \lambda). \tag{4}$$

For λ fixed, the Lagrangian $L(.,\lambda)$ is convex in the first variable and a minimiser u of $L(.,\lambda)$ satisfies the following equation

$$Ku - f + B^\top \lambda = 0. \tag{5}$$

Eqn (5) has a solution if and only if $f - B^\top \lambda$ belongs to the range of K and therefore the following relationship holds

$$R^\top (f - B^\top \lambda) = 0. \tag{6}$$

R denotes the full rank matrix with columns spanning the kernel of K. The kernels of the sub-domains are known and can be assembled directly.

It is necessary to eliminate the primal variable u from eqn (5). It may be easily verified that if u is a solution to eqn (5), then there exists a vector α such that

$$u = K^\dagger (f - B^\top \lambda) + R\alpha \tag{7}$$

where $K^\dagger$ is any symmetric positive definite matrix satisfying $KK^\dagger K = K$.

Substituting eqn (7) into eqn (4), we get the following minimisation problem

$$\min \frac{1}{2} \lambda^\top B K^\dagger B^\top \lambda - \lambda^\top B K^\dagger f, \quad \text{s.t.} \quad R^\top (f - B^\top \lambda) = 0. \tag{8}$$

Let us introduce notations

$$F = BK^\dagger B^\top, \quad G = R^\top B^\top, \quad e = R^\top f, \quad d = BK^\dagger f, \tag{9}$$

so that the problem (8) reads

$$\min \frac{1}{2} \lambda^\top F \lambda - \lambda^\top d \quad \text{s.t.} \quad G\lambda = 0. \tag{10}$$

The final step stems from observation that the problem (10) is equivalent to

$$\min \frac{1}{2} \lambda^\top PFP\lambda - \lambda^\top Pd \quad \text{s.t.} \quad G\lambda = 0, \tag{11}$$

where

$$P = I - Q \quad \text{and} \quad Q = G^\top (GG^\top)^{-1} G \tag{12}$$

stand for the orthogonal projectors on the kernel of G and the image space of $G^\top$, respectively.

The problem (11) may be solved efficiently by the conjugate gradient method because the estimate of the spectral condition number for the FETI method also holds for the TFETI method [5].

It was shown that application of the TFETI methodology to the contact problems converts the original problem to the quadratic programming one with simple bounds and equality constraints. This problem can be further transformed by Semi-Monotonic Augmented Lagrangians with Bound and Equality constraints (SMALBE) method to the sequence of simply bounded quadratic programming problems. These auxiliary problems may be solved efficiently by the Modified Proportioning with Reduced Gradient Projection (MPRGP) method. The detailed descriptions of SMALBE and MPRGP are beyond the scope of this paper and can be found in Dostál [9]. It was proved in Dostál [10] that application of combination of both these methods to solution to contact problems benefits the numerical and parallel scalabilities.

4 Non-linear algorithm

Herein we extend application of the FETI and TFETI methods to problems with the geometric and material non-linearities. The algorithm based solely on the assumptions and relationships presented in the previous Sections is directly applicable to solution to the contact problems, but with other conditions linear. Any additional non-linear effect necessitates employment of the nested iteration strategy, where the inner loop accounts for TFETI based contact solver while the outer loop is concerned with the material and geometric non-linear effects, contact geometry update and equilibrium iterations.

The state of equilibrium is characterised by condition that the internal forces equal the total external forces, i.e., the residual equals zero

$$res = f_{ext} - f_{int} = 0. \tag{13}$$

The total external forces consist of the applied external forces and the contact ones

$$f_{ext} = f - B^{\top}\lambda. \tag{14}$$

The internal forces can be expressed as follows

$$f_{int} = \sum_{nelem} \int_{V_e} B_s{}^{\top}(\epsilon)\, \sigma(\epsilon)\, dV \tag{15}$$

where B_s is an appropriate element matrix relating the element strain tensor ϵ with nodal displacements while considering the geometric non-linearities, and σ is the element stress tensor arising in general from non-linear material behaviour. We sum over the total number of elements $nelem$.

The solution algorithm is shown in the following flowchart.

Initial step: Assemble stiffness matrix $K = diag\{K_1, ..., K_p\}$ and B^E ;
Set $i = 0$; $\mathrm{u}^0 = 0$, $\lambda^0 = 0$, $f^0_{int} = 0$;
Step 1: Evaluate contact conditions B^{Ii} ;
Step 2: Solve contact problem for $\Delta\lambda \rightarrow \Delta\mathrm{u}$;
Step 3: $\lambda^i = \lambda^{i-1} + \Delta\lambda$, $\mathrm{u}^i = \mathrm{u}^{i-1} + \Delta\mathrm{u}$;
$f^i_{int} = \sum_{nelem} \int_{V_e} B_s{}^{\top}(\epsilon^i)\, \sigma(\epsilon^i)\, dV$
Assemble residual load vector $res^i = f - B^{i\top}\lambda^i - f^i_{int}$;
Check on convergence criteria $\frac{\|\Delta\mathrm{u}\|}{\|\mathrm{u}^i\|} < \eta_1$, $\frac{\|res^i\|}{\|f^i_{ext}\|} < \eta_2$;

If fulfilled then STOP,
otherwise set $i \leftarrow i + 1$ and go to **Step 1**

Step 2 represents the inner iteration loop for evaluation of the Lagrange multipliers enforcing compatibility between the sub-domains along both fictitious and contact interfaces and at nodes with the applied Dirichlet conditions.

5 Numerical experiments

To demonstrate the ability of our algorithms to solve contact problems, we show results of two numerical experiments. The first case is concerned with contact problem of two cylinders, and the second one with contact problem of the pin in hole with small clearance.

5.1 Contact problem of two cylinders

Consider contact of two cylinders with parallel axes. We can consider only one half of the problem due to its symmetry. The diameter of the upper cylinder $R_u = 1\ m$ and of the lower one $R_l = \infty$. In spite of the fact that it is the 2D problem, it is modelled with 3D continuum tri-linear elements with two layers of them along the axis of symmetry of the upper cylinder. The model consists of 8904 elements and 12765 nodes. The boundary conditions are imposed in such a way that they generate, from the physical point of view, the plane strain problem. The material properties are as follows: Young's modulus $E = 2.0 \times 10^{11}\ Pa$ and Poisson's ratio $\nu = 0.3$.

First, the upper cylinder is loaded by $40\ MN/m$ along its upper line and the problem is considered linearly elastic and linearly geometric. Fig. 3(a) shows solution in terms of the deformed mesh.

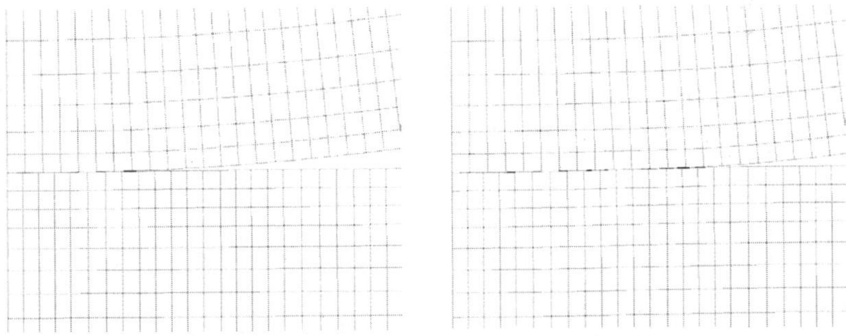

(a) Linear problem. (b) Non-linear problem.

Figure 3: The problem of two cylinders, deformed meshes.

WIT Transactions on Engineering Sciences, Vol 55, © 2007 WIT Press
www.witpress.com, ISSN 1743-3533 (on-line)

Next, the problem was computed on the same mesh with the same loading, but we considered the linearly–elastic–perfectly–plastic material model with the yield stress $\sigma_Y = 800\ MPa$. We also considered the geometric non-linearity. The deformed mesh is depicted in fig. 3(b).

In the latter case we iterated in the outer loop in the sense of the solution algorithm described in Section 4. The number of the outer iterations was 15. The number of iterations of the MPRGP algorithm for contact evaluation at each cycle of the outer loop is shown in fig. 4.

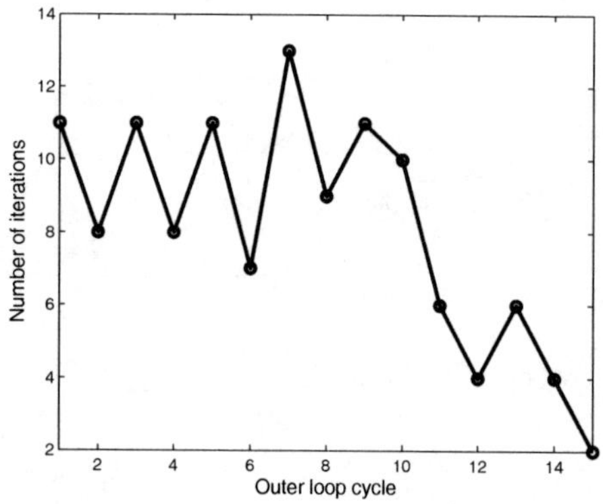

Figure 4: Number of MPRGP iterations at each cycle of the outer loop.

5.2 Pin-in-hole contact problem

Consider a problem of the circular pin in circular hole with small clearance. The radius of the hole is 1 m and the pin has its radius 1% smaller. Again, the 2D problem is modelled with 3D elements. The model consists of 15844 tri-linear elements and 28828 nodes. The pin is loaded along its centre line by 133 MN/m. The geometric non-linearity was considered. The material properties are the same as in the previous case.

Fig. 5 shows the normal contact stress distribution along surface of the pin from the plane of symmetry. The distribution of this stress along the hole is practically identical.

6 Conclusion

Application of a new variant of the FETI domain decomposition method to solution to contact problems with additional non-linear effects was presented. It is called the Total FETI method, and its basic idea, in comparison with the FETI method, consists in replacement of the Dirichlet boundary conditions by the Lagrange multipliers with physical meaning of forces in this context. This feature

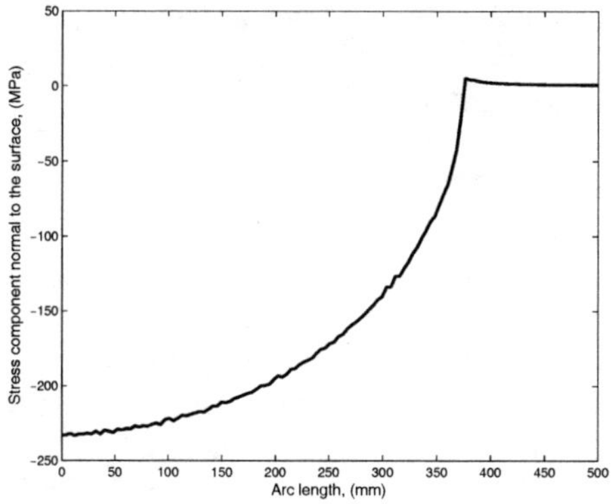

Figure 5: The pin-in-hole problem, normal contact stress on the pin, geometrically non-linear case.

is of great importance from the computational point of view, because the magnitudes of defects of stiffness matrices of all the sub-domains are the same and their magnitude is known beforehand. The numerical experiments show that algorithm stemming from the TFETI method is applicable to solution to contact problems accompanied by other non-linearities.

Acknowledgement

The authors would like to acknowledge the support of GA CR through grant number 101/05/0423 and AS CR through grant number AV0Z20760514.

References

[1] Farhat Ch. & Roux F.X., A method of finite element tearing and interconnecting and its parallel solution algorithm. *International Journal for Numerical Methods in Engineering*, **32(12)**, pp. 1205–1227, 1991.
[2] Farhat Ch., Mandel J. & Roux F.X., Optimal convergence properties of the FETI domain decomposition method. *Computer Methods in Applied Mechanics and Engineering*, **115(5)**, pp. 365–385, 1994.
[3] Dostál Z., Horák D., Kučera R., Vondrák V., Haslinger J., Dobiáš J. & Pták S., FETI based algorithms for contact problems: scalability, large displacements and 3D Coulomb friction. *Computer Methods in Applied Mechanics and Engineering*, **194(2–5)**, pp. 395–409, 2005.
[4] Dostál Z. & Horák D., Scalable FETI with optimal dual penalty for a variational inequality. *Numerical Linear Algebra with Applications*, **11(6)**, pp. 455– 472, 2004.

[5] Dostál Z., Horák D. & Kučera R., Total FETI - an easier implementable variant of the FETI method for numerical solution of elliptic PDE. To be published in *Communications in Numerical Methods in Engineering*.
[6] PMD manuals, www.it.cas.cz/manual/pmd
[7] Laursen T.A., *Computational contact and impact mechanics*, Springer-Verlag: Berlin, pp. 8–17, 2002.
[8] Farhat Ch. & Géradin M., On the general solution by a direct method of a large-scale singular system of linear equations: application to the analysis of floating structures. *International Journal for Numerical Methods in Engineering*, **41(7)**, pp. 675–696, 1998.
[9] Dostál Z. & Schöberl J., Minimizing quadratic functions over non-negative cone with the rate of convergence and finite termination. *Computational Optimization and Application*, **30(1)**, pp. 23–43, 2005.
[10] Dostál Z., Inexact semi-monotonic augmented Lagrangians with optimal feasibility convergence for convex bound and equality constrained quadratic programming. *SIAM Journal on Numerical Analysis*, **43(2)**, pp. 96–115, 2005.

Dimensional reduction for fast simulations of contact problems

T. Geike & V. L. Popov
Institute of Mechanics, Technical University of Berlin, Germany

Abstract

This paper explains a method of how to reduce the dimension of a contact problem under study. In particular it is shown how the three-dimensional adhesive contact can be simulated on the basis of a one-dimensional model. Single contacts as well as multi-asperity contacts with adhesion are discussed in some detail. If lubricants are present it is necessary to solve simultaneously for elastic deformations of contacting bodies and fluid flow between the bodies. We show that the problem can be considerably reduced in the case when the lubrication layer is so thin that the main contribution to the contact interaction comes from a small part of micro contacts with a distance much smaller than the average distance between the bodies. In this case, it is possible to model the dynamics of lubrication by non-conservative forces between surface elements depending both on the distance and relative velocity. The presented reduction method is currently used for the simulation of chemical–mechanical polishing.
Keywords: elastic contacts, computational contact mechanics, friction, adhesion, lubrication, chemical–mechanical polishing.

1 Introduction

Contact and friction play an important role in many technical applications ranging from traditional applications like bearings, clutches and brakes [1] over manufacturing technologies [2, 3] to modern applications like micro-electro-mechanical systems [4]. Further technological progress in these fields requires a better understanding of the friction phenomenon and the development of appropriate simulation tools.

Many tribological systems belong to the class of fractal systems: in friction processes, both the microscopic and macroscopic scales may play an essential

WIT Transactions on Engineering Sciences, Vol 55, © 2007 WIT Press
www.witpress.com, ISSN 1743-3533 (on-line)
doi:10.2495/SECM070211

role [4, 5]. The multi-scale nature of friction processes, however, makes the simulation of such systems difficult. The need of including many scales and physical processes in a simulation model leads to the development of reduced simulation methods.

One of the possibilities to reduce the computation time is to use hierarchical simulation methods [6]. In the present paper, another strategy is discussed: substitution of three-dimensional systems by one-dimensional systems. The basis of the reduction are the following two ideas [7]: (1) The elastic contact energy is a local quantity which only depends on the configuration and deformations in the vicinity of a micro contact, but does not depend on the size and the form of the body as a whole. (2) The interaction between micro contacts is of minor importance. Then the dimensionality of the multi-contact system plays no role as long as the behaviour of a single asperity and the statistics of asperities are modelled correctly.

The one-dimensional model is currently used to study the chemical–mechanical polishing [3]. In this application, elasticity, plasticity, lubrication, adhesion and the surface topography are considered in the simulation tool. Because of this complexity and the need to study a sufficiently huge piece of the specimen to be polished, simulating the evolution of the surface topography in the course of the polishing process is only possible with the proposed model.

In this paper the elastic contact is discussed in detail. Much attention is paid on how to set up the parameters of the one-dimensional model (elastic properties, surface topography). Subsequently two extensions are discussed: adhesion and lubrication.

2 Elastic contact

2.1 Single contact

The first main idea of the proposed reduced description is the following [7, 8]: consider the three-dimensional contact problem with relative radius of curvature R_3 and elastic modulus E^*. The relation between normal force F_3 and approach d reads [9]

$$F_3\left(d\right) = \frac{4}{3} E^* \sqrt{R_3 d^3} \quad , \tag{1}$$

while the relation between normal force F_3 and radius of contact a is

$$F_3\left(a\right) = \frac{4E^*}{3R_3} a^3 \quad . \tag{2}$$

Now consider the one-dimensional contact problem, depicted in figure 1. The respective relations are

$$F_1\left(d\right) = \frac{4\sqrt{2}c_n}{3} \sqrt{R_1 d^3} \quad , \tag{3}$$

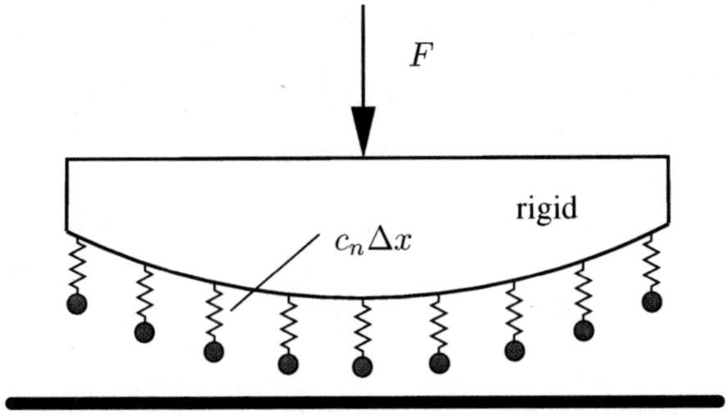

Figure 1: Contact between a rigid plate and a rigid cylinder with elastic layer.

$$F_1(a) = \frac{2c_n}{3R_1} a^3 \quad , \tag{4}$$

where c_n is the stiffness per unit length. Note that the distance Δx between particles is small compared to the size of the contact. The macroscopic relations between force and approach and force and radius of contact will be identical for the three-dimensional and one-dimensional problem if

$$R_1 = \frac{1}{2} R_3 \quad , \quad c_n = E^* \quad . \tag{5}$$

Hence, the three-dimensional contact problem can be reduced to a one-dimensional problem for arbitrary radius of curvature. For the local force in the one-dimensional problem one gets

$$f(x) \propto \left(a^2 - x^2\right) \quad , \tag{6}$$

which is different from the well-known result for the contact pressure in the three-dimensional problem [9]. Introducing a stress according to

$$\sigma(x) = \frac{f(x)}{b\sqrt{\delta(x) R_1}} \quad , \tag{7}$$

where $\delta(x)$ is the local deformation and b is the effective width, yields the desired relation

$$\sigma(x) \propto \sqrt{1 - \frac{x^2}{a^2}} \quad . \tag{8}$$

By choosing the effective width b appropriately, the stress according to eqn (7) is identical to the three-dimensional result. When simulating problems with plasticity the yield criteria should depend on the stress σ and not on the local force f.

2.2 Multi-asperity contact

The second important idea of the proposed 3D to 1D mapping is that the interaction between neighbouring asperities is of no importance for the contact problem as far as the size of micro contacts is much smaller than the distance between them – and this is the case in a typical macroscopic tribological system. It is rather the statistics of heights and radii of curvature which is important for the contact problem [10, 11]. The statistics of micro contacts determines on one hand the normal forces between bodies. On the other hand, it determines the real area of contact and thus the tangential friction forces. The distribution of normal and tangential forces as well as the distribution of contact areas of micro contacts are the most important quantities for the understanding and the qualitative characterisation of tribological systems on the microscale. As we have shown that a single three-dimensional asperity can be equivalently substituted by a one-dimensional asperity – independent of the radius of curvature – the next step is to create a one-dimensional surface with the same statistical properties of the distributions of height and curvature as the two-dimensional surface of the three-dimensional body. It will then have the same contact properties as the inital three-dimensional body. In the present section we study the question whether it is possible to create such an equivalent one-dimensional surface (line) and if possible, how it has to be done correctly.

For simplicity, we assume here that a two dimensional surface topography (of the three-dimensional body) can be characterized by its surface roughness power spectra $C_{2\mathrm{D}}(q)$ defined by

$$C_{2\mathrm{D}}(q) = \frac{1}{(2\pi)^2} \int \langle h(\mathbf{x})\, h(0) \rangle\, e^{-i\mathbf{q}\cdot\mathbf{x}} d^2x \quad , \tag{9}$$

where $h(\mathbf{x})$ is the height measured from the average plane defined so that $\langle h \rangle = 0$ and $\langle . \rangle$ stands for ensemble averaging. Since it is assumed that the statistical properties of the surface topography are translationally invariant and isotropic, the surface roughness power spectra $C_{2\mathrm{D}}(q)$ only depends on the magnitude q of the wave vector $\mathbf{q}$ [12–14].

Similarly a surface roughness power spectra $C_{1\mathrm{D}}(q)$ can be introduced for a one-dimensional surface topography according to

$$C_{1\mathrm{D}}(q) = \frac{1}{2\pi} \int \langle h(x)\, h(0) \rangle\, e^{-iqx} dx \quad . \tag{10}$$

For generating a one-dimensional surface equivalent to the initial two-dimensional surface, the appropriate surface roughness power spectra $C_{1\mathrm{D}}(q)$ must be defined. The qualitative arguments for the choice of the proper one-dimensional spectral density are the following: The height distribution of asperities of a fractal surface has generally the same order of magnitude as the mean square root value of the height $h(x)$ of the profile. The mean curvature of the asperities has the same order of magnitude as the mean square root value of the curvature $\kappa = \partial^2 h(x)/\partial x^2$.

The mean-square values of height for two- and one-dimensional systems

$$\left\langle h^2 \right\rangle_{2D} = 2\pi \int_0^\infty q C_{2D}(q)\, dq \quad , \tag{11}$$

$$\left\langle h^2 \right\rangle_{1D} = 2 \int_0^\infty C_{1D}(q)\, dq \quad , \tag{12}$$

will be equal, if we take

$$C_{1D}(q) = \pi q C_{2D}(q) \quad . \tag{13}$$

It is important to note, that the mean square curvatures $\left\langle \kappa^2 \right\rangle$ will then be equal as well. We now show, that not only the average values of the heights and curvatures but also their distributions will be almost equal for these systems. Thus, the transformation eqn (13) defines the rule for generating equivalent one-dimensional surfaces with the same contact properties as for three-dimensional bodies. For this sake, we study how the statistics of asperities of both two- and one-dimensional systems are related to each other. We generated one-dimensional and two-dimensional surface topographies and determined the statistics of their heights and curvatures numerically. The surface topography is calculated from the surface roughness power spectra according to

$$h(\mathbf{x}) = \sum_{\mathbf{q}} B_{2D}(\mathbf{q}) \exp\left(i\left(\mathbf{q}\cdot\mathbf{x} + \phi(\mathbf{q})\right)\right) \quad , \tag{14}$$

where $\phi(\mathbf{q}) = -\phi(-\mathbf{q})$ are randomly distributed in $[0, 2\pi)$ and

$$B_{2D}(\mathbf{q}) = \frac{2\pi}{L}\sqrt{C_{2D}(\mathbf{q})} = \bar{B}_{2D}(-\mathbf{q}) \quad .$$

For the one-dimensional case the respective equations are

$$h(x) = \sum_{q} B_{1D}(q) \exp\left(i\left(qx + \phi(q)\right)\right) \quad , \tag{15}$$

$$B_{1D}(q) = \sqrt{\frac{2\pi}{L} C_{1D}(q)} = \bar{B}_{1D}(-q) \quad . \tag{16}$$

Numerical generation of surfaces is based on the FFT algorithm rather then on directly calculating the sums in eqns (14) and (15). For each generated surface topography the statistics of asperities is calculated. We introduce the following ratios ϕ_1, ϕ_2 and ϕ_3 which relate the asperity statistics (index p) to the profile statistics.

$$\phi_1 = \sqrt{\frac{\left\langle h_p^2 \right\rangle}{\left\langle h^2 \right\rangle}}, \quad \phi_2 = \frac{\left\langle \kappa_p \right\rangle}{\sqrt{\left\langle \kappa^2 \right\rangle}}, \quad \phi_3 = \sqrt{\frac{\left\langle \kappa_p^2 \right\rangle}{\left\langle \kappa^2 \right\rangle}} \, . \tag{17}$$

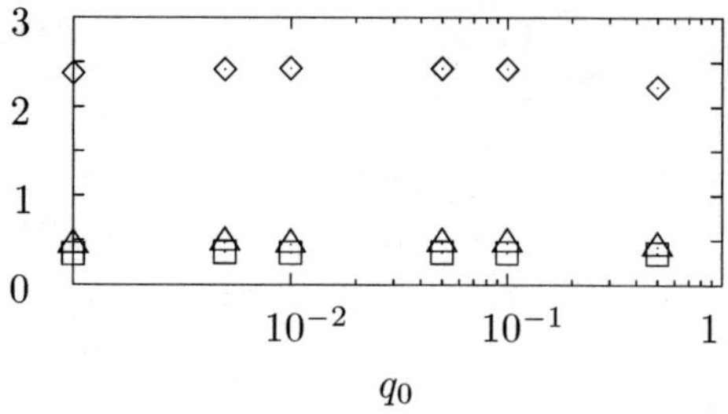

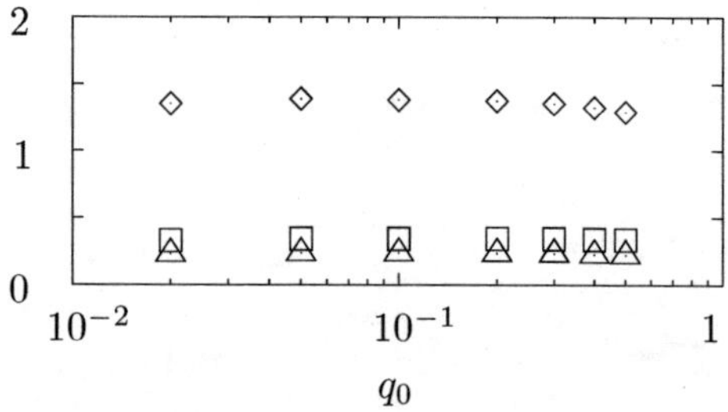

Figure 2: Ratios ϕ_1, ϕ_2 and ϕ_3 according to eqn (17) for the 1D surface (top) and for the 2D surface (bottom), $q_1 = 2q_0$, $\square\, \phi_1$, $\diamond\, \phi_2$, $\triangle\ \phi_3$.

Figure 2 shows ϕ_1, ϕ_2 and ϕ_3 for one-dimensional surfaces (top) and for two-dimensional surfaces (bottom), for a constant power spectrum with cutting wave vectors $q_1 = 2q_0$

$$C_{\rm 2D} = \begin{cases} c & \text{for} \quad q_0 \le q \le q_1 \\ 0 & \qquad \text{otherwise} \end{cases}$$

and $C_{\rm 1D}$ according to eqn (13). ϕ_1, ϕ_2 and ϕ_3 hardly depend on the wave number q_0. Further numerical experiments with $2 \le q_1/q_0 \le 10$ show that this feature is also present for $q_1 \ne 2q_0$. From numerical studies with generated one- and two-dimensional surface topographies ($2 \le q_1/q_0 \le 10$) the following important conclusion regarding the statistics of asperities (index p) can finally be drawn: if the surface roughness power spectra are transformed according to eqn (13) the statistics of asperities will transform according to

$$\begin{aligned} \langle h_p^2 \rangle_{\rm 1D} &\approx \langle h_p^2 \rangle_{\rm 2D} \quad , \\ \langle \kappa_p \rangle_{\rm 1D} &\approx 1.8 \langle \kappa_p \rangle_{\rm 2D} \quad , \\ \langle \kappa_p^2 \rangle_{\rm 1D} &\approx 2.0 \langle \kappa_p^2 \rangle_{\rm 2D} \quad . \end{aligned}$$

Note that according to eqn (5) the relation for the average curvature of asperties should preferably be $\langle \kappa_p \rangle_{\rm 1D} = 2 \langle \kappa_p \rangle_{\rm 2D}$. Choosing the stiffness c_n to get the correct $F(d)$ relation, the contact radius a will not be exactly equal in the two models. In the case at hand the relation $\langle \kappa_p \rangle_{\rm 1D} \approx 1.8 \langle \kappa_p \rangle_{\rm 2D}$ leads to about 5% error in the radius of contact.

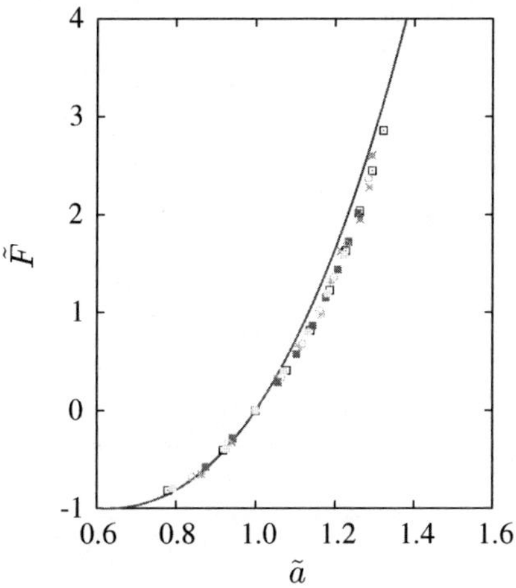

Figure 3: Non-dimensional normal force $\tilde{F} = F/F_A$ vs. non-dimensional contact size $\tilde{a} = a/a_0$, with adhesion force F_A and contact radius a_0 at zero normal force, JKR result (solid line), simulation with the one-dimensional model (points).

3 Adhesion and lubrication

3.1 Adhesion

For many contact and friction problems it is important to take adhesion into account. In particular this is important because adhesion may increase the real area of contact significantly even if no adhesion force is recognised in a pull-off experiment [15].

In the model discussed above adhesion is not considered, thus the interaction forces in this model (see figure 1) are repulsive forces only. Adhesion requires also attractive interaction forces. This can be done by using a Lennard–Jones-type interaction potential for the interactions between particles of opposing bodies.

Numerical experiments and analytical calculations with the one-dimensional model yield for the adhesion force $F_A \propto \sqrt{R_1}$. JKR theory [16] gives for the three-dimensional problem $F_A \propto R_3$. However for any given but fixed radius of curvature, the one-dimensional model gives the correct relation between normal force F and radius of contact a. Figure 3 shows the JKR result (solid line) and results from numerical experiments with five different values for the radius of curvature. Thus the three-dimensional adhesive contact problem can be simulated on the basis of the proposed one-dimensional model under the restriction of having an arbitrary but fixed radius of curvature. Note that this restriction does not apply for the elastic contact without adhesion.

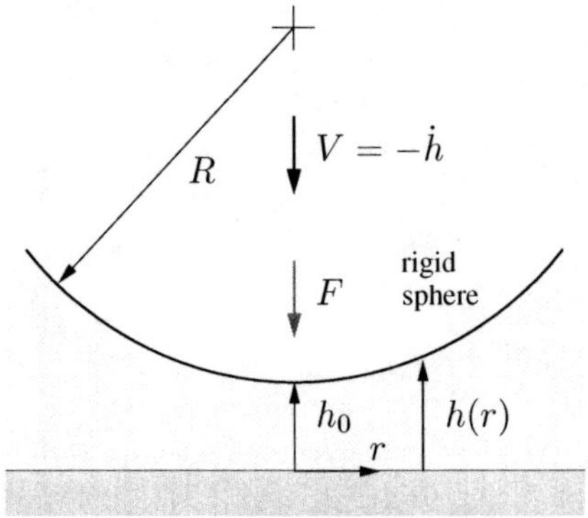

Figure 4: Normal approach of a rigid sphere.

Numerical results for the adhesive contact between rough surfaces especially the dependence of pull-off force on the roughness are presented in [17].

3.2 Lubricated contacts

The reduced description can also be extend to lubricated contacts under the condition that only asperities that are very close to asperities of the opposing body contribute significantly to the total force between the two macroscopic bodies. The reduction is based on the idea that mixed lubrication can be modelled by non-conservative forces between surface elements of the contacting bodies, instead of modelling fluid particles explicitly. Consider the normal approach of a rigid sphere and a rigid plate separated by a classical Newtonian fluid with constant viscosity η (figure 4).

The normal force F acting on the sphere can be calculated from the Reynolds equation and is

$$F = -\frac{6\pi\eta R^2 \dot{h}}{h_0} \quad , \tag{18}$$

where $-\dot{h}$ is the velocity of approach. Under the assumptions made the main contribution to the force comes from the immediate vicinity of the mirco contact. The details of the flow far away form the contact region do not influence the asperity-asperity interactions. The macroscopic result eqn (18) can actually be obtained in numerical simulations by introducing an interaction force

$$F_{pp} = 4.635 \frac{\eta R^{3/2} v}{r^{5/2}} dr_1 dr_2 \quad , \tag{19}$$

where r is the distance of the two interacting particles and v is the projection of the relative velocity onto the direction between them. Note that the interaction force F_{pp} between surface elements depends on the distance r according to a simple power law and is proportional to the relative velocity.

If cavitation is relevant in the application under study a reduced description is build on a kinetic equation for the interaction forces [18].

4 Summary

Starting from the idea that for many applications with contacts between randomly rough surfaces fast simulation tools are required we studied the problem whether it is possible to reduce the dimension of systems from three to one leaving the essential contact properties invariant. We have shown that it is indeed possible as long as the contact area is much smaller then the apparent (macroscopic) contact area. The reduction of the dimension means a huge reduction of computation time, allowing the simulation of multi-scale systems within one model.

The interaction forces between particles depend on the relative distance between particles and in the case of lubricated contacts on the relative velocity as well. Simulations of contacts between randomly rough surfaces also require a conversion of the surface roughness power spectra form two-dimensional to one-dimensional.

A numerical implementation of the described model is currently used to study the chemical–mechanical polishing [3]. Extensive studies with three-dimensional and one-dimensional models are in progress that will give further information on the quality of the one-dimensional model.

References

[1] Ostermeyer, G.P. & Müller, M., Dynamic interaction of friction and surface topography in brake systems. *Tribology International*, **39(5)**, pp. 370–380, 2006.

[2] Lovell, M.R. & Deng, Z., Experimental investigation of sliding friction between hard and deformable surfaces with application to manufacturing processes. *Wear*, **236**, pp. 117–127, 1999.

[3] Popov, V.L. & Filippov, A.E., Modelling of mechanical polishing with lubrication. *Technical Physics Letters*, **31(9)**, pp. 788–792, 2005.

[4] Persson, B.N.J., *Sliding Friction*. Springer, 2nd edition, 2000.

[5] Persson, B.N.J., Bucher, F. & Chiaia, B., Elastic contact between randomly rough surfaces: comparison of theory with numerical results. *Physical Review B*, **70(18)**, p. 184106, 2002.

[6] Yang, C., Tartaglino, U. & Persson, B.N.J., A multiscale molecular dynamics approach to contact mechanics. *European Physical Journal*, **19(1)**, pp. 47–58, 2006.

[7] Popov, V.L. & Psakhie, S.G., Numerical simulation methods in tribology. *Tribology International*, **40(6)**, pp. 916–923, 2007.

[8] Geike, T. & Popov, V.L., Reduction of three-dimensional contact problems to one-dimensional ones. *Tribology International*, **40(6)**, pp. 924–929, 2007.

[9] Johnson, K.L., *Contact mechanics*. Cambridge University Press, 6th edition, 2001.

[10] Greenwood, J.A. & Williamson, J.B., Contact of nominally flat surfaces. *Proceedings of the Royal Society of London Series A*, **295(1442)**, p. 300, 1966.

[11] Bush, A.W. & Gibson, R.D., The elastic contact of a rough surface. *Wear*, **35**, pp. 87–111, 1975.

[12] Persson, B.N.J., Elastic contact between randomly rough surfaces. *Physical Review Letters*, **87(11)**, p. 116101, 2001.

[13] Persson, B.N.J., Adhesion between elastic bodies with randomly rough surfaces. *Physical Review Letters*, **89(24)**, p. 245502, 2002.

[14] Palasantzas G. & De Hosson J. Th. M., Influence of surface roughness on the adhesion of elastic films. *Physical Review E*, **67**, p. 21604, 2003.

[15] Persson, B.N.J., Albohr, O., Tartaglino, U., Volokitin, A.I. & Tosatti, E., On the nature of surface roughness with application to contact mechanics, sealing, rubber friction and adhesion. *Journal of Physics: Condensed Matter*, **17(1)**, pp. R1–R62, 2005.

[16] Johnson, K.L., Kendall, K. & Roberts, A.D., Surface energy and contact of elastic solids. *Proc Roy Soc London A*, **324(1558)**, p. 301, 1971.

[17] Geike, T. & Popov, V.L., Reduzierte Beschreibung des 3D adhäsiven Kontaktes zwischen rauen Oberflächen. *Tribologie und Schmierungstechnik*, **53(3)**, pp. 5–9, 2006.

[18] Geike, T. & Popov, V.L., Reduced description of lubricated contacts with cavitation. *Submitted to Tribology International.*

Determination of potential function in contact problems

F. Sharafbafi & S. Adibnazari
Aerospace Engineering Department, Sharif University of Technology, Tehran, Iran

Abstract

In this paper, a relation is introduced that simplifies the determination of the Muskhelishvili's potential function in plane contact problems. The relation is $\Phi(z)' = 1/2[p(z) - iq(z)]$, which is correct for all uncoupled-elastic contact problems. This relation is proved in a mathematical way and utilized to obtain the potential function in several contact problems. A complete agreement has been observed between our results and the potential functions that have been obtained from complicated methods in the past. Utilization of the relation simplifies the solution of contact problems and analytical calculation of the stress and displacement fields, which is helpful in the analytical studies of contact mechanics. It also may lead to the design of better cutting tools or fretting fatigue test pads.
Keywords: contact mechanics, uncoupled-elastic plane contacts, analytical solution, Muskhelishvili's potential function, a simple relation.

1 Introduction

A routine procedure for solving analytically plane contact problems is the following: find pressure and shear distributions from the contact fundamental equations and substitute those into the Muskhelishvili's integral equation to obtain the contact potential function. The contact stress and displacement fields are achievable through the determined potential function [1–3]. Since the Muskhelishvili's integral equation is a complex singular integral equation, calculation of the potential function, in this way, is not so simple. Therefore, the closed-form potential function of a few contact problems has been obtained. For a contact problem with complicated geometry, there are other procedures to

WIT Transactions on Engineering Sciences, Vol 55, © 2007 WIT Press
www.witpress.com, ISSN 1743-3533 (on-line)
doi:10.2495/SECM070221

obtain the potential function. The most common procedure is that of achieving the Chebyshev expansion of the pressure and shear distributions and putting them into the integral equation to obtain the expanded potential function. However, this method does not give the closed-form potential function because the Chebyshev series are not bounded [4]. The convergence of the Legendre series is Legendre expansion of the traction distributions instead of the Chebyshev expansion [5]. The other procedure to obtain the potential function is employing the Bertrand–Poincare's lemma to reverse the order of integration of the pressure and shear distributions and the Muskhelishvili's integral in order to simplify the integration procedure [6, 7].

Recently, a new method for finding the pressure distribution functions has been introduced. By arranging the flat punches along the contact profile and superposing the pressure distribution functions of the flat punches, for any contact profile, the pressure distribution function can be obtained in terms of a hyper-geometric function. The related potential function can be obtained through the superposing of the potential functions of flat punches along the contact profile [8, 9]. Additionally, Finite Elements Modelling (FEM) has been utilized to solve contact problems and obtaining the potential function [10].

In this paper, it is proved that the potential function of an uncoupled contact problem can be obtained directly through the pressure and shear distributions by utilizing a simple relation such as $\Phi(z) = 1/2[p(z) - iq(z)]$. This relation bypasses the Muskhelishvili's integral equation and simplifies the procedure of analytical solution of the contact problems. Since the relation is correct for all uncoupled contact problems such as complete, incomplete, multi-region and rough contacts, it is necessary to define a generalized contact problem before proving the relation. The generalized contact problem is defined in the "Theory" section of this paper. The relation is proved in the "Proof of the Relation" section of this paper. In the "Case Studies" section, several important contact problems are considered and for each of them, the potential function is obtained through the simple relation. The results were compared with the potential functions calculated by others. A complete agreement has been observed between them.

2 Theory

In this section, a generalized contact problem is considered. Although it might be repetitive, it is helpful to rewrite it in order to expose assumptions and equations essential for proving the simple relation.

Consider two elastic bodies S_1 and S_2, which are in contact with each other. According to figure 1, the body S_1 occupies the upper half-plane and the body S_2 occupies the lower half-plane. Along the boundaries of the bodies, there might be several contact segments namely $\sum_{i=1}^{n}[a_i, b_i]$. For the considered contact problem, the following usual assumptions have been made:

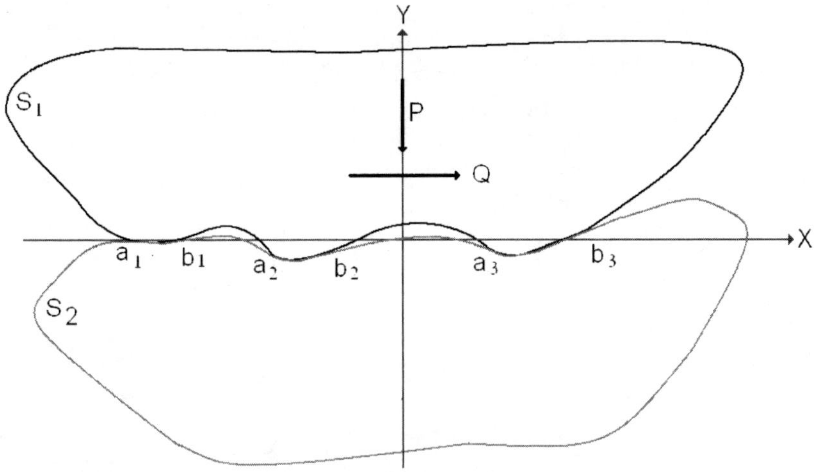

Figure 1: Contact of two bodies with several contact segments.

1- Each of the bodies can be approximated with a half-plane to achieve non-conformal contact problem.
2- Each contact length $L_i = [a_i, b_i]$ is very small in comparison with the dimensions of the bodies.
3- There is no rotation of contact bodies.
4- The stresses in S_1 and S_2 vanish at infinity.
5- The profiles of the boundaries are known before applying the external forces.

The external normal and shear forces produce pressure and shear tractions on the contact surfaces. For the lower body S_2, the pressure distribution function $p(x)$ and the shear distribution function $q(x)$ are defined as

$$p(x) = \sigma_{yy}(x,0) \qquad \& \qquad q(x) = \tau_{xy}(x,0) \tag{1}$$

The positive direction of the pressure distribution function is the same as the positive y-axis [4]. It is assumed that, $p(x)$ and $q(x)$ satisfy the Hölder condition on the contact length and vanish at infinity [1]. The Muskhelishvili's potential function is defined as

$$\Phi(z) = \frac{1}{2\pi i}\int_L \frac{p(x) - iq(x)}{x - z}dx \tag{2}$$

The potential function $\Phi(z)$ is a holomorphic complex function throughout the plane, except in the contact segments of the x-axis. Therefore, along the contact zones, Plemelj formulas can be written as

$$\Phi^+(x) - \Phi^-(x) = p(x) - iq(x) \tag{3}$$

$$\Phi^+(x) + \Phi^-(x) = \frac{1}{\pi i}\int_L \frac{p(\xi) - iq(\xi)}{\xi - x}d\xi \tag{4}$$

It is well known that in plane contact problems, the fundamental equations are coupled singular integral equations, which relate the relative surface vertical overlap function $h(x)$ and the horizontal overlap function $g(x)$, to the contact pressure and shear distribution functions as follows:

$$\frac{1}{A}\frac{\partial h(x)}{\partial x}=\frac{1}{\pi}\int_L \frac{p(\xi)}{(x-\xi)}d\xi-\beta q(x) \tag{5}$$

$$\frac{1}{A}\frac{\partial g(x)}{\partial x}=\frac{1}{\pi}\int_L \frac{q(\xi)}{(x-\xi)}d\xi+\beta p(x) \tag{6}$$

where

$$A=\frac{\kappa_1+1}{4\mu_1}+\frac{\kappa_2+1}{4\mu_2} \tag{7}$$

$$\beta=\frac{\mu_2(\kappa_1-1)-\mu_1(\kappa_2-1)}{\mu_2(\kappa_1+1)+\mu_1(\kappa_2+1)} \tag{8}$$

in which $\kappa=(3-\upsilon)/(1+\upsilon)$ is for plane stress and $\kappa=(3-4\upsilon)$ is for plane strain conditions. The overlap functions $h(x)$ and $g(x)$ can be presented as follows:

$$h(x)=v_1-v_2 \tag{9}$$

$$g(x)=u_1-u_2 \tag{10}$$

where v_1 and v_2 are the vertical and u_1 and u_2 are the horizontal components of the displacement of the contacting surfaces of the two bodies.

If Dundur's constant β is zero ($\beta=0$), that is to say if the materials of the bodies are similar or the relation $\frac{\mu_2}{\mu_1}=\frac{k_2-1}{k_1-1}$ holds, the contact equations become decoupled.

Therefore, eqns (5), (6) reduce to

$$\frac{1}{A}h'(x)=\frac{1}{\pi}\int_L \frac{p(\xi)d\xi}{x-\xi} \tag{11}$$

$$\frac{1}{A}g'(x)=\frac{1}{\pi}\int_L \frac{q(\xi)d\xi}{x-\xi} \tag{12}$$

Equations (11) and (12) are the fundamental equations in the uncoupled contact problems.

3 Proof of the relation

For a generalized and uncoupled contact problem, another boundary condition is obtained through the fundamental contact equations. Multiplying eqn (11) by i and adding it to the eqn (12), yields

$$\frac{1}{A}\left[g'(x)+ih'(x)\right]=\frac{1}{\pi i}\int_{L}\frac{p(\xi)-iq(\xi)}{\xi-x}d\xi \tag{13}$$

which is a Cauchy singular integral equation of the first kind. The solution (also known as inversion) of eqn (13) is

$$p(x)-iq(x)=\frac{\omega(x)}{\pi A}\int_{L}\frac{h'(\xi)-ig'(\xi)}{\omega(\xi)(\xi-x)}d\xi+\Omega(x)\omega(x) \tag{14}$$

where $\omega(x)$ is the weight function and $\Omega(x)$ is a polynomial of up to an order less than n, where n is the number of the contact zones. In accordance with the contact geometry and the behavior of surface tractions at the ends of each contact zone, $\omega(x)$ and $\Omega(x)$ are then determined [1, 2]. The results are summarized in table 1 for multiple contact zones.

Table 1: Weight function $\omega(x)$ and the polynomial function $\Omega(x)$ for different contact boundary conditions

Boundary conditions	**Weight function $\omega(x)$**	$\Omega(x)$
Singular at all end points	$\omega(x)=\left(\prod_{i=1}^{n}(x-a_i)(b_i-x)\right)^{-1/2}$	$\Omega(x)=\sum_{i=0}^{n-1}c_i x^i$
Non-singular at $x=a_j$	$\omega(x)=\left(\prod_{j=1}^{m}(x-a_j)\right)\left(\prod_{i=1}^{n}(x-a_i)(b_i-x)\right)^{-1/2}$	$\Omega(x)=\sum_{i=0}^{n-1}c_i x^i$
Non-singular at $x=b_j$	$\omega(x)=\left(\prod_{j=1}^{m}(b_j-x)\right)\left(\prod_{i=1}^{n}(x-a_i)(b_i-x)\right)^{-1/2}$	$\Omega(x)=\sum_{i=0}^{n-1}c_i x^i$
Non-singular at all end points	$\omega(x)=\left(\prod_{i=1}^{n}(x-a_i)(b_i-x)\right)^{1/2}$ where $\int_{a_i}^{b_i}\frac{h'(\xi)-ig'(\xi)d\xi}{\omega(\xi)}=0$	$\Omega(x)=\sum_{i=0}^{n-1}c_i x^i$

On the other hand, comparing equations (4) and (13) results in:

$$\Phi^{+}(x)+\Phi^{-}(x)=\frac{1}{A}\left[g'(x)+ih'(x)\right] \tag{15}$$

This equation is a non-homogeneous Hilbert–Reimann linear problem and its solution is:

$$\Phi(z)=\frac{\omega^{*}(z)}{2\pi A}\int_{L}\frac{h'(\xi)-ig'(\xi)}{\omega^{*}(\xi)(\xi-z)}d\xi+\frac{\Omega^{*}(z)\omega^{*}(z)}{2} \tag{16}$$

where $\omega^{*}(z)$ is a weight function and $\Omega^{*}(z)$ is a polynomial of up to an order less than n, where n is the number of the contact zones. Since both of the weight functions $\omega(x)$ and $\omega^*(z)$ are dependent on the contact geometry and the behavior of surface tractions at the ends of each contact zone, it can be shown that $\omega^*(x) = \omega(x)$. Additionally, the order of both polynomials $\Omega(x)$ and $\Omega^*(z)$ are the same. Therefore it is possible to consider $\Omega^*(z)$ such that the relation $\Omega^*(x) = \Omega(x)$ holds. Comparison of eqn (14) and (16) yields to:

$$\Phi(z) = 1/2[p(z) - iq(z)] \tag{17}$$

In other words, if the closed form of the pressure and shear distribution functions, $p(x)$ and $q(x)$, are known, for determination of the potential function of the contact problem, it is enough to replace z for x in the pressure and shear distribution functions, $p(z)$ and $q(z)$, and utilize eqn (17) to obtain the potential function. Due to the fact that for all contact profiles, $\omega^*(x) = \omega(x)$ and $\Omega^*(x) = \Omega(x)$, eqn (17) holds true for all contact profiles: be it Hertzian, non-Hertzian, with or without singularities or even rough contacts. It is noteworthy that the imaginary part of the pressure or shear distribution functions should not be omitted since, although the imaginary parts have no meaning in the pressure and shear distributions, they are required for computing the potential function through the eqn (17). Additionally, if the traction distribution functions are defined in the normalized plane, the potential function obtained through eqn (17) is consequently in the normalized complex plane.

By comparing eqns (2) and (17), we can rewrite the Muskhelishvili's integral relation as:

$$\Phi(z) = \frac{1}{2\pi\ i} \int_L \frac{p(\xi) - iq(\xi)}{\xi - z} d\xi = \frac{p(z) - iq(z)}{2} \tag{18}$$

Of course, this relation is correct for uncoupled-elastic contact problems.

Determining the potential function in closed form makes it possible to obtain the closed form relations for the displacement and stress components through the following equations:

$$\sigma_{xx} + \sigma_{yy} = 4(\mathrm{Re}(\Phi(z))) \tag{19}$$

$$\sigma_{yy} - \sigma_{xx} + 2i\sigma_{xy} = 2[(\bar{z} - z)\Phi'(z) - \overline{\Phi}(z) - \Phi(z)] \tag{20}$$

$$2\mu\left(\frac{\partial u}{\partial x} + i\frac{\partial v}{\partial y}\right) = (\bar{z} - z)\overline{\Phi'}(\bar{z}) + \Phi(\bar{z}) + \kappa\Phi(z) \tag{21}$$

4 Case studies

Although the simple relation, eqn (17), is proved in mathematics, it is important to find the conformity of the potential function obtained through the relation and those introduced in the literature. It is noteworthy that the closed form of the potential function has been achieved only for a few simple contact problems.

In this section, four important contact problems, which had analytical solutions, are considered such as 1- the classic Hertz contact problem, 2- contact of a flat punch with half plane, 3- contact problem of a symmetrical wedge with half plane and 4- contact of a punch generating uniform pressure. Figure 2 shows profile of the indenters and pressure distribution of each contact problem. For each mentioned contact problem, the potential function has been extracted from eqn (17) and compared with the existing results in the literature. Agreement between these two potential functions, confirm the accuracy of the simple relation.

The asterisk sign on the parameters shows that the related equation had been achieved before.

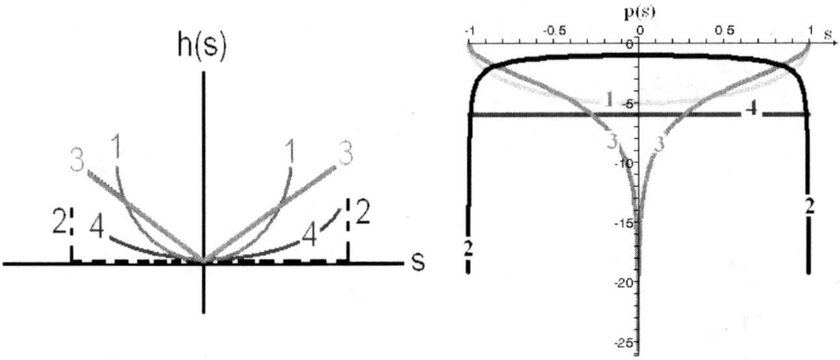

Figure 2: Indenters profiles (left) and pressure distributions (right) of the considered contact problems.

4.1 Hertz contact problem

In 1882, Hertz introduced and solved contact problem of two cylinders. The pressure distribution of the famous Hertz contact problem is:

$$p^*(s) = \frac{-ka}{A}\sqrt{1-s^2} \qquad s \in [-1,1] \tag{22}$$

where a is half of contact length and s is the normalized axis along the contact surface. The potential function for this contact problem was calculated by integrating the Cauchy integral [4, table 3.5]:

$$\Phi^*(z) = \int_{-1}^{1} \frac{p(\xi)}{\xi - z} d\xi = \left(-\frac{kai}{2A}\right)\left[z - \sqrt{z^2 - 1}\right] \tag{23}$$

It is seen that substitution of the achieved pressure distribution eqn (22) into the eqn (17) does not yield the same result as eqn (23) and is a major nonconformity. However, this discrepancy must have occurred because of the imaginary part of the pressure distribution. To obtain the imaginary part of the pressure distribution, the contact problem is solved again and the complete form of the pressure distribution is obtained as follows

$$p(s) = -\left(\frac{ka}{A}\right)\left[\sqrt{1-s^2} + is\right] \tag{24}$$

It is obvious that the real part of the pressure distribution function eqn (24) is the same as the pressure distribution function in eqn (22). Substituting the complete pressure distribution function eqn (24) into the eqn (17) gives:

$$\Phi(z) = -\left(\frac{ka}{2A}\right)\left[\sqrt{1-z^2} + iz\right] = \left(-\frac{kai}{2A}\right)\left[z - \sqrt{z^2-1}\right] \tag{25}$$

The potential function achieved through the eqn (17) is exactly the same as the previously obtained potential function eqn (23). The obtained potential function is plotted as a two dimensional function in figure 3-1.

4.2 Contact of a flat punch with a half plane

As a second example, a flat punch in contact with a half plane with no shear force or friction is considered. Muskhelishvili [1] has extracted pressure distribution and potential function as follows:

$$p^*(s) = \frac{-P}{a\pi\sqrt{1-s^2}} \tag{26}$$

$$\Phi^*(z) = \frac{-P}{2a\pi\sqrt{1-z^2}} \tag{27}$$

It is obvious that substitution of the pressure distribution function eqn (26) into the eqn (17) yields the same potential function as eqn (27). The obtained potential function is plotted as a two dimensional function in figure 3-2.

4.3 Contact of a symmetrical wedge with a half plane

Complete form of the pressure distribution of contact problem of a symmetrical wedge with a half plane can be obtained as:

$$p(s) = \frac{-2\phi}{\pi A}\left(\cosh^{-1}\left(\frac{1}{|s|}\right) + \frac{\pi i}{2}\right) \tag{28}$$

Substitution of the pressure distribution in eqn (17), gives the potential function:

$$\Phi(z) = \frac{-p_0}{2}\left[\cosh^{-1}\left(\frac{1}{|z|}\right) + \frac{\pi i}{2}\right] \tag{29}$$

Simplifying the above equation leads to:

$$\Phi^*(z) = \frac{-i\,p_0}{2}\sin^{-1}\left(\frac{1}{z}\right) \tag{30}$$

The latter form of the potential function is the same as the closed potential function obtained by Truman and Sackfield [11] for this contact problem. The obtained potential function is plotted as a two dimensional function in figure 3-3.

4.4 Contact of a punch generating uniform pressure

Hills et al [4] obtained the profile of a punch which generates uniform pressure over the contacting half plane. Due to the specific profile of the punch, the pressure distribution is uniform and equal to:

$$p(s) = -p_0 = \frac{P}{2a} \qquad -1 \le s \le 1 \tag{31}$$

Substituting the pressure distribution in the Muskhelishvili's integral equation gives the potential function as:

$$\Phi(z) = \frac{-p_0}{2\pi i}\int_{-1}^{1}\frac{ds}{s-z} = \frac{-p_0}{2\pi i}\ln\left(\frac{1-z}{-1-z}\right) = \frac{p_0}{2} + i\frac{p_0}{2\pi}\ln\left(\frac{1-z}{1+z}\right) \tag{32}$$

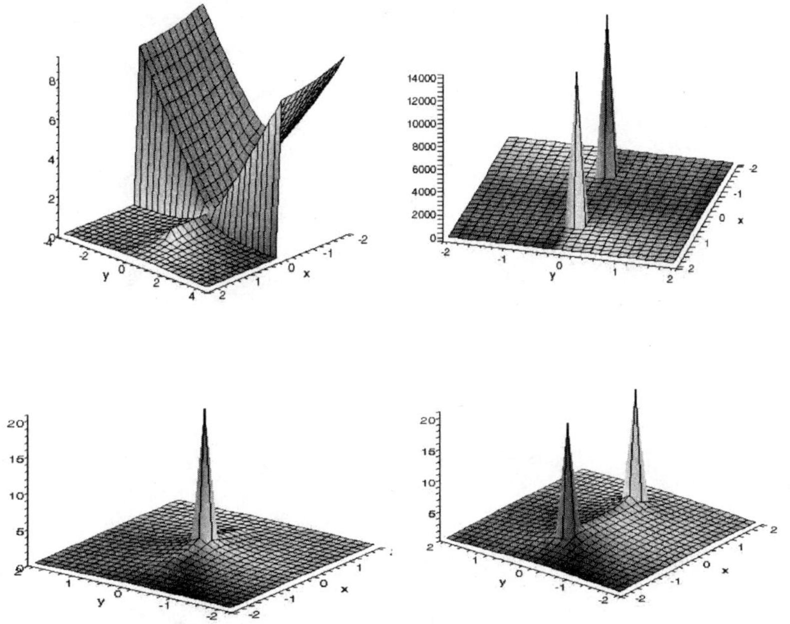

Figure 3: The potential functions of the related contact problems 3-1 Hertz contact (top- left), 3-2 Flat punch (top-right), 3-3 symmetrical wedge (bottom-left) & 3-4 Punch generating uniform pressure (bottom- right).

This form of the potential helps to determine the imaginary part of the complete pressure distribution function. According the eqn (18), the complete form of the pressure distribution function will be equal to:

$$p(z) = 2\left\{\frac{p_0}{2} + i\frac{p_0}{2\pi}\ln\left(\frac{1-z}{1+z}\right)\right\} = p_0 + i\frac{p_0}{\pi}\ln\left(\frac{1-z}{1+z}\right) \tag{33}$$

The real part of $p(z)$ is the same as the defined pressure distribution in eqn (31) and this is another confirmation of the eqn (17). The imaginary part of the uniform distribution is obtained as $\frac{p_0}{\pi}\ln\left(\frac{1-s}{1+s}\right)$. The obtained potential function is plotted as a two dimensional function in figure 3-4. The singularity of the potential function shown in the figure is very interesting.

5 Conclusion

Conformity of the potential function obtained through the simple relation with those obtained by complicated methods in the literature, shows the accuracy and simplicity of the relation. Therefore, in order to determine the potential function of an uncoupled contact problem, it is recommended to find the complete form of the pressure and shear distribution functions and utilize the simple relation $\Phi(z) = 1/2[p(z) - iq(z)]$ to achieve the potential function.

References

[1] Muskhelishvili, N.I., Some Basic Problems of the Mathematical Theory of Elasticity, translated by Radok, JRM, Noordhoff Ltd, Groningen, Netherlands, 1963.

[2] Gladwell, G.M.L., Contact Problems in the Classical Theory of Elasticity, Sijthoff & Noordhoff, Rockville, MD, 1980.

[3] Johnson, K.L., Contact Mechanics, Cambridge University Press, UK, 1987.

[4] Hills, D.A., Nowell, D. & Sackfield, A. Mechanics of Elastic Contacts, Oxford Press, Butterworth-Heinemann, 1993.

[5] Badr, A.A., On the Numerical Treatment of the Contact Problem, Int. J. of Mathematics & Materials Science, 23, No. 12, pp. 865–871, 2000.

[6] Sackfield, A., Dini, D. & Hills, D.A., The Finite and Semi-infinite Tilted Flat but Rounded Punch, Int. J. of Solids and Structures, 42, pp. 4988-5009, 2005.

[7] Sackfield, A., Dini, D. & Hills, D.A., The Tilted Shallow Wedge Problem, European Journal of Mechanics A/ Solids, 24, pp. 919-928, 2005.

[8] Jäger, J., A New Principle in Contact Mechanics, Journal of Tribology, 120, pp. 677- 683, 1998.

[9] Jäger, J., New Analytical and Numerical Results for Two-Dimensional Contact Profiles, International Journal of Solids and Structures, 39, pp. 959-972, 2002.

[10] Mackerle, J., Contact Mechanics: Finite Element and Boundary Elements Approach, a Bibliography (1995-1997) Finite Element Analysis, Dec 29, pp. 275-285, 1998.

[11] Truman, C.E. & Sackfield, A., Closed-Form Solution for the Stress Fields Induced by Blunt Wedge Shaped Indenters in Elastic Half Planes, Int. J. Applied Mech., 68, pp. 817-819, 2001.

Inverse problems of plane elasticity for the determination of contact stresses

A. N. Galybin
Wessex Institute of Technology, Southampton, UK

Abstract

This article presents some results of numerical experiments on the reconstruction of contact stresses by using synthetic data on displacement monitoring on a traction-free surface. The corresponding boundary value problem is ill-posed therefore a regularisation procedure based on the SVD decomposition is employed. Statistical analysis of results has been performed.
Keywords: contact problem, plane elasticity, boundary value problems, inverse problems, SVD regularisation.

1 Introduction

There are many applications where data on stresses or displacements may not be available on the entire boundary of a body (including internal boundaries). Such problems appear in strain-stress measurements, interferometry, rock mechanics, monitoring the fracture development in strength tests etc. They require the consideration of a specific boundary value problem, BVP, which is overspecified on a part of the boundary and underspecified on the rest of it.

The theory of classical BVPs of plane elastostatics is fully presented in the classical monograph by Muskhelishvili [1], it assumes that two scalar boundary conditions are given on the entire boundary of a domain. In contact mechanics the BVPs are usually formulated as the mixed type problems when displacements are given in the contact zone and tractions on the rest of the surface. Other formulations of contact conditions are discussed in detail in Johnson [2]. In all these cases the boundary value problem is well posed, therefore it possesses a unique and stable solution.

Despite classical boundary conditions describe a wide class of mechanical phenomena there is still the necessity to use additional assumptions in

WIT Transactions on Engineering Sciences, Vol 55, © 2007 WIT Press
www.witpress.com, ISSN 1743-3533 (on-line)
doi:10.2495/SECM070231

formulation of the boundary conditions. This can lead to paradoxical results, as for instance, in the problem for a flat stamp indented into the boundary of a half-plane without sliding. The exact solution shows oscillation of the contact stresses in a small zone near the stamp edges, which does not obey the assumption that the stamp and the half-plane are in contact everywhere. This effect should not be overlooked if one intends to investigate the stress distribution under the stamp edges, which presents the case of interest, for instance, for fracture development. Thus, corrections of the boundary conditions may be required to provide the consistence of obtained results. However this presumes that some new assumptions will be needed for describing the contact conditions.

Alternatively the BVPs of this type can be solved with the use of the displacement monitoring data over the free boundary outside the contact zone. In this case both the displacements and the contact stresses remain unknown under the stamp and the BVP is underdetermined on a part of the boundary and overdetermined on the rest of it. Perhaps, the first comprehensive analysis of solvability of these problems has been reported by Shvab [3] for an isotropic elastic domain with the following boundary conditions: displacement vector is given on a part of the boundary simultaneously with the stress vector; the rest of the boundary has no conditions posed. This problem can be viewed as consecutive problems for holomorphic vectors, on which the proof of uniqueness can also be based, e.g. [4]. Methods involving complex variables for investigation of this problem in 2D have also been applied [5,6]. The problem can be referred to as conditionally ill-posed [3], one can rarely find analytical solutions for it (with exceptions for simple domains, e.g. for wedge-like domains [5]), therefore the development of stable numerical methods has been the main focus during the last years. The considerable progress has recently been achieved by researches from University of Leeds (UK) in the development of regularisation techniques, iterative methods and algorithms for solving non-classical BVPs of this type, see for instance [7-10]. In particular, it has been shown that methods based on the Tikhonov regularisation provide stable solutions in elastostatics, [8,9]. Other studies, e.g. [10-12], confirm this conclusion, in particular, it has been found [10,12] that the use of the SVD regularisation presents a valuable computational tool in elastostatics. In this study the SVD regularisation is applied for stress identification in contact mechanics problems.

2 Mathematical formulation

2.1 Inverse BVP: displacements are known on traction-free surfaces

Let Γ be a closed contour separating the complex plane into interior Ω^+ and exterior Ω^- domains. In contact problems one of these domains can be associated with one contacting (plane) body, say, Ω^+ if one considers finite bodies or Ω^- if infinite. We further consider the stress state of only one of the contacting bodies. The action of another body is replaced by unknown stresses distributed over the contact zone, therefore the shape of this body is unimportant in the present formulation and without loss of generality it is assumed that it occupies the complement of Ω^+(or Ω^-) with respect to the whole plane. For the same reason it is assumed that elastic moduli of contacting bodies are the same. Stress states in both domains can be expressed through sectionally holomorphic functions (complex potentials) $\varphi(z)$ and $\psi(z)$ of complex variable $z=x+iy$ by the Kolosov–Muskhelishvili solution [1]. In particular on the boundary one has the following expressions for tractions, $\mathbf{t}=(t_1,t_2)$, and displacements, $\mathbf{u}=(u_1,u_2)$

$$F(\zeta)=\varphi(\zeta)+\overline{\mathrm{T}(\zeta)},\quad W(\zeta)=\kappa\varphi(\zeta)-\overline{\mathrm{T}(\zeta)} \tag{1}$$

Hereafter $F=t_1+it_2$; $W=2G(u_1+iu_2)$; $\zeta\epsilon\Gamma$ is a point on the boundary; G is shear modulus, elastic constant $\kappa=3\text{-}4\nu$ to for plane strain and $\kappa=(3-\nu)(1+\nu)^{-1}$ for plane stress, ν is Poisson's ratio; $\mathrm{T}(\zeta)$ is boundary value of bi-holomorphic function

$$\mathrm{T}(z,\bar{z})=\bar{z}\varphi'(z)+\psi(z) \tag{2}$$

From (1) it is evident that

$$(1+\kappa)\varphi(\zeta)=F(\zeta)+W(\zeta),\quad (1+\kappa)\overline{\mathrm{T}(\zeta)}=\kappa F(\zeta)-W(\zeta),\quad (1+\kappa)\mathrm{T}(\zeta)=\kappa\overline{F(\zeta)}-\overline{W(\zeta)} \tag{3}$$

If tractions and displacements on the entire boundary are know it is easy to determine complex potentials everywhere inside the domain considered by using the integral Cauchy formula as follows

$$\begin{aligned}\varphi(z)&=\pm\frac{1}{2(1+\kappa)\pi i}\int_\Gamma\frac{F(t)+W(t)}{t-z}dt,\quad \varphi(\infty)=0\\ \mathrm{T}(z,\bar{z})&=\pm\frac{1}{2(1+\kappa)\pi i}\int_\Gamma\frac{\kappa\overline{F(t)}-\overline{W(t)}-(\bar{t}-\bar{z})\left(F'(t)+W'(t)\right)}{t-z}dt,\quad T(\infty)=0\end{aligned} \tag{4}$$

Hereafter upper/lower sings refer to domains $\Omega^\pm$ respectively; Γ is traversed in counterclockwise direction.

It should be noted that the functions $F(t)$ and $W(t)$ are dependent on the entire boundary, however they can be chosen independently on a part of the boundary, which leads to the boundary value problem depicted in Figure 1.

Let us represent this problem as follows

$$F(\zeta)=\begin{cases}0 & \zeta\in\Gamma_1\\ X(\zeta) & \zeta\in\Gamma_2\end{cases},\quad W(\zeta)=\begin{cases}w(\zeta) & \zeta\in\Gamma_1\\ Y(\zeta) & \zeta\in\Gamma_2\end{cases} \tag{5}$$

where $X(\zeta)$ and $Y(\zeta)$ are unknown functions; it is also assumed that displacements are monitored on a part of the traction-free surface and therefore $w(\zeta)$ is known there. The existence of traction-free surfaces does not narrow the formulation because due to linearity one can superimpose a solution for known tractions on the solution of the problem (5).

The problem can be reduced to the consequent determination of holomorphic functions by analytic continuation. It is shown, e.g. [3-5], that this problem has a unique solution. Here we derive an integral equation for the determination of unknown functions $X(\zeta)$ and $Y(\zeta)$ on the basis of representations (4).

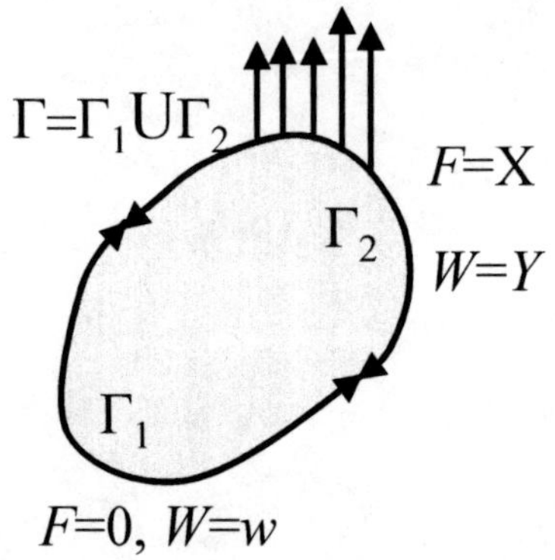

Figure 1: The problem.

2.2 Integral equations

Let us introduce the following integral operators

$$\mathbf{S}^k g = \frac{1}{\pi i}\int_{\Gamma_k}\frac{g(t)}{t-\zeta}dt, \quad k=1,2; \quad \mathbf{S}g = \mathbf{S}^1 g + \mathbf{S}^2 g, \quad \mathbf{S}(\mathbf{S}g) = g \tag{6}$$

$$\begin{aligned} \mathbf{R}_1^k g &= \frac{1}{\pi i}\int_{\Gamma_k}\frac{\bar{t}-\bar{\zeta}}{t-\zeta}g'(t)\,dt = \frac{1}{\pi i}\int_{\Gamma_k}\left(\frac{d\bar{t}}{dt}-\frac{\bar{t}-\bar{\zeta}}{t-\zeta}\right)\frac{g(t)}{t-\zeta}dt \\ \mathbf{R}_2^k g &= \frac{1}{\pi i}\int_{\Gamma_k}\left(\frac{t-\zeta}{\bar{t}-\bar{\zeta}}\frac{d\bar{t}}{dt}-1\right)\frac{g(t)}{t-\zeta}dt, \quad \mathbf{R}_j g = \mathbf{R}_j^1 g + \mathbf{R}_j^2 g, \quad k,j=1,2 \end{aligned} \tag{7}$$

where **S**g is singular and $\mathbf{R}_j g$ are regular operators. The latter are used in the representations for bi-holomorphic function and complex conjugation of **S**g

$$\mathrm{ST} = \bar{\zeta}\mathbf{S}\varphi' + \mathbf{S}\psi + \mathbf{R}_1\varphi' \tag{8}$$

$$\overline{\mathbf{S}g} = -\mathbf{S}\bar{g} - \mathbf{R}_2\bar{g} \tag{9}$$

With the use of these operators one can present the boundary values of the functions in (3) in the following form

$$\varphi = \pm\mathbf{S}\varphi, \quad \mathrm{T} = \pm\mathbf{S}\mathrm{T} \mp \mathbf{R}_1\varphi, \quad \overline{\mathrm{T}} = \pm\overline{\mathbf{S}\mathrm{T}} \mp \overline{\mathbf{R}_1\varphi} = \mp\mathbf{S}\overline{\mathrm{T}} \mp \overline{\mathbf{R}_1\varphi} \mp \mathbf{R}_2\overline{\mathrm{T}} \tag{10}$$

Substitution of (3) into (10) results in

$$\mathbf{S}(F+W)=\pm(F+W)$$
$$(\mathbf{S}+\mathbf{R}_2)(\kappa F\mp W)+\overline{\mathbf{R}_1(F+W)}=\mp(\kappa F+W) \tag{11}$$

Now one can derive a system of integral equations on the traction-free boundary Γ_2 by decomposition of the left hand sides in (11) and applying boundary conditions (5) on different parts of the entire boundary.

$$\begin{cases}\mathbf{S}^2(X+Y)=\pm w-\mathbf{S}^1 w \\ (\mathbf{S}^2+\mathbf{R}_2^2)(\kappa X-Y)+\overline{\mathbf{R}_1^2(X+Y)}=\pm w+(\mathbf{S}^1+\mathbf{R}_2^1)w-\overline{\mathbf{R}_1^1 w}\end{cases} \tag{12}$$

It should be noted that operators $\mathbf{S}^2$ are not singular on Γ_2, with possible exception of the ends of the contour (fixed singularities), therefore the system (12) is a system of integral equations of the Fredholm type. Both equations in (12) are conditionally ill-posed because they are of the first kind [13], therefore there exists a unique solution of this system but it is unstable, which requires regularisation. One can formally introduce two stable operators **P** and **Q** that represent regularised inverse operators of $\mathbf{S}^2$ and $\mathbf{S}^2+\mathbf{R}_2^2$ respectively, i.e.

$$\mathbf{S}^2 g\approx\mathbf{P}^{-1}g,\quad (\mathbf{S}^2+\mathbf{R}_2^2)g\approx\mathbf{Q}^{-1}g \tag{13}$$

By applying these operators to equations in (12) one obtains an approximate solution of the system in the following form

$$\begin{pmatrix}X\\Y\end{pmatrix}=\frac{1}{1+\kappa}\begin{pmatrix}\mathbf{P} & \mathbf{Q}\\ \kappa\mathbf{P} & -\mathbf{Q}\end{pmatrix}\begin{pmatrix}f_1\\ f_2-\overline{\mathbf{R}_1^2(\mathbf{P}f_1)}\end{pmatrix} \tag{14}$$

where the right hand sides in equations (12) are denoted as

$$f_1=\pm w-\mathbf{S}^1 w,\quad f_2=\pm w+(\mathbf{S}^1+\mathbf{R}_2^1)w-\overline{\mathbf{R}_1^1 w} \tag{15}$$

As soon as tractions and displacements on Γ_1 are found one can determine complex potentials by formulae (3)-(5). It should be emphasised that solution (14) is stable with respect to small perturbations in monitored displacements w that appear due to experimental errors or limited resolution in measurements. Construction of the inverse operators **P** and **Q** is, obviously, an important part in engineering applications. The next section presents an example employing the SVD regularisation, which, as shown in [10,12], is an effective tool for this type of problems.

3 Numerical analysis

3.1 Integral equation for half-plane

Let us consider the lower half-plane under a symmetric load distributed on (-1,1) while displacements are monitored on (1,1+L) as shown in Figure 2. In this case the boundaries are $\Gamma_1=(1,1+L)$ and $\Gamma_2=(-\infty,1)\cup(1+L,\infty)$; it is also evident that the regular operators $\mathbf{R}_1=\mathbf{R}_2=0$. Hence the system (12) assume the following form

$$\begin{cases} \mathbf{S}^2(X+Y) = -w - \mathbf{S}^1 w \\ \mathbf{S}^2(\kappa X - Y) = -w + \mathbf{S}^1 w \end{cases} \tag{16}$$

Here the lower sign is chosen since the lower half-plane is associated with Ω^-. The sum of equations in (16) results in $\mathbf{S}^2X=-2(1+\kappa)^{-1}w$, which is the following integral equation for the determination of unknown tractions under the stump

$$\frac{1}{\pi i}\int_{-1}^{1}\frac{X(t)}{t-x}dt = -\frac{2}{1+\kappa}w(x), \quad 1<x<1+L \tag{17}$$

After differentiation of (17) with respect to the contour variable one obtains the relationships between normal stresses and derivatives of displacements that are often used in contact problems for half-plane, e.g. [2].

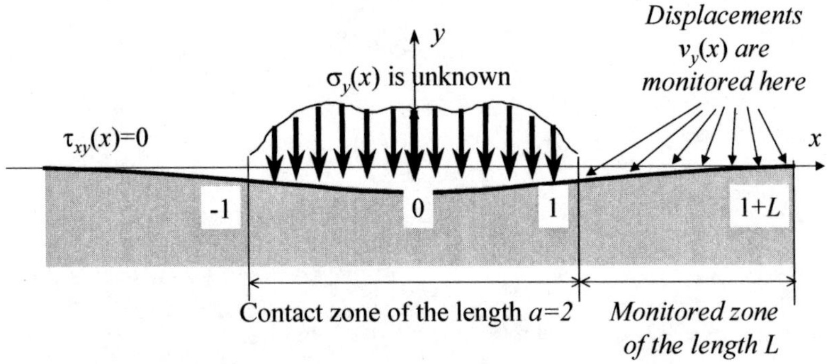

Figure 2: Scheme of displacements monitoring.

It should be noted that it is possible to derive an analytical solution of (17) if $L=\infty$, see [5], however no analytical solutions for the case of finite L are known. We further apply a numerical approach similar to [12], which seems also to be suitable for more complex geometries.

Firstly one can rewrite (17) for the case of symmetrical polynomial load and for simplicity to neglect shear stresses as shown in Figure 2. This leads to the following integral equation

$$\int_0^1\frac{N(t)}{t^2-x^2}dt = g(x), \quad 1<x<1+L, \quad g(x)=\frac{2G}{1+\kappa}\frac{\pi}{x}u_2(x) \tag{18}$$

Here $N(t)$ stands for the normal component of the resultant vector acting under the stump, the right hand side $g(x)$ depends on normal displacement only due to the absence of shear stresses.

Equation (18) is further reduced by the collocation method to a system of linear algebraic equations **AC=G** which solution (it depends on L) is found by the SVD method as follows

$$\mathbf{C} = \mathbf{V}^T \mathbf{D}''\mathbf{U}\mathbf{G}, \quad \mathbf{C} = \mathbf{C}(L) \tag{19}$$

Here **U** (mxn) and **V** (nxn) are orthogonal matrices in the SDV decomposition $\mathbf{A}=\mathbf{U}\mathbf{D}\mathbf{V}^{\mathrm{T}}$ and **D** (nxn) is a diagonal matrix formed by the singular values, d_j, placed in descending order, $d_1 \geq d_2 \geq \ldots \geq d_n$. the matrix **D′** is the regularised inverse of **D** that has the rank k: $\mathbf{D}'=diag\{d_1^{-1}, d_2^{-1}, \ldots d_k^{-1}, 0 \ldots 0\}$.

3.2 Statistical analysis

This subsection presents the results of numerical experiments with synthetic data on displacement monitoring.

3.2.1 Synthetic data

To model displacement monitoring we introduce synthetic data as follows.

a)"True" stresses have been specified in the contact zone as polynomials of the fourth degree $N(x)=c_0+c_1x+c_2x^2+ c_3x^3+c_4x^4$. In j^{th} test all $c_k=0$ except $c_j=1$, $N^{\text{true}}(t)=(k+1)t^k$, which provides unit resultant force acting under the stump.

b) Ideal right hand sides corresponding to the loads above have been evaluated analytically on (1, 1+L) and than computed at collocation points (the number of which was set as n= 40 for all L varying within 0.01-10).

c) Distortion has been generated at each collocation point by introducing an independently generated random error normally distributed within ±5%.

Therefore in calculation the vector **G** had the components

$$g_m = \sum_{k=0}^{4} (k+1) I_k(x_m, L)(1+\xi_m), \quad I_{k,m}(x, L) = \int_0^1 \frac{t^k \, dt}{t^2 - x^2} \tag{20}$$

where ξ_m, m=1,..,n are Gaussian errors. The integrals $I_{k,m}$ have been evaluated exactly, which provides exact representation of the matrix **A**, so the errors are only associated with the left hand side of the linear algebraic system of equations.

3.2.2 Examples and statistics

We study the following arrays of results in the set of 200 numerical experiments (k=1…200) for the resultant force

$$P_k = \int_0^1 N^{\text{recovered in } k\text{-test}}(x)\,dx, \quad \mathbf{P} = \{P_k\} \tag{21}$$

and the divergence of the recovered and true contact stresses

$$\delta_k = \left[\int_0^1 \left(N^{\text{recovered in } k\text{-test}}(x) - N^{\text{true}}(x) \right)^2 dx \right]^{\frac{1}{2}}, \quad \boldsymbol{\Delta} = \{\delta_k\} \tag{22}$$

Examples of contact stress recovery are presented in Figure 3 for the following cases

(a) – constant load with maximum and minimum errors of Err_{max}=0.039, Err_{min}=-0.041 normally distributed with Mean=$1.2\cdot 10^{-3}$, <u>*StDev*</u>=0.03, recovered resultant force is P_{rec}=0.999;

(b) – linear load with Err_{max}=0.039, Err_{min}=-0.041 *Mean*=$1.9\cdot 10^{-3}$, *StDev*=0.07; P_{rec}=1.002;

(c) – parabolic load with Err_{max} =0.047, Err_{min}=-0.03 *Mean*=$-6.0\cdot 10^{-3}$, StDev=0.06; P_{rec} =1.006;

(d) – cubic load with Err_{max}=0.038, Err_{min}=-0.028 Mean=$4.8\cdot 10^{-3}$, StDev=0.086; P_{rec} =1.005

In all these examples the length of the monitored zones was L=1.

Statistical properties of (21) and (22) are presented in Figure 4 as functions of L.

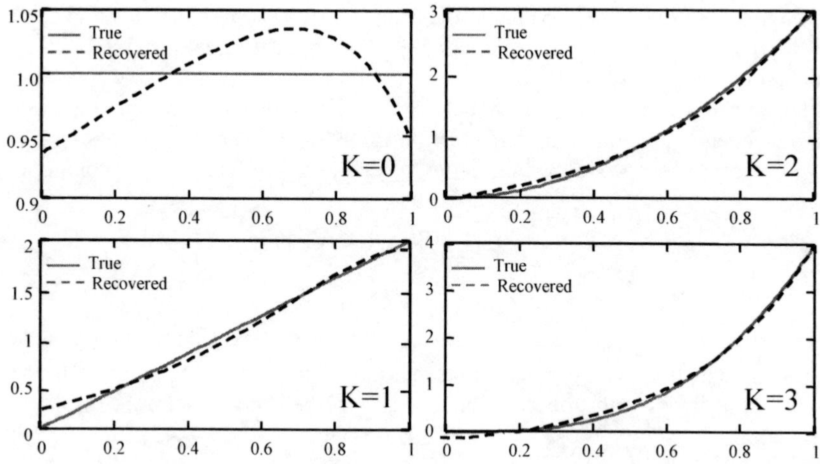

Figure 3: Examples of contact stresses recovering.

4 Conclusions

Solution of non-classical contact problem (5) for plane bodies is presented in operator form (14). The results of numerical experiments on the reconstruction of contact stresses demonstrate the SVD regularisation is an effective tool that provides stable numerical solutions. Statistical analysis of the results has been performed for the case of half-plane loaded by polynomial loads. It shows that resultant forces and traction distributions are better reconstructed for longer monitoring zones (Figure 4).

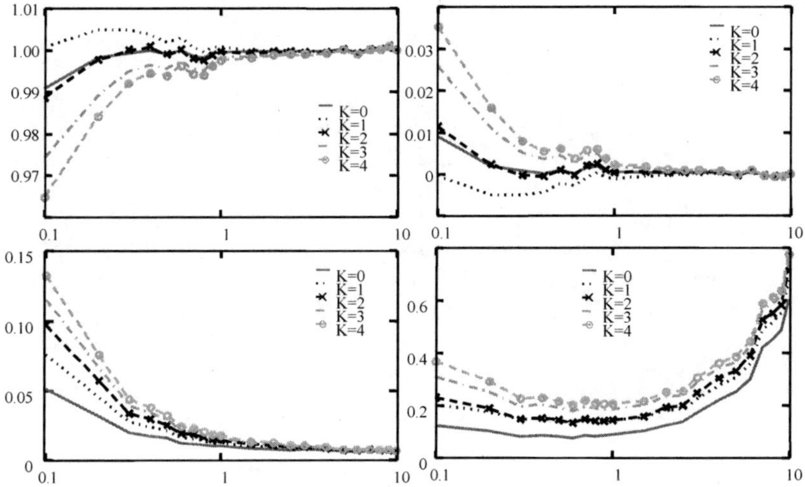

Figure 4: Means and standard deviations: **P** (left) and **Δ** (right).

References

[1] Muskhelishvili, N.I. Some basic problems of the mathematical theory of elasticity, P. Noordhoff, Groningen: the Netherlands, 1963.

[2] Johnson, K.L. Contact mechanics. Cambridge University Press, 1985.

[3] Shvab, A.A. 1989. Incorrectly posed static problems of elasticity, Mechanics of Solids. 24 (6), 98-106.

[4] Schwab, A. A. 1994. The inverse problem of elasticity theory: Application of the boundary integral equation for the holomorphic vector. Physics Of The Solid Earth. 30 (4), 342-348.

[5] Galybin, A.N., 1999. A non-classical plane elastic boundary value problem. Moving Boundaries V. Computational Modelling of Free and Moving Boundary Problems (Eds Sarler et al). WIT Press, Southampton, UK, 59-68.

[6] Tsvelodub, I. Yu. 2000. An inverse problem for an elastic] medium containing a physically non-linear inclusion. J. Appl. Maths Mechs. 64 (3), 407-412.

[7] Marin, L., L. Elliott, D.B. Ingham and D. Lesnic, 2001. Boundary element method for the Cauchy problem in linear elasticity, Engineering Analysis with Boundary Elements. 25 (9), 783-793.

[8] Marin, L., D.N. Hào and D. Lesnic. 2002. Conjugate gradient-boundary element method for the Cauchy problem in elasticity. Quarterly Journal of Mechanics and Applied Mathematics. 55 (2), 227-247.
[9] Marin, L. and D. Lesnic. 2002. Regularized boundary element solution for an inverse boundary value problem in linear elasticity. Communications in Numerical Methods in Engineering. 18 (11), 817-825.
[10] Marin, L. and D. Lesnic. 2002. Boundary element solution for the Cauchy problem in linear elasticity using singular value decomposition. Computer Methods in Applied Mechanics and Engineering. 191 (6),3257-3270.
[11] Galybin, A.N., 2001. A Method for determination of stress distributions in the process zone ahead of a 2D crack. "Moving Boundaries VI" Computational Modelling of Free and Moving Boundary Problems (Eds B. Sarler and C. A. Brebbia). WIT Press, Southampton, UK 243-252.
[12] Galybin, A.N., 2002. Determination of softening law by measuring crack opening displacements. Structural Integrity and Fracture (Eds A.V. Dyskin et al). Swets & Zeitlinger B.V., Lisse, The Netherlands, 35-41.
[13] Tikhonov, A.N. and Arsenin, V.Y. Solution of Ill-Posed Problems, New York: Winston, Wiley, 1977.

Contact problems in geomechanics focused on bumps occurrence

V. Doležel[1] & P. Procházka[2]
[1]*University of Pardubice, Czech Republic*
[2]*Czech Technical University in Prague, Czech Republic*

Abstract

The phenomenon known as bumps or rock bursts are one of the greatest dangers to human health during mining operations. They are caused by the accumulation of energy of forces (stresses) in close vicinity to a face of excavations of seams. This causes a sudden release of energy, which is strongly dependent on a development of material properties in both rock overburden and coal. In this case the coal material behavior can be neglected as for the time development. Hereditary problems have to be taken into account, as velocity of excavation is one of the biggest problems in this field of research. The faster the velocity is, the shorter time for redistribution of forces can be reached. This causes a more dangerous situation, as for the bumps initiation. These ideas are valid for pure openings, but not for wall reinforcement during mining, support of the ceiling of mines, etc. These additional elements also have to be taken into account, but not in this paper. We are concentrating only on time dependency of redistribution of stresses in the vicinity of seam face.
Keywords: rock bumps, coupled modeling, certain time-dependent problems, bumps on dislocations, physical scale modeling, numerical solution.

1 Introduction

In the paper, a time-dependent contact problem is formulated and solved. The main application is focused on the field of geomechanics, namely on bumps occurrence in deep mines. The time factor, which plays decisive role in the solution of the problem of bumps, depends on wide scale of circumstances. One of the most important items describing the sudden failure of the rock, i.e. influencing the real behavior of the rock, and consequently leads to the bumps

WIT Transactions on Engineering Sciences, Vol 55, © 2007 WIT Press
www.witpress.com, ISSN 1743-3533 (on-line)
doi:10.2495/SECM070241

occurrence is the velocity of excavation of mines in extreme depths. In the deep mines the way of depositing of packs and its mechanical properties are also decisive. Their mutual coupling of these items can principally influence the safety against bumps. For correct understanding of behavior of the rock aggregate (coal seam vs. overburden) rheology has to be reliably described, i.e. creep and relaxation is necessary to formulate in a proper way. Before formulate the above stated problem, large extent of laboratory experiments and "in situ" measurements have to be conducted. Since the on site tests are too expensive, and virtually impossible, laboratory models are prepared and the results from these tests are transmitted to mathematical models.

Contact conditions between particles of the rock are very important in the formulation of distinct element methods, which is a powerful tool for solving cracking in materials and its deterioration, softening, damage and other hereditary problems, i.e. time-dependent behavior. The discrete methods have been solved in many papers of the second author for statical case. Either lagrangian multipliers or penalty formulation were used. New formulation has to be submitted in terms of extended lagranigian principle. Creep is involved into formulation instead of eigenparameters, which can be substituted for quantities describing both hereditary development of the change of material and plastic behavior of the compacted rock. As we concentrate our attention on bumps occurring in coal deep mines, brittle behavior of coal is considered and the plasticity and time-dependent development of stiffness in rock material is involve only in surrounding rock. From couple of examples it was shown in static case that the behavior at the face of longwall mining is close to that near the crack tip, and the differences in material properties of coal and overburden are also not negligible. These factors are also expected to be important in the case of dynamical problems. Some examples will show applications of the procedures proposed.

2 Physical modeling

Principles of a new projecting method of underground construction in soft rocks can be formulated on the basis of results from extensive tests on physical models which were carried out in experimental department of Pardubice University.

Properties of these rocks differ from site to site. Rocks are frequently separated by discontinuity surfaces. These surfaces and the weakening zones cause disintegration or susceptibility to disintegration of the rock mass into structural units of various forms, size and properties. Their properties change also with the stress mode and depend on the stressing force, to which the rock was exposed in the past.

It results from the above-mentioned facts that the properties of the rock environment cannot be measured either on small rock samples or by isolated sporadic tests in situ. In the first case, we cannot evaluate the effect of weakening planes, and in the other one an unpleasant dispersion variance of material properties is caused by inhomogeneities of the rock mass. In both cases, the necessary conditions of the physical similitude are usually not observed. These conditions would require measuring the rock properties under stress conditions equal to stressing force, to which they are exposed during construction and after completion of construction works. These requirements can very easily be observed on physical models from equivalent materials.

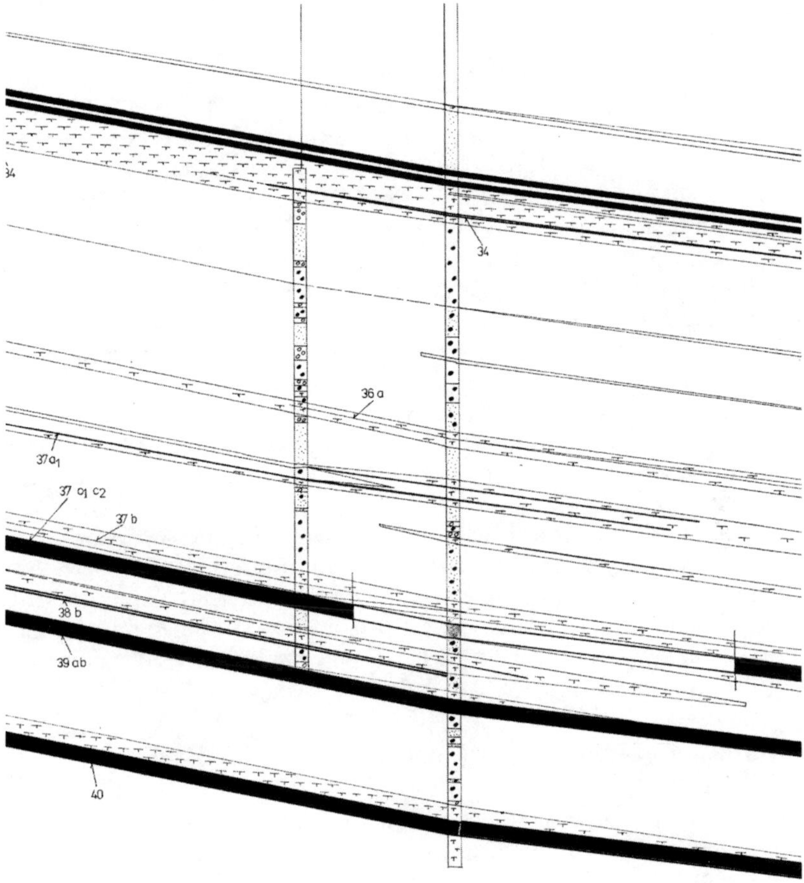

Figure 1: Side of the physical model.

Results from laboratory tests can be used directly as impute data for mathematical solution provided the constitutional relation between the stress tensor and strain (deformation) tensor is linear and provided the relative homogeneity and isotropy (e.g. of solid rocks) are supposed. This is not our case and large simulations in laboratory have to be carried out.

On the other hand the stresses are difficult to obtain from the physical modeling. This is the moment when the mathematical model can help and coupling of both models can approximate the real state of the rock continuum and the structure.

A very important conclusion results from some selected rock mechanics problems using the method of physical and mathematical modeling, and from their mutual comparison. It is possible, by means of a test on a physical model, to establish conditions, which are of primary interest to us (e.g. when the resistance of internal forces against failure is optimal). For conditions thus established, the deformation and stress of the rock environment in the neighborhood of the underground opening are determined by the finite element method, boundary element method, or with the aid of the combination of both. A part of problem can therefore be resolved by tests on a physical model and the results obtained can be used as input values for final solution by some numerical methods. In this way, both procedures can be adequately combined and completed, advantages and drawbacks of both of them acting complementarily.

With respect to the extension of the modeling equipment, material properties of physical material, measurement tools, and time factor it has been selected geometric ratio 1:200 and time scale 1:120. The mutual relation between real parameters and model properties of the material is given by adequate rules of similarity. The modeling stand (basin with glassed front side reinforced by steel frame was 6 m high, 2 m long and 1 m wide). The rear side has been constructed from plastic plates with longitudinal openings in the shape of distinguished seams. Modeling equipment enables researchers to load the terrain by pneumatic pillows. Front view of the model in depicted in Fig. 1. The progress of the failure is visible in Fig. 2.

The results of movements of the equivalent material in the front plane of the stand are determined for vertical and horizontal displacements and have been measured by photogrammetric method. This method enables one to get stereo couples of photographic pictures; one of them is in the starting (virgin) position. From the stereo couples of the model states the movement vectors at selected points for given stages of physical model are derived.

3 Solution of underground continuum by the BEM

In this part we briefly describe an implementation of boundary element method to the solution of specific problems of underground continuum, for which the numerical method appears to be extraordinary advantageous.

The method, among others, reduces the problem by one. Further good application of the boundary element method is the optimization and/or contact problems which concern the boundary only. Then, in spite of the finite element method it suffices to study a change of location of boundary elements only.

The problem is solved as two-dimensional, i.e. a possible dislocation is long enough, and narrow seam is considered. Moreover, the nonlinear behavior is considered in the region, which is sufficiently close to the dislocation, according to Mises theory. Suspicious dislocation is given from the experimental model from physically equivalent materials.

In our following consideration we will concentrate to the physically nonlinear problems (nonlinear evolution is also included in boundary conditions). Let us solve the problem in domain Ω. We originate from the Cauchy equations:

Figure 2: View of the experimental stands with the physical model failure.

$$(\alpha + \mu)\frac{\partial}{\partial x_i} div\ u + \mu \Delta u_i + X_i + \sum_j \frac{\partial \sigma_{ij}^0}{\partial x_j} = 0, \quad i=1,\dots,3 \qquad (1)$$

where

$$div\ u = \frac{\partial u_1}{\partial x_1} + \frac{\partial u_2}{\partial u_2}\ , \qquad \Delta = \frac{\partial}{\partial x_1} + \frac{\partial}{\partial x_2}$$

and $u = (u_1, u_2)$ is the displacement field, (X_1, X_2) are components of the volume weight and σij0 are components of the tensor of initial stress.

These equations will be solved in coordinate system $0x_1{,}x_2$. In the sense of BEM, (1) may be reformulated in an equivalent form:

$$c_{k1}(\xi)u_1(\xi) = \left(\varepsilon_{ijk}^*, \sigma_{ij}\right) + \left[p_{ik}, u_i\right] - \left\lfloor u_{ik}^*, p_i \right\rfloor - \left(u_{ik}^*, x_i\right) \qquad (2)$$

where [.] are boundary integrals, (.) are plane integrals, c_{kl} *is* the matrix of coefficients depending on a position of ξ, p is the vector of external forces, ε is the tensor of deformations and quantities with asterisk denote the relevant quantities of fundamental solution. Such the function was derived by Melan and can be found in publication [1], for instance.

Now the trick starting with the polarization tensor used in [2] is applied for describing the nonlinear behavior of the whole massif. In the polarization tensor also initial stress can be involved. From the Cauchy equations (respecting shearing stresses to be zero) we have well known relations for virgin state to get initial stresses:

$$u_1^o = -\frac{1-2v}{4\mu(1-v)} X_1 x_1^2 + const., \qquad u_2^o = 0$$

$$\sigma_{1,1}^o = -X_1 x_1, \sigma_{1,2}^o = 0, \sigma_{2,2}^o = -\frac{v}{1-v} X_1 x_1 \tag{3}$$

4 Contact problem

Before we start the analysis preliminary considerations will be introduce. In order to explain the process of computation two-dimensional problem will be treated. The three-dimensional problems are solved similarly.

Let the problem be described from experimental study. The field of vertical displacements is depicted in Fig. 3, vertical stress in Fig. 4. In numerical version the problem is illustrated in Fig. 5 by domain Ω, Γ_c is a part of boundary splitting the plane into left Ω^l and right half plane Ω^r, along Γ_p the distribution of given surface forces is done and Γ_c is fictitious slip surface (dislocation) either prescribed (this is our case, the location is estimated from the physical model), or the location is to be searched by an enlarged numerical process. Ω^{seam} is the domain of the seam, for which the dislocation and the bearing capacity is to be assessed.

After discretization of (2) in the sense of boundary element method the problem leads to the system of algebraic equations:

$$\begin{bmatrix} H_{11} & H_{12} & H_{13} \\ H_{21} & H_{22} & A_{23} \\ H_{31} & H_{32} & A_{33} \end{bmatrix} \begin{Bmatrix} u \\ u_c^- \\ u_c^+ \end{Bmatrix} - \begin{bmatrix} G_{11} & G_{12} & G_{13} \\ G_{21} & G_{22} & G_{23} \\ G_{31} & G_{32} & G_{33} \end{bmatrix} + G_{11}^g \begin{Bmatrix} g \\ g_c^- \\ g_c^+ \end{Bmatrix} = \begin{Bmatrix} F \\ F_c^- \\ F_c^+ \end{Bmatrix} \tag{4}$$

where the upper index - denotes "from the left" and + denotes "from the right", g is the vector of prescribed surface forces along the boundaries Γ and Γ_p, p_c is the vector of surface forces on fictitious contact Γ_c and F includes the effect of volume weight.

As the vector g contains known quantities we can rearrange the previous equations to obtain:

$$\begin{bmatrix} H_{11} & H_{12} & H_{13} \\ H_{21} & H_{22} & A_{23} \\ H_{31} & H_{32} & A_{33} \end{bmatrix} \begin{Bmatrix} u \\ u_c^- \\ u_c^+ \end{Bmatrix} - \begin{bmatrix} G_{12}p_c^- + G_{13}p_c^+ \\ G_{21}p_c^- + G_{23}p_c^+ \\ G_{31}p_c^- + G_{33}p_c^+ \end{bmatrix} = \begin{Bmatrix} F + G_{11}^g \\ F_c^- + G_{21}^g \\ F_c^+ + G_{31}^g \end{Bmatrix} \tag{5}$$

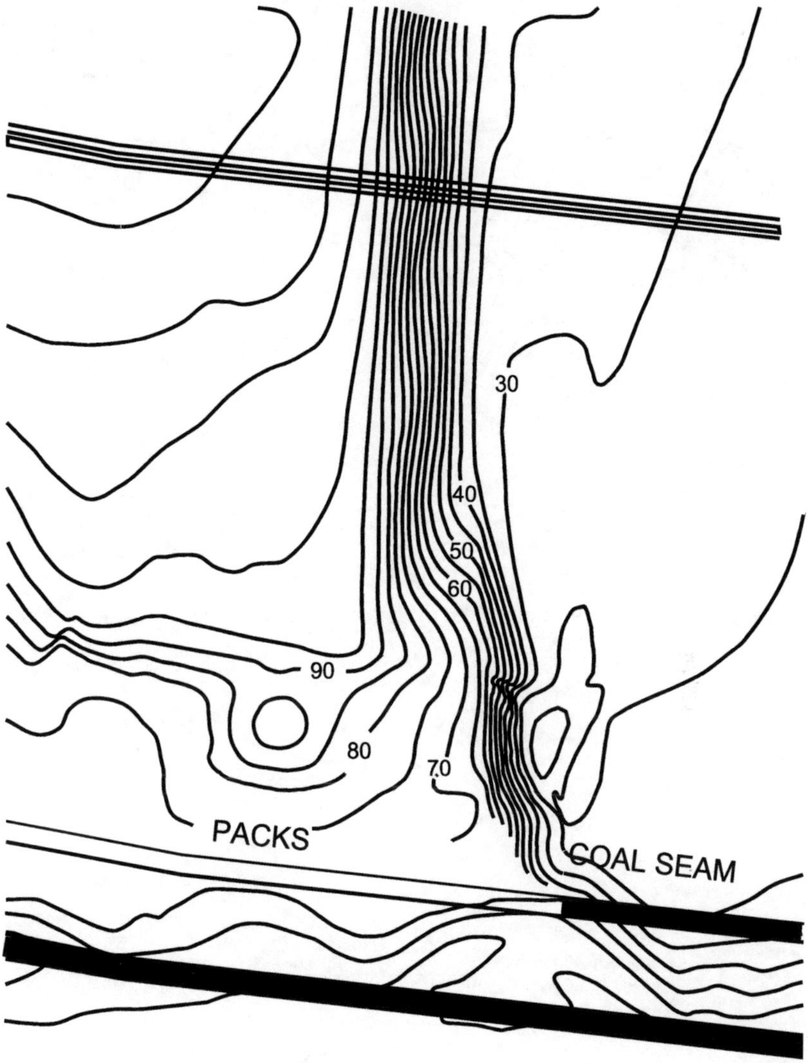

Figure 3: Cross section of the massif with vertical displacements (physical model).

Suppose now that for example u_c^- and u_c^+ is known. Then the problem is uniquely solvable, so that the matrix H_{11} is regular. For the similar reason the matrices H_{kk} are regular, too. Also, the same assertion holds for the matrices $G_{11,}$, i = 1,2. This is the general result of solvability of linear problems of elasticity by boundary element method. We can conclude that the matrix H is singular, but the last submatrices are regular matrices. This is why it is possible to rearrange the system in the sense of matrix canonical transformations (in algorithm we use Gaussian elimination) to obtain:

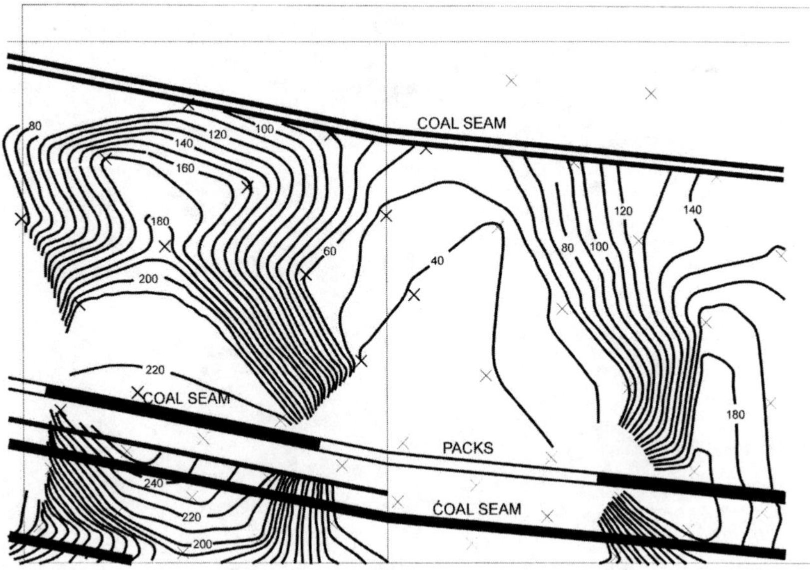

Figure 4: Cross section of the massif with vertical stress (physical model).

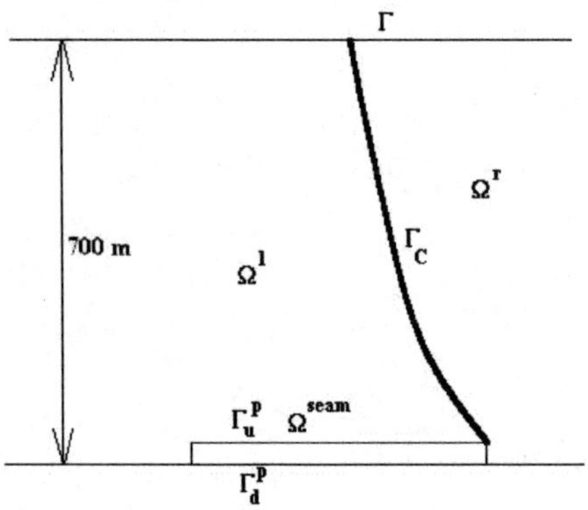

Figure 5: Domain and denotation of the example under study.

$$\begin{bmatrix} H_{11} & H_{12} & H_{13} \\ 0 & A_{22} & A_{23} \\ 0 & A_{32} & A_{33} \end{bmatrix} \begin{Bmatrix} u \\ u_c^- \\ u_c^+ \end{Bmatrix} - \begin{Bmatrix} B_{11} \\ B_{21} \\ B_{31} \end{Bmatrix} \{P_c\} = \begin{Bmatrix} C_{11} \\ C_{21} \\ C_{31} \end{Bmatrix} \tag{6}$$

where the balance condition

$$p_c = p_c^- = -p_c^+ \tag{7}$$

was employ. The matrices are known while the vectors u and p remain unknown. From the last form the reducibility follows and we can employ the following system of equations:

$$\begin{aligned} A_{32}u_c^- + A_{33}u_c^+ - B_{31}p_c &= c_{21} \\ A_{22}u_c^- + A_{23}u_c^+ - B_{21}p_c &= c_{22} \end{aligned} \tag{8}$$

Generally, along the contact line only balance condition holds and the compatibility is prescribed with the aid of more complicated relations. For example, suppose that at each nodal point along the contact line holds:

$$\begin{aligned} [u]_n = u_n^1 - u_n^2 \geq 0 \\ |p_t| \leq Tp_n + c \\ |p_t| \leq Tp_n + c \Rightarrow E\lambda > 0, [u]_n = -\lambda p_t \end{aligned} \tag{9}$$

where T and c are prescribed coefficients (they may very along the contact), p_n and p_t are projections of tractions to the normal and tangential direction with respect to the contact line, respectively. Then Uzawa's algorithm can be applied to the contact problem – see, e.g., [3].

5 Conclusions

In this paper a complex study based on coupled numerical end experimental studies has been proposed and applied to a real structure of underground work. It is aimed to assess the bearing capacity of dislocation, its position has been detected by experimental studies on scale models, and the results involving positioning of the dislocation have then been taken as input data of a great importance to observing stability (particularly to bumps) in the coal seams.

Many others studies have been carried out and in the future we intend to test with the aid of this procedure another approaches to improve the information about the three-dimensional behavior of the system of dislocations in the rock mass. This will lead to much more realistic information and helps to designers of underground works to prepare better plans.

The approach is very close to a back analysis, which is widely used in other parts of civil and underground structures.

Acknowledgements

The first author has been financially supported by GACR, project No. 103/05/0679; the second author has been supported by GACR, project No. 103/05/0334.

References

[1] Brebbia, C.A., Teles, J.C.F. & Wrobel, L.C., Boundary element techniques. Springer Verlag, Berlin, Heidelberg, NYC, 1984.

[2] Prochazka, & P. Šejnoha, J., Behavior of composites on bounded domain. BE Communications, Vol. 7, No. 1, pp. 6-8, 1996.

[3] Prochazka, P. & Sejnoha, M., Development of debond region of lag model. Computers and Structures, Vol. 55, No. 2, pp. 249-260, 1995,

[4] Dolezel, V., Experimental measurement on physical models of the deep mine in Ostrava. Research report, CSAV, Prague, 1991.

Section 5
Material surfaces in contact

In situ measurement of contact area in coated surfaces

J.-H. Sick & G.-P. Ostermeyer
Institute of Dynamics and Vibrations,
Technical University of Braunschweig, Germany

Abstract

Few attempts to directly measure the contact area of solid bodies have been made. In fact, it shapes up as a complex problem, because there usually is no direct access to the relevant surface of the experimental setup. Promising work in electrical measurement of the contact area by relating it to the contact resistance has been made by Bowden, Tabor and Holm from the 1930s on. This method has some restrictions due to the mutability of contamination layers involved in the most sensitive part of the setup. It is suggested to reintroduce observation of the contact area by electrical measurement, taking advantage of recent progress made in thin film coatings. By use of low conductive coating materials like oxides or carbon based films the contact resistance is increased by several orders of magnitude. The measured value of contact resistance in low conductive coated samples is within a range to be measured nearly independent of parasitic inductions like usual contaminations. Current work is on quasi-statical experiments and calibration. The first results joining earlier efforts are reported and the experimental setup, preparation method, advantages and known limitations are put up for discussion.
Keywords: real contact area, contact model, friction, wear, contact resistance.

1 Introduction

The question of how to measure the real contact area of solid bodies directly leads to the problem of defining the real contact area. Relationship of load and electrical contact resistance is investigated, thus observing effects influenced by parameters on a subatomic scale, leading to results perceivable in macroscopic dimensions. To define what to address as the real contact area is not quite simple. In the experimental setup the contact of a solid body dipped into a conductive

WIT Transactions on Engineering Sciences, Vol 55, © 2007 WIT Press
www.witpress.com, ISSN 1743-3533 (on-line)
doi:10.2495/SECM070251

liquid is used to aim for a reproducible holohedral electric contact. This standard of comparison has several imperfections like the dependence to wettability of the surface, roughness, pressure, atmospheric composition etc. Nevertheless it can be useful to compare the immersion standard measurements with results from solid body contacts.

The contacts investigated by Bowden and Tabor were cleaned metal surfaces, for instance silver, steel, brass. Measured contact resistances vary from about $100 \cdot 10^{-6}\Omega$ for a load of 5 N and $50 \cdot 10^{-6}\Omega$ for a load of 5 kN [3]. The setup of Bowden and Tabor involves several precautions to obtain reliable results. This includes a cleaning procedure, four wire connection because of the contact resistance being low compared to the leads, and gentle movement of the surfaces before settling of the measured value. Results show sensitivity to the presence of contamination layers of any kind, as well.

Today one may choose from several facilities to prepare surfaces with well researched thin film coatings, raising the contact resistance by several orders of magnitude, providing a considerable improvement to the usability of this technique. The coating itself might change the character of the surface drastically, as found in tribological, rather nearly all physical and chemical properties. Nevertheless, a better insight to friction dynamics of tribological boundary layers may be anticipated. In micro hardness testing with nano indenters, used for thin film coatings, it is difficult to measure the mechanical properties of the coating itself and not the substrate. The all over mechanical behaviour of several coatings is within given limits similar to a flexible skin, that permits and follows deformations of the substrate before delamination happens. Even though the coating might be harder than the substrate itself, the macroscopic load carrying capacity of the composite is not increased, unless edge layer of the substrate is strongly biased like in nitrided or kolsterised steel (cf. Bach [2] et al).

2 Experimental setup

2.1 Equipment

For electrical measurement a low noise current source Keithley 6221 is used in combination with a Nanovoltmeter Keithley 2182A and a digital transient recorder Yokogawa DL750 with 16 bit resolution inputs. The Nanovoltmeter has an input biasing current < 5 nA and an input resistance > 10 GΩ to be suitable for signal source resistances up to some Megaohm. For comparison of bandwidth, noise and load affects a Microdul HIP101 high impedance probe is used. This is configured as an input buffer with an input current of typical 0.2 pA (FET leakage current at 293 K), what is less than about 4 ppm of the supply current used. All measurements are carried out in DC mode at low currents to achieve ohmic proportionality of voltage and current density. The motion speed is kept at low rates with regard to the bandwidth and resolution performance of the measurement instrumentation. Temperature is controlled with a drift of less than $\pm$ 0.1 K per hour.

The probing force transducer is a Kistler type 9207, connected to a charge amplifier type 5019B140. A small drift typical to this device is compensated calculational retrospectively to data acquisition.

To measure displacement of the sample holder a Micro-Epsilon laser position encoder LD1605-20 is used.

Resistances in the measuring chain are small for all wiring, substrates, immersion, constriction resistance and current source resistance with the exception of the film resistance, boosting the contact resistance, and even more for the instrument input stages. For quasi-statical constitutional changes as in this case noise is kept at a low level by low pass filtering with a cut-off frequency of 10 Hz in Nanovoltmeter Voltage measurement, 100 Hz in force measurement, 400 Hz acquisition bandwidth limit and post acquisition cut-off to 10 Hz in high impedance probe voltage measurement (phase-neutral filter).

2.2 Preparation

Sample substrates are made of common stainless steel EN 1.4301/ASTM 304, polished sheet metal. With a Brinell hardness of about 170 and a Young's modulus of 200 Gigapascal this material grade is good machineable with the benefit of good corrosion resistance, what eases handling and storage of samples. Coated samples have a carbon based coating of 4 µm thickness, available as diamond like carbon (DLC) for anti-wear purposes in machine parts. For flat-ball contact bearing balls made of EN 1.3505/ASTM A-295 are used, which is remarkably harder than the coated substrate, as well as the coating is. It is important to compare samples from the same coating lot each, because obtained thickness and conductance may scatter significantly. Samples are kept untouched after coating, stored in an exsiccator, contaminated with oxygen and moisture just due to exposition whilst handled in the laboratory.

2.3 Contact area standard of comparison

To align measured values of contact resistance to a known relation of resistance to contact area there is need for a reproducible standard to refer to. This involves definition of the contact area as itself to be able to give some absolute values. At present this definition appears to be only partly finished by metrologic institutes. In the first instance a standard experiment is used with a simple, practical definition of the contact area. The sample is immersed into mercury, which is liquid at lab temperature of 294 K. The electrical resistivity of mercury is $95.9 \cdot 10^{-8}$ Ωm at 294 K (calculated from the CRC Handbook of Chemistry and Physics [5]), what is a high value for a metal (mercury is a relative weak conductor). The transition metals iron, nickel, chromium, manganese are hardly soluble in mercury, they do not form intermetallic phases with mercury as in amalgames like silver and copper in mercury. No change in the properties of the steel substrate material as well as the carbon film is to be due to contact with mercury, over and above the solid steel is not wettable to mercury in general. Aluminium for example amalgamates, but is usually not wettable due to its tight mercury-repellent oxide layer, investigated by [13].

In the experiment a coated electrode sheet is attached to an electrical insulating holder. By lowering the holder with a remote controlled, vibration-free motor drive the electrode is immersed to and extricated from a bath of mercury (Fluka, Hg puriss.). The mercury is kept in a glass vessel and is attached by a set of two wire electrodes of stainless steel. A schematic representation of the setup is given in Figure 1. The setup is built onto a vibration-isolated table to keep the mercury surface even. The level of the holder is measured by a laser triangulometer synchronously. Change in the mercury level due to displacement caused by the dipped electrodes is compensated upon data interpretation.

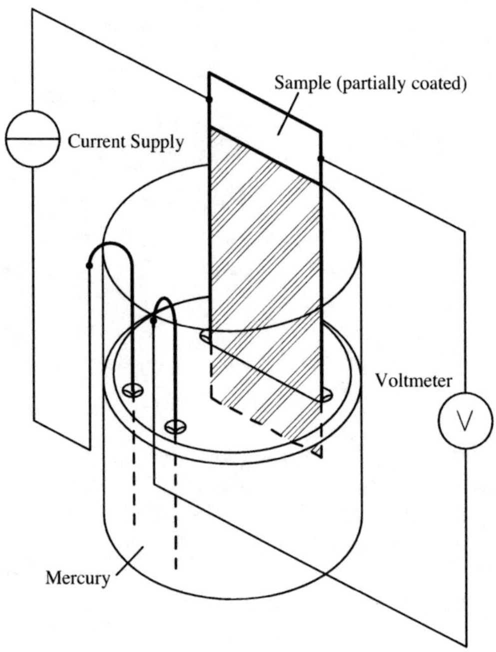

Figure 1: Schematic representation of the coated sample immersed into mercury to obtain holohedral contact.

2.4 Static loaded solid-solid contacts

Aim of the investigation is to obtain data about the real contact area depending on acting load by monitoring the contact resistance. The concept of "contact resistance" shall be used in terms of Holm, i.e. a resistance at the conduction area of the contact, no matter if there is a pure constriction resistance at a clean surface or a film contributing to the resistance [9].

To allow precision measurement, quasi-statical load-unload cycles are performed without lateral relative movement of the samples. The coated flat sample is mounted upside down to a perpendicularly driven rack. It is the same holder as used for the immersion experiment. The ball with a diameter of 10 mm is held in a countersink coaxial on the tip of a force transducer. The assembly compensates angular misalignment and keeps any electrical junctions symmetrical, except for the contact under test. The wires are clamped directly to the ball and the flat sample respectively, supply and measuring wires separately (4 wire connection). A schematic representation of the setup is given in Figure 2 (ball on flat sample). Current is kept constant at 50 nA, what turns out to be a value suiting well the ranges of the involved supply and meters, avoiding unwanted heating of the contact. Voltage measurement is accomplished in a fixed range of 12 V for the Nanovoltmeter and 9 V for the high impedance probe. Load effects are small for the Nanovoltmeter and negligible for the high impedance probe. They achieve results close to each other. The supply voltage is limited to 12 V, thus lowering electrical power loss to a maximum of 0.12 µW. The integration time is set to 1 power line cycle (20 ms) for suppression of high frequency noise and mains hum.

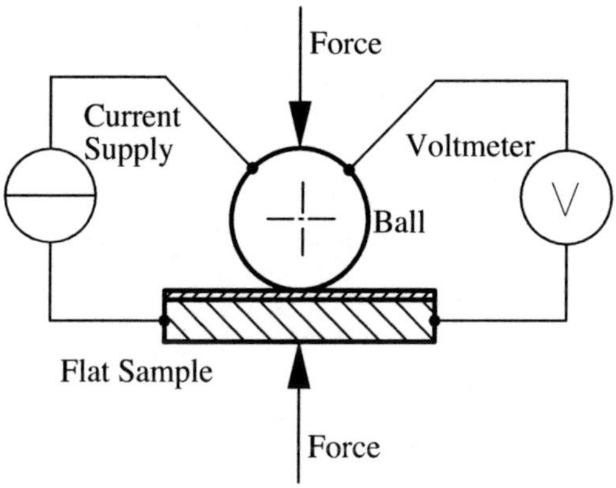

Figure 2: Schematic representation of the ball on flat contact.

2.5 Known issues

2.5.1 Temperature

Electric conductivity is temperature dependent. In case of metals like mercury and iron we have a small positive temperature coefficient of resistance (TCR). Semiconductive carbon based films show a strong temperature dependency. Robertson reports carbon films with a negative TCR in the range of 1 % per Kelvin at 294 K [12]. Even small temperature changes or gradients within the sample will have a noticeable influence to the conductivity. The contact angle of mercury to several substrates is reported to be temperature dependent, as accounted from Ellison et al [7].

To avoid relevant uncertainties of temperature all experiments are arranged in a full-time thermocontrolled laboratory at a temperature of about 294 K, performed one day of settling time after each handling of samples.

2.5.2 Capacitance

High contact resistances may have a noticeable capacitance, thus producing a delay in voltage changes due to current control. They might be part of the cause for a hysteresis found in several measurements.

2.5.3 Square resistance

The lateral electric conduction will cause an additional non-linearity to the ideally reciprocal dependence of immersion depth and contact resistance. The effect decreases for a rising contact area.

2.5.4 Semiconductive properties

Robertson reports carbon based films as semiconductors with a typical dependency of current density and resistance [12]. Thus the current has to be kept very low or the voltage across the junction has to be kept constant, to avoid a non-linear response to changes of junction's cross sectional area that is involved in transport. Some films have piezoresistive properties, as shown from Tibrewala [14]. The polarity dependence found in measured results is small.

2.5.5 Wettability, adhesion

According to de Wet, there is no dissolving of mercury to the steel sample itself to be expected [16]. Wright reports little adherence of mercury to steel [17] and Cuthrell even less to carbon [6], if not particularly plasma treated as reported from Sudarshan et al [13].

A common method to measure the open porosity of solids non-wettable by mercury is known as mercury porosimetry. Mercury porosimetry characterises a material's porosity by applying various levels of pressure to a sample immersed in mercury. The pressure required to intrude mercury into the sample's pores is inversely proportional to the size of the pores, as described at Micromeritics [11]. Thus a pressure dependency in contact resistance of rough surfaces is most likely. Probably a computational correction for hydrostatic pressure should be applied. This is to be investigated separately.

2.5.6 Contamination layers

Both sides of the contact, solid and liquid, may be contaminated with layers of oxides and adsorbed moisture. These layers affect the contact directly by adding an accessory resistance as well as indirectly by altering the wettability, the latter reported from Sudarshan et al [13]. The accessory resistance of thin contamination layers themselves is very small in relation to the resistive coating, but relevant on blank surfaces, as Holm found already [9].

2.5.7 Thermoelectric forces

There will be a typical thermoelectrical force induced voltage at any junction in the measurement chain. The setup is elaborated to provide symmetry of electrical junctions, thus compensating for thermoelectric forces (except for the contact under test itself). Measuring current is kept at low values to avoid heating of the samples. As long as the temperature of antipodal junctions is kept equal they will add just some noise to the result (Low Level Measurement Handbook [10]).

For the coated samples measured voltages are up to several volts, so the influence of thermoelectric forces in the range of millivolts may be neglected.

2.5.8 Reference sensors

Sensors used as reference for load and displacement have certain non-linearity and hysteresis, adding up to a relevant measurement uncertainty of up to 1.1% of the particular range.

3 Results

3.1 Immersion of coated sample in mercury

To measure electric properties of thin films they usually are deposited on a non-conductive substrate, hence electrical measurement proceeds with lateral direction of carrier stream in the film. In the present case the setup for measurement of electric conductivity shall be at the best accordance to the solid body contact to be compared with. The metal mercury, molten at room temperature, permits to make a setup with carrier stream normal to the film, resembling a lot of the electronic properties of the solid body contact.

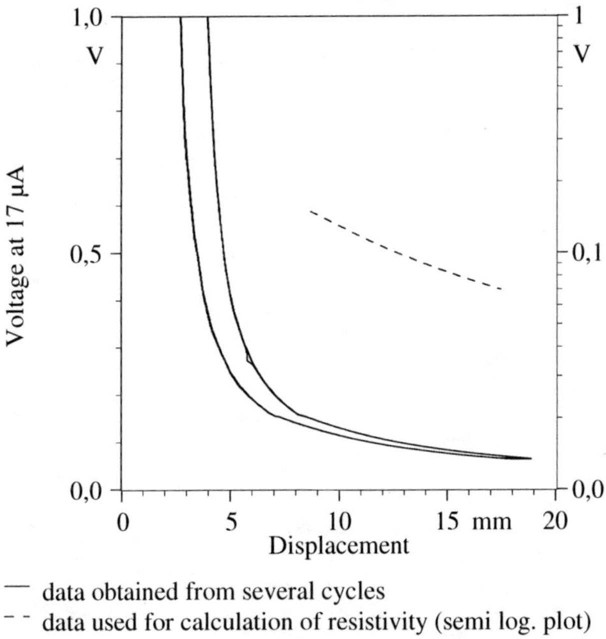

Figure 3: Measured voltage vs. vertical displacement of sample.

The conductivity calculated from the obtained data is about $3.5 \cdot 10^{-4}\Omega^{-1}m^{-1}$, what is within a possible range as reported from Robertson, Holiday [12, 8]. The specific electrical resistance is well within typical ranges for various amorphous carbon films, as portrayed in VDI guideline 2840 [15].

3.2 Flat-ball contact under normal load

The contacts are tested in normal load cycles without lateral movement from 0–10 N up-and-down.

For comparison the contact of a bearing ball and a blank metal sheet is investigated, too. Low voltages in the millivolt range have to be measured to keep the current low. A current of 50 µA turns out to be suitable for the given contact to measure within a range of 10 mV (except for the first few Micronewton of load, what is beyond access due to a voltage compliance to limit the electrical power loss). This supply current is a thousand times the current used for the resistive coated samples. The result is shown in Figure 4 and 5 in plots of voltage and conductance vs. load. Note the very distinctive hysteresis with higher voltage for increasing load, which was found similar in several runs.

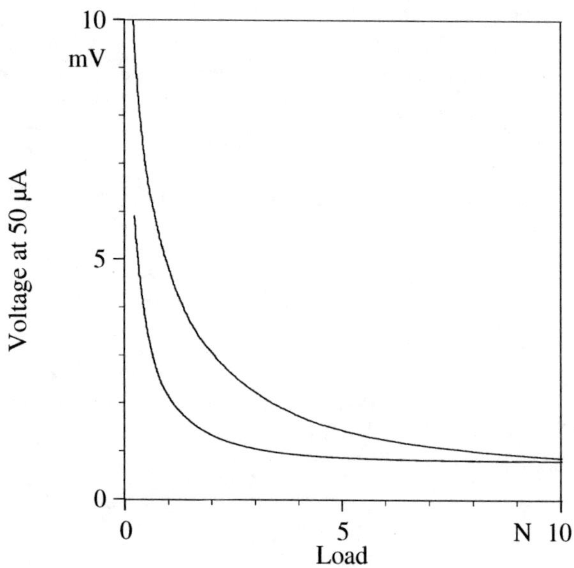

Figure 4: Measured voltage vs. load for blank ball on flat contact.

A resistive coated sample of the same substrate is contacted by a blank bearing ball. Due to the raised resistance the measured voltage is within a plainly gaugeable range of 10 V for a current as small as 50 nA. To afford measurement on a high impedance signal source like this contact an instrumentation amplifier with a field effect transistor input stage is useful, otherwise the electrical load on the DUT might corrupt the result.

Results measured with a high impedance probe are shown in Figure 6. From the measured voltage the conductance is calculated to produce the common logarithmical plot of conductance vs. load given in Figure 7.

Combined with the area resistance proportion from the immersion experiment the conductance is plotted as real contact area vs. load in Figure 8.

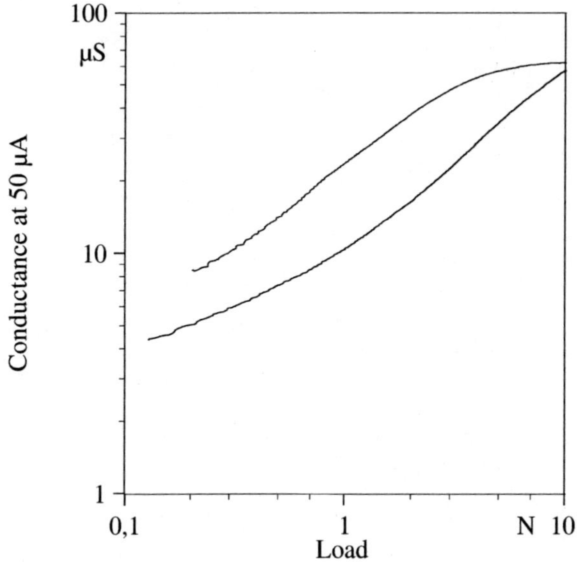

Figure 5: Calculated conductance vs. load for blank ball on flat sample.

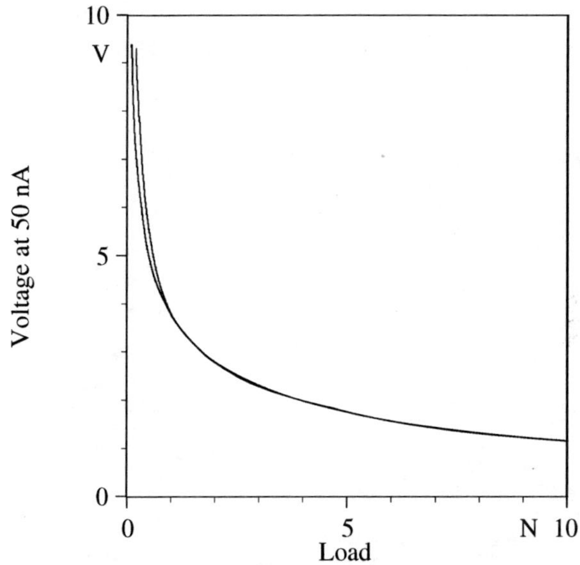

Figure 6: Measured voltage vs. load for ball on coated flat sample.

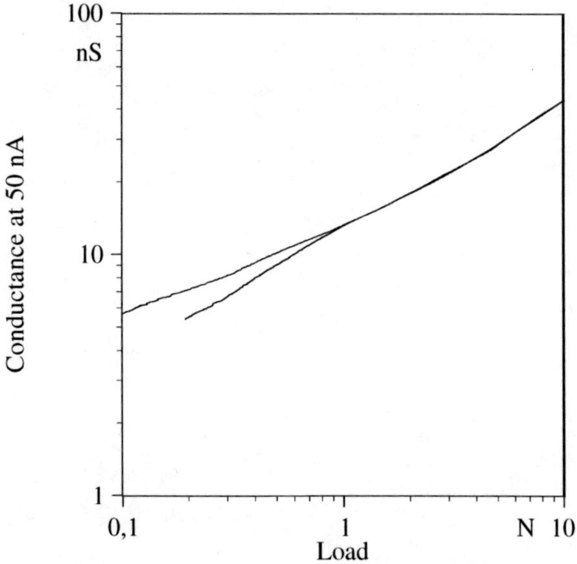

Figure 7: Calculated conductance vs. load for ball on coated flat sample.

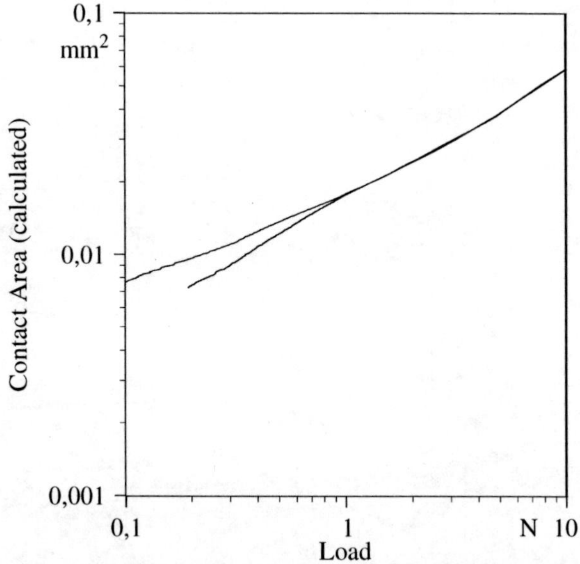

Figure 8: Calculated contact area vs. load for ball on coated flat sample.

Note the linear dependence of load and contact area. This is similar to recent predictions based on numerical calculations from Almqvist as well as from Ciavarella [1, 4].

4 Discussion and conclusions

It is shown that a configuration of a resistive coated steel sample and a blank counterpart permits convenient measurement of contact resistances. The dependence of measured conductance to load is almost linear already for small load, as expected from theory. The contact area is quantified for the given setup. Several issues mentioned above might cause a non-ideal contact load response of electric contact properties. Each of this will be subject to further investigations.

An alternative to the proposed immersion method with isostatic solid-liquid contact would be to build solid conductive electrodes, either made of electrical conductive gluing compound or of vapour deposited metal films (as used in preparation of samples for electron microscopy). Such electrodes on top of the resistive layer should obtain full tangency on a measurable portion of the surface. It might result some metal doping of the resistive layer however, leading to a resistivity different from the not treated area. Accurate measurement of conductivity and thickness (using known relationship of film thickness, specific resistance and square resistance as well as eddy current testing and profilometry), and investigation of anisotropic features of individual coatings could help to adjust another reference value for the relation of contact area and contact resistance.

Acknowledgement

This work was supported by the German Research Foundation (DFG), under contract number OS 166/2-1.

References

[1] Almqvist, A., Sahlin, F., Larsson, R., Glavatskih, S., On the dry elasto-plastic contact of nominally flat surfaces, Tribology International 40, 574–579, 2007.

[2] Bach, F.-W., Moehwald, K., Laarmann, A., Wenz, T. (editors), Moderne Beschichtungsverfahren, Wiley-VCH, Weinheim, 2004.

[3] Bowden, F. P., Tabor, D., Reibung und Schmierung fester Koerper. Springer-Verlag, Berlin, 32, 1959.

[4] Ciavarella, M., Delfine, V., Demelio, G., A “re-vitalized” Greenwood and Williamson model of elastic contact between fractal surfaces, Journal of the Mechanics and Physics of Solids 54, Elsevier, 2569–2591, 2006.

[5] CRC Handbook of Chemistry and Physics, Chemical Rubber Publishing Company, New York, 79th edition 1998-1999.

[6] Cuthrell, R. E., Evaluation of electrical contact materials for mercury switches designed to detect angular rotation, Journal of Materials Science, Chapman and Hall Ltd., 21, 2119–2123, 1986.

[7] Ellison, A. H., Klemm, R. D., Schwartz, A. M., Grubb, L. S., and Petrash, D. A., "Contact angles of mercury on various surfaces and the effect of temperature," J. Chem. Eng. Data 12, No. 4, 607, 1967.

[8] Holiday, P., Dehbi-Alaoui, A., Matthews, A., Physical properties of carbon films produced using a hybrid vapour deposition technique, Surface and Coatings Technology 47, Elsevier, 315–326, 1991.

[9] Holm, R., Electric Contacts, 4th edition, Springer-Verlag, Berlin, 1967.

[10] Low Level Measurements Handbook, 6th Edition, Keithley Instruments Inc., Cleveland, Ohio, 2004.

[11] Mercury Porosimetry Using AutoPore Porisimeters From Micromeritics published on AZoM at http://www.azom.com/details.asp?ArticleID=3227

[12] Robertson, J., Diamond-like amorphous carbon, Materials Science and Engineering, Elsevier Science B.V., R 37, 236, 242, 2002.

[13] Sudarshan, T. S., Lim, M. H., Hefley, P. L., Thompson, J. E., Wetting of aluminium electrodes with mercury, Journal of Applied Physics, American Institute of Physics, 56 (8), 2236–2240, 1984.

[14] Tibrewala, A., Piezoresistive effect in diamond-like carbon films (dissertation, Technical University of Braunschweig, Germany), Cuvillier Verlag, Goettingen, 2006.

[15] VDI Verein Deutscher Ingenieure (The Association of Engineers) (Editor), guideline VDI 2840 "Carbon coatings - Basic knowledge, coating types and properties", Beuth Verlag GmbH, Berlin, Germany, 2005.

[16] de Wet, J. F., Haul, R. A. W., Zur Loeslichkeit einiger Uebergangsmetalle in Quecksilber, Zeitschrift fuer anorganische und allgemeine Chemie, WILEY-VCH Verlag GmbH & Co. KGaA, Weinheim, Volume 277, Issue 12, 96–112, 1954.

[17] Wright, D. J., Hysteresis of the Angle of Contact of Mercury against Steel, Proc. Phys. Soc. B 68 297–303, 1955.

Wear assessment of tin and tin alloy coatings

W. P.-W. Lam[1], K. Mao[2], C. Kerr[1] & T. A. Stolarski[2]
[1]ITRI Ltd., UK
[2]Department of Mechanical Engineering, Brunel University, UK

Abstract

Tin and tin alloys have traditionally been used as coating materials for a variety of functions, e.g. decorative, solderable and corrosion resistant, but rarely in applications that require physical durability, owing to the soft nature of tin. Although it has been reported that some alloys of tin are comparatively harder, these materials have traditionally been used as corrosion resistant coatings and there has been little research conducted on exploring their physical properties. This paper aims to demonstrate the range of durability augmentation achieved from alloying tin with copper and nickel in coatings electrodeposited onto steel by assessing the degree of coating removal caused by wear from a sliding ceramic ball contact on the coating surface. The degree of wear is measured by the level of iron detected (i.e., where coating removal has resulted in the exposure of the underlying steel), via EDX, and this is correlated with 1) different loads (10 – 50 N at ten Newton increments) at which the ceramic ball is applied; 2) the thickness of the tin/tin alloy deposits (2, 5 & 10 microns) and 3) the wear testing duration (1 – 8 hours at one hour increments). The results are then presented in the form of a matrix. The study confirmed that tin performed poorly, and revealed significant improvements when tin is alloyed with nickel or copper.
Keywords: sliding wear, coating, SEM/EDX, tin, tin-nickel, tin-copper.

1 Introduction

This paper assesses the possibility of using tin and tin alloy coatings in an application not previously considered before. The suitability of the materials in question is evaluated using a novel method, where the degree of coating removal is quantified. Concluding comments and suggestions for further work are presented at the end of the paper.

WIT Transactions on Engineering Sciences, Vol 55, © 2007 WIT Press
www.witpress.com, ISSN 1743-3533 (on-line)
doi:10.2495/SECM070261

1.1 General applications of tin and tin alloy coatings

Uses of tin and tin alloys in packaging and electronics industry comprise around 60 % of the tin usage market [1-3]. Other uses include decorative or corrosion resistant coatings [4].

Tin and tin alloys, however, have not traditionally been considered as wear resistant coating materials, owing to its physical properties. When compared with other materials, metals and alloys, tin is a soft and malleable metal [1] that is often dismissed as a candidate material for wear applications that have traditionally required hard and robust properties [5].

Previous research has already reported on the corrosion resistant properties of tin-copper and tin-nickel alloys [6-7], and the intention is to assess these two alloys, with tin, in a sliding wear environment. A coating with a dual resistance to both corrosion and wear would have clear implications for industry.

1.2 Selecting the technique for assessing wear resistance

Although coatings can be assessed in a number of ways, including pull-off tests for adhesion and four-point bend tests for inducing coating delamination [5], it was decided that the testing in this research should introduce a form of deformation of the coating material that is representative of in-service wear.

The most common technique for assessing wear resistance is by measuring the coefficient of friction or by weight loss of the material [8], however, both of these methods have not been considered for the following reasons:

- The effectiveness of soft coatings is due to the solid lubrication properties they offer via reduction of friction. Measuring of this reduction in friction coefficient may provide some indication of when the coating has been breached, but it does not give any quantitative indication of the degree of coating depletion.
- The weight loss incurred by the coatings in this research are negligible and initial trials using a thick coating, large load and long test duration have indicated that there was insufficient material loss for detection.

A coating that offers protection to the underlying substrate is only performing its role while it is still adhered. It is therefore decided that the retention of the coating is to be assessed in this research and the continual monitoring of the removal of the coating is carried out by inspection of the wear scar surface. The use of a surface sensitive analytical technique, namely EDX, is considered for the examination of elemental composition in the wear scar location, enabling the ratio between detected coating and substrate material to be evaluated.

1.3 Coating deposition and characterisation

1.3.1 Electrodeposition

Using a combination of Hull Cell testing and available electroplating literature [4, 9-12], coatings of tin, tin-nickel and tin-copper were deposited on steel substrates measuring 40 x 22 mm^2.

1.3.2 Characterisation

The electrodeposits were characterised on several counts:

- The electroplated tin and tin alloy deposits were examined using a Topcon SM300 Scanning Electron Microscope (SEM) to show that, in all cases, a uniform coating and complete coverage was deposited.

- Using an Oxford Instruments INCA Energy Dispersive X-ray (EDX) system, it was possible to determine that the elemental compositions of deposits have been electroplated. It is also important to confirm the absence of iron from the surfaces looked at, indicating that the underlying steel substrates have been completely obscured.

- Using conventional metallographic preparation techniques, the deposits were microsectioned in order to determine their thicknesses with respect to the plating duration. In this way, it was possible to ascertain the exact plating times for producing 2, 5 and 10 micron coatings for the three deposits.

2 Methodology

The wear assessment is conducted on the tin, tin-nickel and tin-copper coated coupons. This Section presents the method for determining wear resistance, which has been divided into two parts.

2.1 Wear testing

The wear coupons, which have been coated with three different finishes, each at three different thicknesses, are tested using a TE70 micro-friction machine. This setup involves the sliding contact of a hard ceramic ball over the surface of the wear specimen (Figure 1).

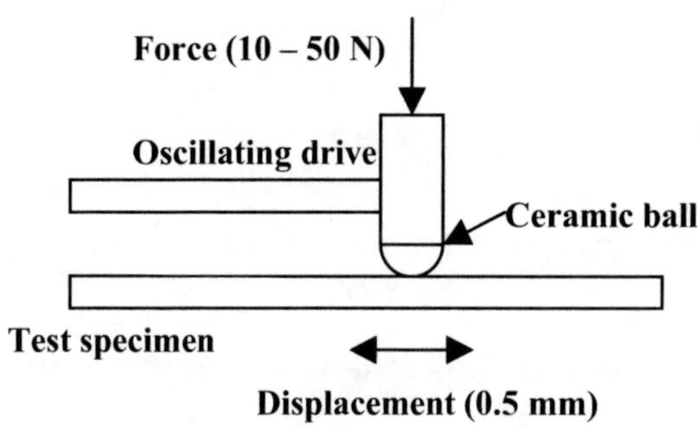

Figure 1: Schematic of TE70 machine setup used for wear testing.

The TE70 default settings of frequency (10 Hz) and displacement (0.5 mm) are maintained. These determine, respectively, the speed at which the ball slides and the distance of the sliding action.

An adjustable load carrier directly over the ceramic ball enables the test to be carried out under different loads. The load carrier applies a force of 10 N, whilst its capacity allows up to four 10 N weights to be added for a maximum total of 50 N. In this study, wear testing was conducted at 10, 20, 30, 40 and 50 N to investigate the effect of an increasing load.

The wear testing was carried out at 1, 2, 3, 4, 5, 6, 7, and 8 hour durations to investigate the effect of an increasing wear time.

2.2 Wear scar assessment

The resulting wear scars are examined using SEM and EDX in order to determine the extent of damage introduced by contact with the sliding ceramic ball. This is achieved by measuring the amount of detectable Fe within the wear scar. The amount of Fe detected for 0 hour samples is negligible; therefore, any Fe detected within the wear scar will be from the underlying steel substrate, which has been exposed by removal of the coating.

EDX is used to evaluate the approximate level of exposed Fe by acquiring from a rectangular area within the wear scar. Figure 2 illustrates the rectangle drawn within the wear scar, designating the area where the EDX acquisition is to take place.

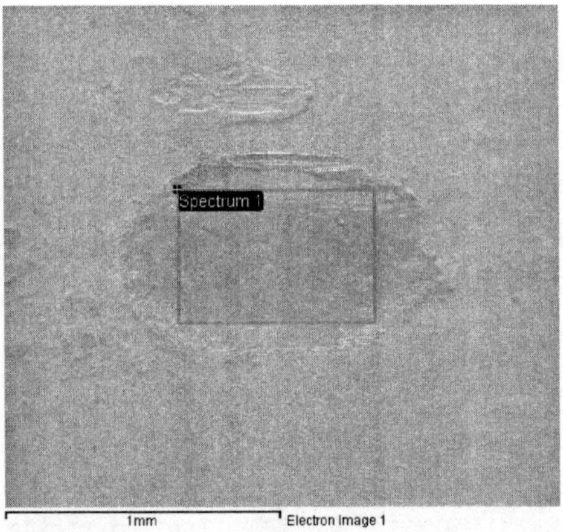

Figure 2: Area within wear scar selected for EDX analysis.

The Fe content detected is presented as a weight percentage, and recorded with reference to the coating material, coating thickness, load used and the duration of the wear test.

For the benefit of presentation and data analysis, the degree of coating removal has been separated and graded into five different categories:

- Negligible: 0 – 20 wt % Fe detected
- Moderate: 20 – 40 wt % Fe detected
- Significant: 40 – 60 wt % Fe detected
- Critical: 60 – 80 wt % Fe detected
- Failure: 80 – 100 wt % Fe detected

The results are presented, for each coating, in the form of a matrix, where an individual cell corresponds to a certain set of test parameters. The cells are shaded with respect the degree of coating removal, with a darker shade denoting more iron detected, the key to which is presented by Table 1.

Table 1: Legend defining shading used in Results Matrices for each condition/range of wt % Fe detected.

Negligible	Moderate	Significant	Critical	Failure
0 – 20	20 – 40	40 – 60	60 – 80	80 – 100

3 Results

The results matrix is presented for the coatings tested (Table 2), and is shaded according to the amount of exposed Fe detected.

3.1 Tin coating

Examination of the wear scars showed that there was a negligible degree of tin coating removed (0 – 20 wt % Fe detected) from the conditions tested. Even the testing of the thickest coating (10 micron) with the minimum load (10 N) within the minimum test duration (1 hour) yielded a moderate amount of tin removal (20 – 40 wt % Fe detected).

Testing on the tin coating was conducted up to 30 N, due to the amount of coating removed at the weight. Further increases in load would have yielded little further information.

3.2 Tin-nickel coating

The tin-nickel deposit was investigated with loads from 10 to 50 N. Although the thinnest coating (2 microns) had been fully removed after just 2 hours at 30 N, 50 N was necessary to ensure full removal of the 5 and 10 micron coatings.

3.3 Tin-copper coating

Testing of the tin-copper deposit was also conducted using the full 10 – 50 N load range. 10 N yielded negligible steel substrate exposure, even when testing for 8 hours on the thinnest coating thickness (2 microns). Using the maximum load of 50 N was sufficient to remove all coating from the 2 and 5 micron deposits, but not from the 10 micron samples.

Table 2: Results matrix for testing conducted on tin, tin-nickel and tin-copper coatings.

Testing Force / N	Testing duration / hrs	Tin 2 µm	Tin 5 µm	Tin 10 µm	Tin-Nickel 2 µm	Tin-Nickel 5 µm	Tin-Nickel 10 µm	Tin-Copper 2 µm	Tin-Copper 5 µm	Tin-Copper 10 µm
10	1									
	2									
	3									
	4									
	5									
	6									
	7									
	8									
20	1									
	2									
	3									
	4									
	5									
	6									
	7									
	8									
30	1									
	2									
	3									
	4									
	5									
	6									
	7									
	8									
40	1									
	2									
	3									
	4									
	5									
	6									
	7									
	8									
50	1									
	2									
	3									
	4									
	5									
	6									
	7									
	8									

4 Discussion

Both tin-nickel and tin-copper significantly out-perform tin, as expected, with the alloy coatings being retained for a far longer test duration and resistant to far greater loads.

4.1 Tin-nickel coating

Increasing the tin-nickel coating thickness significantly prolongs the retention of the coating and retards the exposure of the underlying steel substrate by a greater margin than the same increase has for a tin coating.

At 10 N, the level of exposed steel after testing for 6 hours on a 2 micron deposit is nearly 90%, decreasing to ~40% with a 5 micron deposit and further falling to ~0% for a 10 micron deposit. This large improvement is also reflected for tests using 20, 30 and 40 N.

The increasing of the load from 10 to 20 N initiates the trend at an earlier stage. At 4 hours, testing with 20 N, the 2 micron coating had failed, whilst the 5 micron coating was at a critical condition, however, there was negligible damage detected for the 10 micron coating.

At 30 N, the 2 micron coating fails after 2 hours. Under the same load and within the same duration, the 5 micron coating is significantly removed whilst there is negligible Iron found at the surface of the 10 micron coating. The 10 micron coating shows no sign of any removal for testing at 8 hours with loads of up to 30 N.

The 2 micron coating fails after one hour of testing with 40 N, whilst the 5 micron coating suffers critical damage. The 40 N load, however, is not sufficient to cause any notable damage to the 10 micron coating until 7 hours, when moderate coating removal is detected. Even after 8 hours testing, the 10 micron coating only yields ~50 wt % iron at the wear scar.

At 50N, the load appeared to be sufficient, after just 1 hour, to cause a critical condition (60 – 80 wt % Fe) for both the 5 and 10 micron coatings, with failure occurring after 2 hours of testing.

A 5 micron coating of tin-nickel, when tested using a 30 N load, performs comparably with a 2 micron coating of tin-nickel tested at 20 N. In both cases, testing for 1 hour yields a moderate degree of coating removal and a second hour of testing results in a significant case of steel exposure. At 3 hours, the coating removal becomes critical and the coating is considered to have failed after 4 hours.

A 10 micron coating of tin-nickel, when tested using a 50 N load, performs comparably with a 2 micron coating of tin-nickel tested at 30 N. In both cases, testing for 1 hour yields a high level of coating removal and, after a second hour of testing, the coating fails.

4.2 Tin-copper coating

The increase in tin-copper thickness from 2 to 5 microns does not appear to significantly prolong the retention of the coating. This is characterised by the proximity of the two sets of results, particularly at high loads. This is not so apparent at 10 N, where, after 8 hours of testing, less than 40 wt % Fe is detected and negligible damage is introduced to the 5 and 10 micron deposits.

At loads of 20 N and above, the increase from 5 to 10 microns, however, appears to yield a marked increase in performance of the coating at retarding the exposure of the underlying steel substrate.

After increasing the load to 20 N, failure of the thinnest coating is first observed after 7 hours of testing. At this point, the 5 micron coating is in a critical state, however, there is still negligible damage to the 10 micron deposit. There is no significant increase in Fe detection after 8 hours.

With a load of 30 N, failure of both the 2 and 5 micron coatings were discovered after 3 hours, whilst the 10 micron coating had still yet to experience any notable damage. This is similar to observations made at 20 N for the SnNi coating, where a similar set of results were obtained after 4 hours.

One hour of testing using 40 N appears to be sufficient to nearly completely remove the 2 micron coating. At this stage, the 5 micron coating exhibits some degree of removal, whilst the 10 micron coating is showing no evidence of being breached.

A second hour of testing yields significant further damage in the 5 micron coating, which ultimately fails after 3 hours testing. The 10 micron coating, however, is gradually depleted but falls short of the critical status when testing concludes at 8 hours.

At 50 N, the 2 and 5 micron coatings behave very similarly, failing at 1 and 3 hours respectively. The 10 micron coating reaches a critical condition after 6 hours, and maintains this status for the remainder of the 8 hour test.

At low loads, the 2 micron SnCu coating appears to out-perform the 5 micron SnNi deposit.

4.3 Summary

The pure tin coating performs relatively poorly, as expected, owing to its comparatively soft and malleable nature, and it is possible that the metal is so soft that the even the solid lubricating properties are nullified. Testing with the lowest load, 10 N, was sufficient to cause complete coating removal in both the 2 and 5 micron deposits (after 4 and 5 hours, respectively), and critical damage to the 10 micron coating (after the maximum 8 hour test duration). Increasing the load to 20 N caused complete failure of the 10 micron coating (after 3 hours).

Coatings of tin-nickel performed significantly better and testing with 10 N resulted in failure of only the 2 micron coating (after 6 hours). Testing with 20 N yielded failure in the 5 micron coating (after 5 hours), whilst the 10 micron coating was able to endure testing with loads of up to 40 N without critical damage. The 10 micron coating was eventually removed after 2 hours of testing with a maximum load of 50 N.

The tin-copper coatings were also a significant improvement on the tin metal deposits, in most case, even surpassing the performances of the tin-nickel alloy. 10 N appeared to be insufficient for generating failure in any of the tin-copper deposits, with failure of the 2 micron coating first achieved with 20 N (after 7 hours). Failure for the 5 micron thickness was observed at 30 N (after 3 hours), whilst the 10 micron coating was never fully removed, even after testing at the maximum 50 N load for the full 8 hour duration.

5 Conclusions

Three different deposits (tin, tin-copper and tin-nickel) of three set thicknesses (2, 5 and 10 microns) were successfully applied onto steel coupons. Using a combination of analytical techniques, it is possible to confirm deposition of complete coating coverage, and uniformity in composition and thickness. This is important for the generation of consistent results from the subsequent testing and analysis.

The wear resistant properties of the coatings produced are assessed using a novel methodology, where the degree of wear is quantitatively evaluated by analysis of the wear scar with SEM/EDX.

Although tin and tin alloys are not generally considered as candidate coating materials for wear resistant applications, but rather for decorative, corrosion resistant and solderable coatings, the alloying with copper and nickel greatly increases the wear characteristics of tin. For example, a 20 N force, that would yield a wear scar and remove nearly all the tin coating from the surface, could be applied to a similar thickness of tin-nickel or tin-copper for the same length of time, and result in little or no coating removal.

6 Further work

The effect of alloying other metals with tin is currently being considered, with intentions for tin-zinc, tin-cobalt and, possibly, tin-manganese to be investigated on in the near future.

Another avenue that is being looked into is the effect of incorporating particles, such as, PTFE and WC into the tin-alloy coatings, by process of co-deposition, on the wear resistant property.

References

[1] Evans C.J., Tin Handbook, Huthig GmbH, 1994
[2] Klein-Wassink R.J., Soldering in Electronics, Electrochemical Publications, 1989
[3] Hwang J.S., Environment-Friendly Electronics: Lead-Free Technology, Electrochemical Publications, 2001
[4] Chapman A.H., Tin and Tin-Alloy Plating – A Review, ITRI Publication No. 606, 1980
[5] Mellor B.G., Surface Coatings for Protection against Wear, Woodhead Publishing, 2006
[6] Britton S.C., Tin Versus Corrosion, ITRO Publication No. 510, 1975

[7] Warwick M.E., Hampshire W.B., Atmospheric Corrosion of Tin and Tin Alloys, John Wiley & Sons, 1982
[8] Hutchings I.M., Tribology: Friction and Wear of Engineering Materials, 1992
[9] Cuthbertson J.W., The Hull Cell, Metal Industry, 1951, Vol. 79 No. 5, 87
[10] Parkinson N., Electrodeposition of Bright Tin-Nickel Alloy Plate, J. Electrodepositors' Tech. Soc., 1951, 27, 129
[11] Price J.W., Tin and Tin-Alloy Plating, Electrochemical Publications, 1983
[12] Baugh L.M., Processes and Applications for Tin and Tin-Based Alloy Surface Coating Technologies, A Technical Review and Assessment of Recent Developments Compiled for Tin Technology, 2005

Section 6
Fracture and fatigue

Features of fretting fatigue strength/life and its mechanical considerations

T. Hattori, M. Yamashita & N. Nishimura
Department of Mechanical and System Engineering, Gifu University, Gifu, Japan

Abstract

The fretting fatigue process has many features such as early stage crack initiation at the contact edge, very slow crack propagation and fatigue failure after a very long life operation. In a previous paper we presented a new fretting fatigue model which can explain these fretting fatigue features reasonably. In this paper we try to explain many other fretting features such as fretting fatigue strength and life dependence on contact pressure and contact edge shapes. Firstly we try to discuss the dependence of fretting fatigue strength/life on contact pressure. In accordance with the increase of the contact pressure the stress concentration at the contact edge increased and crack initiation stress level decreased. But to open these small cracks initiated at contact edges more wear or more load cycles are needed. So fretting fatigue strength limit decreased in accordance with the increase of contact pressure and fretting fatigue life increased in accordance with the increase of contact pressure. Then we discuss the fretting fatigue strength dependence on the contact edge shape, such as stress release projection or interference of the contact edge with the stress concentration fillet. Experimental results of fretting fatigue strength improvement with stress release projection can be explained analytically. The two-stage S-N curve can be shown in joint structures, in which contact edge is set near the stress concentration fillet. These features can also be explained analytically in this paper.
Keywords: fretting fatigue, fretting wear, contact pressure, contact edge shape, stress singularity parameter, stress intensity factors.

WIT Transactions on Engineering Sciences, Vol 55, © 2007 WIT Press
www.witpress.com, ISSN 1743-3533 (on-line)
doi:10.2495/SECM070271

1 Feature of fretting fatigue

The fretting fatigue process has many features such as early stage crack initiation at contact edge, very slow crack propagation and fatigue failure after very long life operation. For instance 660MW turbogenerator rotor failed in England during the 1970s as a result of fretting fatigue cracking as shown in Fig. 1[3]. In this case the loading cycles in just one year is about 1.6×10^{9} and this trouble was observed after many years operation. Firstly the biggest question for turbogenerator design engineers is that, why the fatigue crack propagates so wide an area in the rotor cross section after very long life operation. The answer is that, the operating stress amplitude is very low less than 10MPa and crack propagate very slowly. Then the next question is that why the crack initiate under these low stress amplitude. In that time there was no enough answer except fretting under residual stress and crack initiation under barring conditions.

Figure 1: Fretting fatigue failure example of turbogenerator rotor. After Lindley and Nix [3].

2 Fretting fatigue mechanisms

I think that above mentioned ultra high cycle fatigue life can't be explained using only initial stress analysis results. We can't neglect the wear of the contact surfaces near contact edge and change of contact pressure in accordance with the progress of wear. Here, in this paper we present fretting fatigue process model as illustrated in Fig. 2. Cracking due to fretting fatigue starts very early in fretting fatigue life. We used stress singularity parameters at the contact edge to estimate the initiation of these cracks [4–6]. During this early period, fretting fatigue cracks tend to close and propagate very slow, due to the high contact pressure acting near this contact edge. But wear on the contact surface reduces the contact pressure near the contact edge, and cracks gradually start to propagate. Hence, fretting fatigue life will be dominated by the propagation of this small cracks initiated at the contact edge. So to estimate the fretting fatigue strength or life, the precise estimation of the fretting wear progress is indispensable. The propagation life in long crack length region can be estimate using ordinal fracture mechanics. In this paper we discuss the estimation method of wear extension on contact surfaces near the contact edge, and present the fretting fatigue crack propagation estimation method considering fretting wear extension.

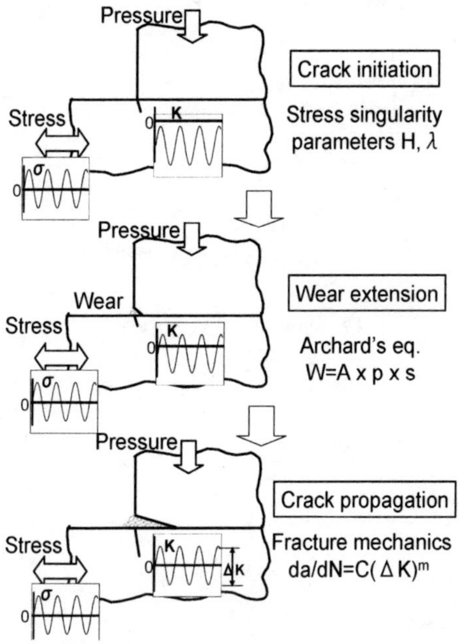

Figure 2: Fretting fatigue mechanisms in various processes.

Then I will show the flow of fretting fatigue life analysis considering the extension of fretting wear In Fig. 3. Firstly the fretting wear amount is estimated using contact pressure and relative slippage on each loading condition. Then the shapes of contact surfaces are modified following the fretting wear amount. And finally fretting crack extension or arrest evaluation is performed using fracture mechanics, if the operating ΔK is higher than the threshold stress intensity factor range ΔK_{th}.

We can estimate this load cycle as fretting life, and if the operating ΔK is lower than the threshold stress intensity factor range ΔK_{th} fretting wear amount is estimated using new contact pressure and new relative slippage and repeat these process until operating ΔK reach to the threshold stress intensity factor range ΔK_{th}.

And using this flow chart I estimated as shown in Fig. 3 by solid line. This estimated S-N curves especially in ultra high cycle region is compared with the experimental results and both results coincide well and this tendency of decrease of fretting fatigue strength especially in ultra high cycle region can explain above mentioned fretting troubles in industrial fields.

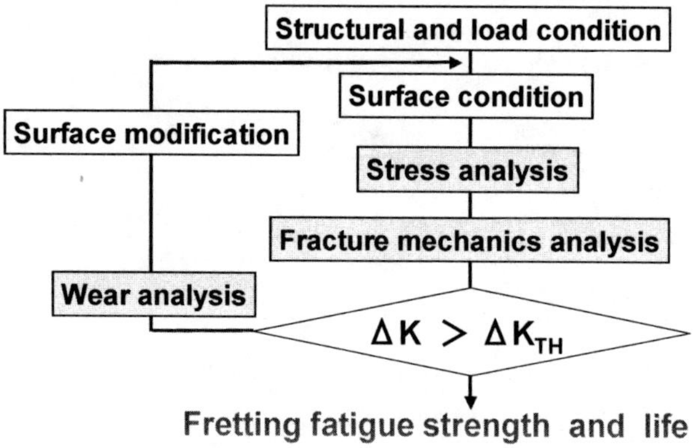

Figure 3: Flow chart of fretting fatigue life analysis.

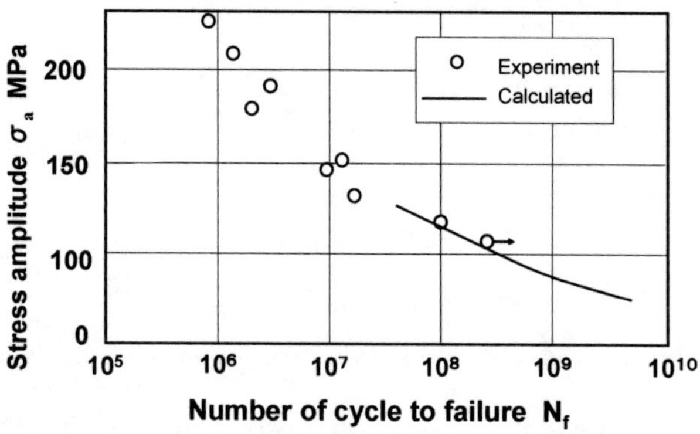

Figure 4: Estimated and experimental fretting fatigue S-N curves.

From these estimated results considering fretting fatigue processes such as crack initiation, wear extension and crack propagation, we can propose the general view of fretting fatigue S-N curve as shown in Fig. 5. The S-N curve in high stress region can be obtained without consideration of fretting wear. But in low stress and high cycle region we must consider the fretting wear extension. To estimate the S-N curve especially in ultra high cycle region more than 10^8, 10^9 we can use the hint that this S-N curve will converge to the crack initiation limit as an asymtote as sown in Fig. 5.

3 Mechanical consideration for some fretting fatigue features

3.1 Fretting fatigue strength / life dependence on contact pressure

In Fig. 6 we show the estimated example of dependence of fretting fatigue strength / life on contact pressure. In accordance with the increase of the contact pressure the stress concentration at contact edge increase and crack initiation stress level decreased. But to open this small cracks initiated at contact edges more wear or more load cycles are needed. So fretting fatigue strength limit decrease in accordance with the increase of contact pressure and fretting fatigue life increase in accordance with the increase of contact pressure. This estimated results coincided well with the experimental results as shown in Fig. 7 [7].

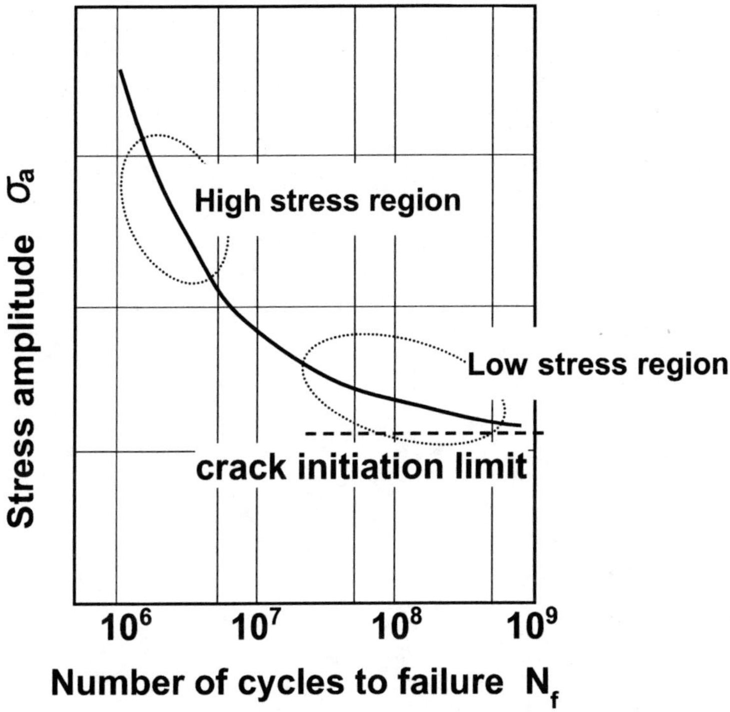

Figure 5: Schematic view of fretting fatigue S-N curve.

3.2 Fretting fatigue strength / life dependence on contact edge shape

To improve the fretting fatigue strength, the stress release projection is sometimes made on contact edge as shown in Fig. 8 [8]. This projection piece reduces the local stiffness and releases the pressure and stress concentration near contact edge. This reduction of stress concentration at contact edge improves the crack initiation limit, and similarly reduction of contact pressure concentration at contact edge decrease the wear rate and so increase the fretting fatigue life. This tendency can be seen in Fig. 8. By making a suitable projection near contact edge, the fretting fatigue strength can be improved about 30% compared with that of plain fretting model.

3.3 Fretting fatigue strength / life estimation considering the interference with stress concentration fillet

In many joint structures we must set contact edge near a fillet as shown in Fig. 9 [8]. In these cases we must consider the interference of stress concentration at contact edge with that at fillet. In the case of Fig. 9 both the stress and pressure concentration at contact edge decrease and fretting fatigue strength / life increase as the crack initiation limit increase and wear rate decrease. But, the most important notice in structural design of these joint is that if we mistake the fillet shape the fatigue at fillet decrease and it regulate the fatigue strength of joint structure. Fig. 9 shows the shrink fitted shaft coupling with fillet. In this case the fretting fatigue strength increase in accordance with the increase of stress concentration at fillet (decrease of fillet radius ρ). From this result we can see that the best choice of fillet radius is near 7mm or more small 6mm. On this condition the fretting fatigue strength at contact edge become same with fatigue strength at fillet. And S-N curve just on near this condition show two-stage curve as shown in Fig. 9. The reason of this feature is the slow propagation behavior of fretting cracks accompanying with the wear extension.

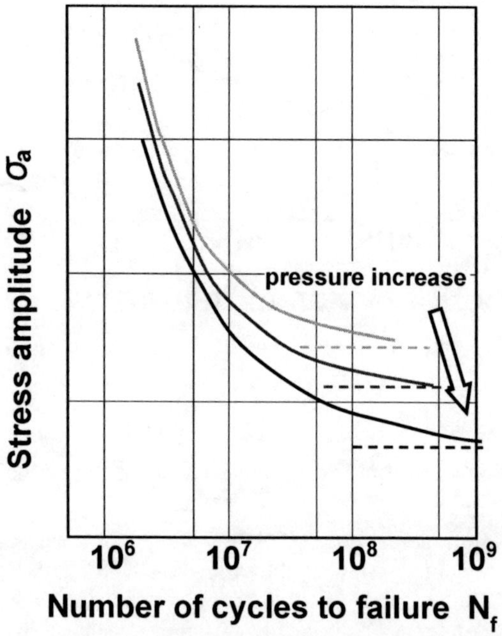

Figure 6: Estimated fretting fatigue strength dependence on contact pressure.

The next example of interference of contact edge with fillet is shown in Fig. 10. Unfortunately in this case the interference of contact edge with fillet increases both stress and contact pressure concentration at contact edge and decreases the fretting fatigue strength. In Japan many trouble happened on the hub structure of trailer truck as shown in this figure. Ordinary we test the fretting fatigue strength of whole parts before delivering these products to confirm the reliability. In these fatigue tests the most important notice is that fretting fatigue strength / life can't be confirmed in the ordinal load cycle number range such as $10^{7.}$ In these load cycle number range we ca only confirm the fatigue strength at fillet. As mentioned above the fretting fatigue failure at contact edge appears after long life with wear extension. So to confirm the reliability of these joint structures we must perform the fatigue test more than $10^{8\text{ or }}10^{9}$ cycles.

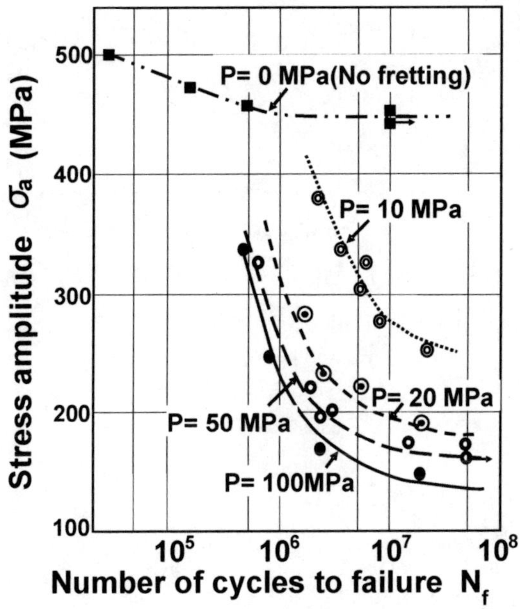

Figure 7: Experimental results of fretting fatigue strength for each contact pressure [7].

4 Conclusions

Fretting fatigue strength / life of several contact conditions are estimated based on the fretting fatigue model, which we presented before, as follows.

1. Fretting fatigue strength and life dependence on contact pressure was estimated and these results coincide well with the experimental results.
2. The interference of contact edge with ordinal fillet structure is analyzed and the existence of two stage S-N curve can be estimated. By using these results we present the methodology for designing optimum fillet shape and for confirming the reliability by fatigue test.

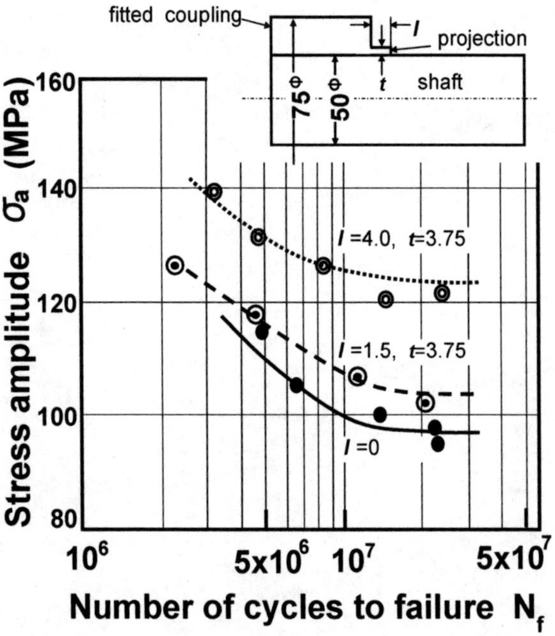

Figure 8: Experimental results of fretting fatigue strength improvement by making the projection at contact edge [8].

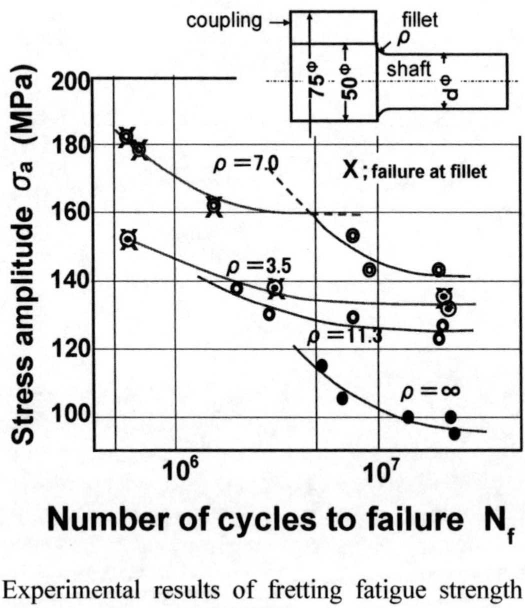

Figure 9: Experimental results of fretting fatigue strength of shrink fitted shaft coupling with fillet [8].

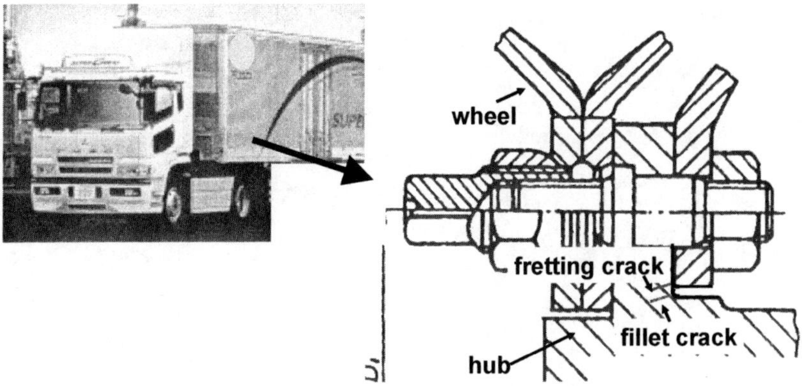

Figure 10: Hub-wheel joint structure in trailer.

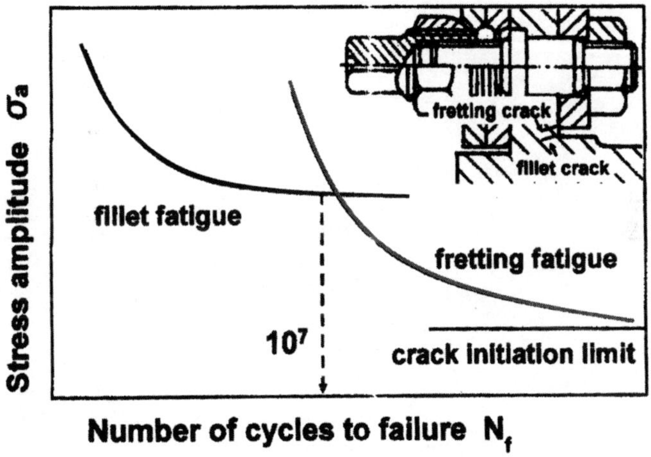

Figure 11: Two stage S-N curve of joint structure with contact edge and fillet.

References

[1] Hattori, Tand Watanabe, T., Fretting wear and fretting fatigue process at the contact edge, *Computational Methods in Contact Mechanics VI, WIT Press,* 2003, p.169-178.

[2] Hattori, T. and Watanabe, T., Fretting fatigue strength estimation considering the fretting wear process, *Tribology International,* 2006, 39, p.1100-1105.

[3] Suresh, S., *Fatigue of Materials 2nd Edition*, Cambridge University Press, 1998, p. 469.

[4] Hattori, T., Sakata, H. and Watanabe, T., A stress singularity parameter approach for evaluating adhesive and fretting strength, *ASME Book No. G00485, MD-vol.6,* 1988, p. 43.

[5] Hattori, T. and Nakamura, N., Fretting fatigue evaluation using stress singularity parameters at contact edges, *Fretting Fatigue,* ESIS Publication 18, 1994, p. 453.
[6] Hattori, T., Nakamura, M. and Watanabe, T., Simulation of fretting fatigue life by using stress singularity parameters and fracture mechanics, *Tribology International,* 2003, 36, p. 87.
[7] Funk, W., *Materialpruf,* 1969, 11, 7, p.221.
[8] Nishioka, K. and Komatsu, H., *Trans. Of JSME (in Japanese),* 1967, 33, 248, p.503-511.

Near surface modification affected by hydrogen/metal interaction

Y. Katz, M. Tymiak & W. W. Gerberich
University of Minnesota, Minneapolis, U.S.A

Abstract

Hydrogen/metal interaction represents a complex topic that attracted a high volume of experimental/theoretical efforts. Thus, for better understanding, ample activities have been conducted that spread over various levels and scales. The current study emphasizes some additional facets manifested mainly by confined and localized information. In this context, nano tests assisted by contact mechanics methodology provided small volume information to be even stretched to other surface related behaviour. This includes implications regarding the quantification of sequential events as related to hydrogen embrittlment (HE) or fracture processes transition. In addition, questions emerged regarding wear aspects or tribological contacts insights. Basically, the study considered free hydrogen (either external or internal) to be an aggressive environmental agent in elastic–plastic crystalline solids. Experimentally based information has been gathered in the bulk, supplemented also on the nano scale. In fact, this comprehensive study with emphasis on the scale effects provided striking insights. For example, in terms of contact tribological effects, normally mechanical driving force is solely considered. However, here, the unique contribution by environmental interaction includes phase stability effects and localized plasticity. Mainly two materials have been focused: first, as a background, AISI 316L metastable austenitic stainless steel, and secondly Ti/Cu, thin film specimens affected by hydrogen. The outcoming effects have been sorted out by mechanical response tracking and morphological visualization. Quantitatively, it was assessed by nano indentation and continuous scratch tests. Besides mechanical testing the information was supplemented by scanning Probe Microscopy (SPM) observations. It became evident that hydrogen affected dramatically the investigated systems with exploration of dislocation nucleation and dynamics. As such, remarkable differences occurred on both plastic localization and micro crack onset. In the thin layers the effective work of adhesion was reduced, indicating significant degradation that could be expressed quantitatively. Finally it was concluded that since fracture is a localized phenomena nano scale information might suggest "critical experiments" so vital for fundamental concepts confirmation.
Keywords: hydrogen embrittlment, metastable austenitic stainless steel, nano data, dislocation emission, wear.

WIT Transactions on Engineering Sciences, Vol 55,
www.witpress.com, ISSN 1743-3533 (on-line)
doi:10.2495/SECM070281

1 Introduction

Hydrogen/metal interactive effects have significant implications on structure integrity aspects due to crack stability transition. Regardless the specific enhancing damage origins, crack nucleation and growth are critical forms of mechanical degradation events. Note that susceptibility to aggressive environment requires special concern regardless the specific structural scale. Attention to small volume circumstances initiated alternative avenues in order to expand the spectrum of insights into hydrogen/deformation interactive effects. The current study selected mainly metastable austenitic stainless steel as well as Ti/Cu thin layers system. Previous studies established already that phase stability aspects in austenitic stainless steel are either enhanced by solely mechanical driving force or by hydrogen interaction. Hydrogen can be provided externally, internally or both. For the sake of the current investigation, beside hydrogen availability due to residuals at microstructural trapping sites, hydrogen was charged intentionally. The charging process can be achieved by electrolytic cathodic charging or by high temperature/pressure gasous charging. Based on previous experimental confirmation it was concluded that regardless the exact procedure of hydrogen charging, austenite decomposition in austenitic stainless steel prevails [1,2]. However, major differences are more associated to the damage evolution caused by a more aggressive high fugacity charging [3, 4].

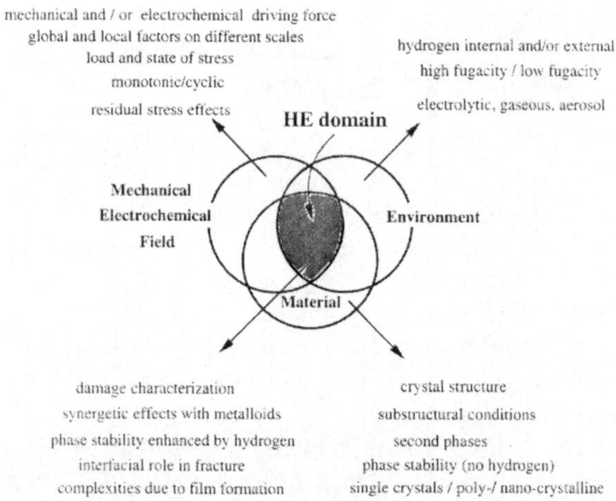

Figure 1: Global description of hydrogen/Metal interaction aspects.

Generally, the following notion remains in order by emphasizing that the present study is engaged only with free hydrogen effects, in contrast to the role of hydride formation that provides other damaging origins. In order to

WIT Transactions on Engineering Sciences, Vol 55, © 2007 WIT Press
www.witpress.com, ISSN 1743-3533 (on-line)

summarize the intense efforts that have been invested in the understanding of the highly complex hydrogen/deformation interactions, some of the variables that dominate this interactive process are depicted in Fig. 1. In the current study, the two selected materials that seem to be hardly related represent at least the small volume approach. The distinction between highly localized resolutions compared to macroscopic contact mechanics damage model, expanded the interpretation capabilities. For example, friction and wear damage evolution have been treated macroscopically from a cumulative damage standpoint [5, 6]. In contrast the nano data provided information that enabled to deal also with point contact (asperities). Moreover, yield excursions alluded to dislocation emission activity and microplasticity analysis. Since the study proposes practical measurements potential, the initial stages of wear - damage might be tackled. The metastable stainless steel is affected by free hydrogen in a typical surface upset beside crystal plasticity habits, phase transformation and micro cracks formation. The thin films emphasize the interfacial strength variability or delamination that clearly affects tribological contact behaviour. These were manifested by ultra fine features that could be measured at the near surface.

2 Experimental procedures

In the bulk surfaces of 304L, 316L and 310 metastable stainless steels, thermal stability was previously established. In this context, even after the immersion into liquid nitrogen and helium the monolithic austenitic phase (γ) was preserved in all three materials. Similar results were obtained after plastic straining at ambient temperature. However, in the 304L stainless steel after very extensive straining at 296K small amount of the body centered austenitic phase was detected. Completely different behavior occurred under load at cryogenic temperatures and straining at 77K in which significant austenite decomposition occurred. The transformation characterization has been assisted by X-ray diffraction and Mossbauer transmission spectra with emphasis on the composition and the reaction sequence. For the latter, the reaction was confirmed to be $\gamma \rightarrow \varepsilon' + \alpha'$ where ε' and α' are the hexagonal close packed and the body centered tetragonal martensitic phases respectively [7]. For example, X-ray diffraction enabled to detect the intermediate ε' hcp martensitic phase at the early stage of 0.02 true strain. More about the austenitic stainless steels have been addressed elsewhere [3, 4]. Here to mention that the 316L stainless steel behaved similarly to the 304 besides revealing always higher thermo-mechanical stability. The exceptional stability behavior of the 310 stainless steel became apparent after extensive straining at 77K. Once again, phase stability degree was a major distinction between the aforementioned austenitic steel types. In contrast to hydrogen-free systems, the case of hydrogen interaction highly depends on the hydrogen charging namely the fugacity degree conditions. Specimens were prepared from flat materials by utilizing spark erosion for the metal working technique; followed by electrolytic polishing. By using such procedures cold working influences were drastically reduced. In most of the specimen series, hydrogen was introduced at ambient temperature by cathodic charging from 1 to

16 hours at a current density of 50-500 mA/cm^2. The electrolytic cell consisted of platinum coil as the anode in 1N H_2SO_4 aqueous solution with 250 mg arsenite per one liter. The hydrogenated specimens were examined periodically for their structure stability confined to the affected hydrogen interfaced layer. The X-ray diffractometry was performed by using shallow penetrating CuK_α radiation, utilizing also a graphite-monochromator that eliminated the high fluorescence background radiation. For resolving the hydrogen affected layer deeper penetrating radiation like MoK_α or CoK_α were supplemented. For the Mossbauer transmission spectroscopy, a constant acceleration computerized spectrometer was used in conjunction with a 25mCi^{57}Co (Pd) source. For this purpose, hydrogen-free specimens were electrolytically thinned to about 25μm while for hydrogen effect studies it was necessary to reduce the specimen thickness to 5μm prior to hydrogen charging. Also here, specimens were prepared by following the aforementioned procedures namely, by utilizing spark erosion for metal working process, followed by electrolytic polishing. The X-ray diffraction and the Mossbauer spectra analysis intended to engage with the following factors. The lattice expansion and contraction, the role of preferred orientation on magnetic polarization, overlapping of the (hkl) reflection peaks, the hydrogen affected layer size and products concentration. Beside others, complementary studies were conducted, i.e., gas chromatography and hydrogen collection measurements combined with microscopic gas release analysis. Here to mention that the quantitative analysis considered also the Debye-Waller factor influences. In addition to the bulk surfaces information, nano scale procedures have been conducted with attention to hydrogen/stainless steel interaction. For this purpose, the selected material was AISI 316L metastable stainless steel a polycrystalline systems that consisted of 50-100 μm grain size. Mechanical response and plasticity behaviour with and with no hydrogen was characterized by contact mechanic methodology including mainly nano indentation and lateral nano scratch techniques. In this case hydrogen was charged also by cathodic charging namely by 1M NaOH under current densities in the range of 10 to 500 mA/cm^2. Similar to the previous post charging studies, sequential events vs. elapsed times were tracked during hydrogen out gaseous. For the nano indentation test, the mentioned microstructure grain size enabled at least ten consecutive tests to be conducted within the same grain. Fine features visualization was carried out for the nano indentation as well as for the scratch trances. These functions were performed by Scanning Electron Microscopy (SEM) and by Atomic Force Microscopy (AFM).

3 Experimental results

3.1 The metastable stainless steel system: global approach

Second phase aspects in austenitic stainless steels considered the austenitic decomposition enhanced by mechanical, hydrogen or both interactions. In fact this whole material class is unstable below the Md temperature even with no hydrogen. Moreover, hydrogen with or with no mechanical field can result in martensitic transformation, mechanical degradation due to localized displacement, delayed cracking and ductility reduction. In this context, few phenomenological observations are cited involving mainly electrolytic charging. Studies in 304, 316 and 310 stainless steel have covered a relatively wide range of phase stability as related to the degree of mechanical degradation. For example, high fugacity charging of 304L with an initial concentration of 0.34 H/M at. resulted in an affected layer of about 1μm. With no mechanical field, about 0.3 volume fractions of transformed austenitic products were observed in contrast to 0.85 volume fraction with an imposed true strain of 0.28. Generally (except for the 310 case) the early work of Holzworth and Louthan [1] indicated that similar phases resulted from plastically strained material compared to those charged with hydrogen. It is emphasized that phase transformation and the related fracture modes have been established regardless a specific charging procedure methods (Fig. 2). The transformation reaction pattern that was already mentioned takes a different form during the transient time, in which expanded ε^* and α^* were confirmed. These were identified as pseudo-phases by X-ray diffraction and internal friction studies [8].

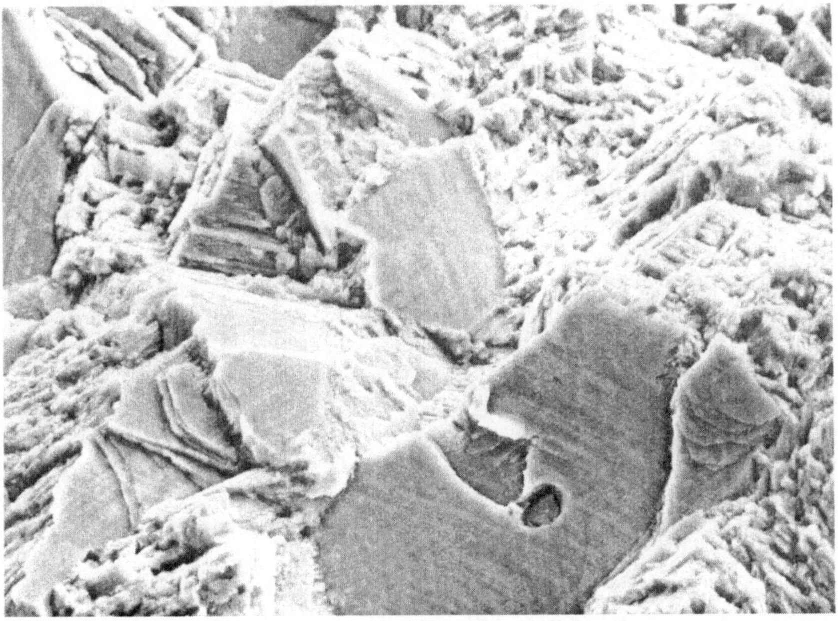

Figure 2: SEM fractograph indicating alternative modes in hydrogenated 316L stainless steel.

3.2 Metastable 316L stainless steel system: local approach

Indentation tests for a prescribed load of 1000μN were carried out with Hysitron nano indentation device utilizing a 90° conical indenter with a 400 nm tip radius of curvature. Tests were performed prior to hydrogen charging, instantly post-charging and one day after charging. Reproducible displacement excursions at an average load of 200 μ*N* were observed for the non-charged samples as show in Fig. 3.

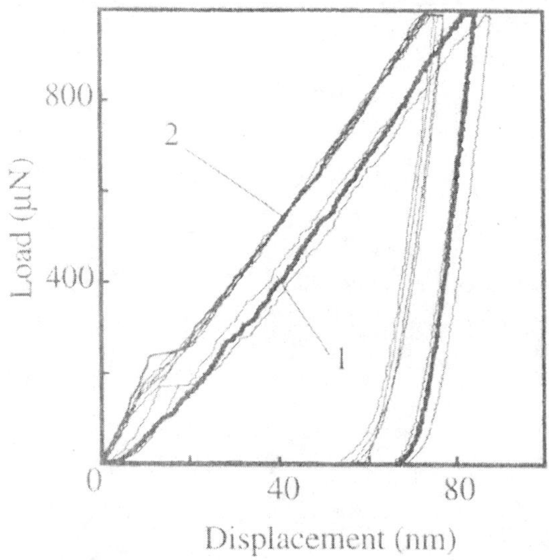

Figure 3: Indentation into non-charged samples. 1 and 2 correspond to arrays of indents in two different areas.

These were attributed to plasticity initiation since unloading just before the excursion load yielded no residual deformation. One example of hydrogen charged specimen is illustrated in the load displacement curve (Fig. 4). Here yield initiation occurred at 650-700 μ*N*. One day after charging the yield points ranged between 300-350 μ*N*. An interesting observation supporting hydrogen charging induced impediment of dislocation motion is that, the load-depth dependence after yield initiation appears to be different from charged samples as compared to non-charged ones. Regarding the scratch test, it appeared that hydrogen increased localized plasticity along a given slip band by as much as a factor of three. In principal, quantitative local strain (λ) arguments based on the surface slip height (h) and spacing (s) could be developed. Similar to what has been estimated as cumulative strain damage in fatigue such measurements of localized slips could be accomplished also here. These in fact are the first estimates of localized slip by probe microscopy as enhanced by hydrogen uptake. Thus, the following can be summarized at this stage. Surface modification due to environmental interaction in metastable austenitic stainless steel result at least from two origins.

First, strain due to phase stability aspects and secondly from hydrogen enhance localized plasticity that could be measured. Clearly, results like microcracking or other damaged sites introduce additional wear variations.

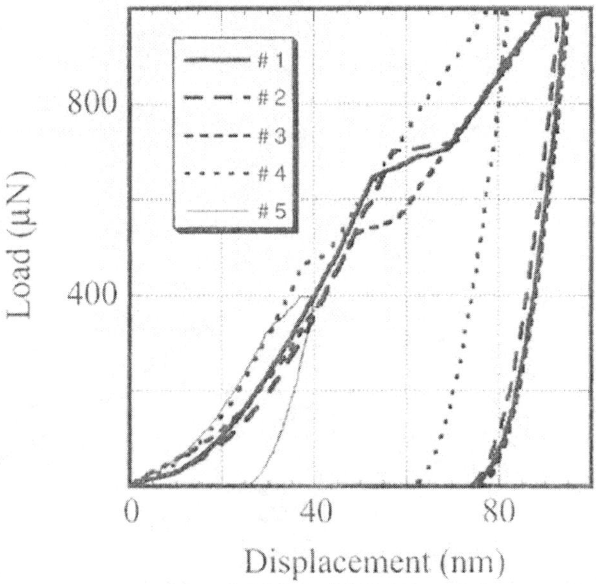

Figure 4: Nano indentation instantly after charging at high hydrogen concentration, time increases from 1 to 4 over 35min time frame.

4 Discussion and conclusion

In this paper the main attention remains in exploring the near surface modification affected by hydrogen/metal interaction. Even so, the connection to hydrogen embrittlment phenomenon activities becomes understandable. Hydrogen interaction often forms a near surface affected layer caused by sequential events that are manifested from localized irreversible displacements up to micro cracking onset. Microscopic observations have revealed consistently that austenite decomposition enhanced also by hydrogen beside mechanical driving force is accompanied by the formation of a characteristic surface relief. In this context, bands in certain crystallographic directions were developed shortly after hydrogen charging. At this stage bands were observed with faint appearance gradually becoming clearer with elapsing time. Further examination under polarized light has revealed the anisotropic nature of the aforementioned bands network. The bands size enabled to evaluate the magnitude of the hydrogen affected layer thickness in the order of 1.5μm. For specific hydrogen charging conditions displacements on one hand and micro cracking onset become important findings in terms of the surface characterization. On the ultra refined scale, two sources of information become quantitatively striking and could be briefly summarized. The nano indentation revealed major changes in the band behavior after hydrogen charging. In this case of 316L stainless steel, after low fugacity hydrogen charging, the initial yield increase is related to the increased hydrogen and its resultant stress gradient. It is still to be determined what the relative magnitude of the residual stress effect is compared to the hydrogen-dislocation interaction at the onset of plasticity. From the other source of the fine features observation namely the scratch test, it appeared that hydrogen increased the localized plasticity along a given slip band by as much as a factor of three. In principal, surface modification is based here on quantitative local plasticity arguments founded on measured localized strain. This was established by measured values of the surface slip height (h) and the slip spacing (s) as given in Table 1.

Table 1: Scratch test findings on plastic deformation.

<table>
<tr><th colspan="2">Surface features</th><th>Non-charged</th><th>charged</th></tr>
<tr><td rowspan="2">Along the pile-up</td><td>Slip step spacing (nm)(s)</td><td>101</td><td>291</td></tr>
<tr><td>Slip step height (nm)(h)</td><td>15</td><td>94</td></tr>
<tr><td rowspan="2">Perpendicular to the pile-up</td><td>Slip step spacing (nm)(s)</td><td>106</td><td>261</td></tr>
<tr><td>Slip step height (nm)(h)</td><td>13</td><td>16</td></tr>
</table>

This in fact is the first estimate of the localized strain ($\lambda \sim h/s$) assisted by micro probe capabilities, providing surface characterization that was modified by hydrogen uptake. By utilizing nano indentation a significant contribution can be achieved not only regarding the localized plasticity behavior but also to the excursions as related to dislocations dynamics bounded to the operating slip systems. The described surface modification clearly has connectivity to the mechanisms affecting the local friction and the wear of tribological contacts. Moreover, some of the experimental procedures alluded to new approaches to be applied in exploring asperity contact effects in an initial stage of wear. For example, Kubota et al [9] addressed the issue of fretting fatigue systems concluding the significant life decrease caused by hydrogen interaction. Accordingly the following is summarized and concluded:

Critical experiments might eventually provide the building blocks for modeling effort in order to simulate hydrogen/metal interaction. It becomes apparent that novel techniques regarding localized contact mechanics and fracture processes can synergistically evaluate modified surfaces caused by complex interaction. Thus:

(1) Macro and micromechanical testing of hydrogen/deformation interaction suggest mutual supplementary information as related to multi-scale modeling development.

(2) Hydrogen enhances near surface modification with implications to structural integrity aspects.
(3) Sorting out the origins for surface modification due to environmental interaction become possible.
(4) Applications that include in service segments in contact depend highly on surface characterization. An implication to fretting fatigue is only a selected example.

References

[1] Holzworth, M.I., & Louthan, M.R., J.Corrosion-NACE, 24, pp110-124, 1968.
[2] Mathias, H., Katz, Y., Nadiv. S., Sec Int. Cong' on hydrogen in metals. 2C11,Paris,pp6-11,1977.
[3] Gilad, I. & Katz. Y., Zeiteschrift fur physikalisch chemie, Genfold Bd, 1645,1989.
[4] Mathias, H., Katz, Y., Nadiv. S., Metal hydrogen sys, Veziroglu, T.N., (ed), Oxford pergamon press, pp 225-249, 1982.
[5] Lim, S.C., & Ashby, M.F., Acta metall, **35**,1,1987.
[6] Blau, P., Friction and wear transition of materials, Park Ridge, NJ, Noyes, 1989.
[7] Mathias, H., Katz, Y., Nadiv. S., Metal Sci, **12**, pp 129-137,1978.
[8] Gavrilijuk, V.G., Hanninen, H., Tarasenko, A.V., Terechenko, A.S., & Ullakko, K., Acta metall mater, 43,559,1995.
[9] Kubota, M., Noyama, N., Fuata, N., Sakae, C., & Kondo, Y., J. of the soc. of mater. Sci., Japan, **54**, pp1231-1236, 2005.

Nanoscratch evaluation of adhesive strength of Cu/PI films

K. Tanaka, K. Gunji & T. Katayama
Department of Mechanical Engineering, Doshisha University, Japan

Abstract

A scratch test has provided a simple, rapid means of assessing the adhesion strengths of thin films on substrates. However, it is not standardized how to detect the interfacial failure. Copper thin films deposited on Polyimide (PI) substrates are used for Flexible Print Circuit (FPC). It is needed to improve and measure the adhesive strength of Cu films on PI substrates to avoid interfacial fractures caused by cyclic bending. In this study, in order to improve the adhesive strength, the surfaces of PI substrates were modified by ion beam using argon or oxygen with a linear ion source and the adhesion strengths of Cu films on PI substrates were measured using nanoscratch tests. Cu thin films were deposed by Physical Vapor Deposition (PVD) using an electron beam. The friction coefficient rapidly decreased and normal displacement rapidly increased at the delaminated point at the same time. The slope of the friction coefficient increased at the delaminated point, whose normal load is called the critical load. The scratched surfaces were observed from films (Cu) side and substrate (PI) side. A transparency of PI substrates allowed us to observe the interfaces between Cu film and PI substrate. The scratched mark observed from substrate side started at just the point where the slope of the friction coefficient changed. As the roughness of the surface of polyimide substrates increased, critical load increased.
Keywords: nanoscratch test, adhesive strength, copper films, polyimide substrates.

1 Introduction

There are a lot of methods to measure the adhesive strength of interfaces between coatings and underlying substrates, which include pull-off test, peel test,

WIT Transactions on Engineering Sciences, Vol 55, © 2007 WIT Press
www.witpress.com, ISSN 1743-3533 (on-line)
doi:10.2495/SECM070291

four point bending test, scratch test and indentation test. The scratch adhesion test has provided a simple, rapid means of assessing the adherence of coatings. The mechanism or stress field of scratch test have been studied by Bull et al [1] and many researchers [2-4]. Scratch test is the test by drawing a diamond indenter along the coated surface. The applied normal load is increased continuously until delamination is occurs. This critical normal load is considered as a semi-quantitative measure of the coating-substrate adhesion. There are several ways of detecting the occurrence of delamination. Mutoh et al [5] used Acoustic Emission, Shibutani et al [6] used microscopes, and Li and Bhushan [7] used the coefficient of friction to detect the delamination. Observation of scratched surface can detect the delamination with fractures of films. In order to detect the initial fracture in interfaces between films and substrates, it is very important to observe the behavior of friction coefficient. Therefore, it is necessary to clarify the relationship between fracture mechanism and the changes of friction coefficient.

Copper (Cu) films prepared on Polyimide (PI) substrates are used in electronics field, such as, for the Flexible Print Circuits. The interfacial fractures caused by cyclic bending have been a serious problem. To obtain the reliable devices, it is needed to improve and measure the adhesion strength of films on substrates.

In this paper, nanoscratch test was used to measure the adhesive strength of Cu thin films prepared on PI substrates by Physical Vapor Deposition (PVD) and the influence of modification on PI surface by ion beams using argon or oxygen with linear ion source was discussed.

2 Experimental details

2.1 Specimens

The substrates used for this study are Kapton® EN films (thickness: 50μm) manufactured by Du Pont. This film is the PMDA-ODA type polyimide, which is optimized for direct coating without adhesive. Prior to the deposition process, surfaces of the substrates were modified by ion beam modification system developed by Ektessabi et al [8]. The system using a linear ion source, which has a couple of electrodes, enables us to modify large surface area. Cu thin films were prepared on as-received and ion-irradiated surfaces of PI substrates by PVD, whose thicknesses were set for 170μm and 350μm. Oxygen and Argon ions were irradiated to the surfaces of PI substrates. The accelerating voltage was between 750V to 1500V. The roughness of as-received and ion-irradiated surfaces was measured by Atomic Force Microscope (AFM, Veeco Instruments, NanoScope3). The sample names, ion accelerating voltage, names of irradiated ions and the RMS roughness of the surfaces of PI substrates are shown in table 1.

2.2 Nanoscratch test

The adhesive strengths of Cu films on PI substrates were measured using nanoscratch tests. The nanoscratch tests were performed by a Triboindenter®

(Hysitron Inc.). A triangular pyramid diamond tip of 90° angles was used for scratch tests in the face forward direction and edge forward direction as shown in Figure 1.

Table 1: Sample name, ion accelerating voltage, ion, RMS roughness.

Sample Name	Ion Accelerating Voltage (V)	Ion	RMS Roughness (nm)
As received	-	-	1.6
Ar1500	1500	Argon	1.3
Ar750	750	Argon	2.6
O1500	1500	Oxygen	0.7
O1250	1250	Oxygen	0.9
O1000	1000	Oxygen	1.2
O750	750	Oxygen	2.8

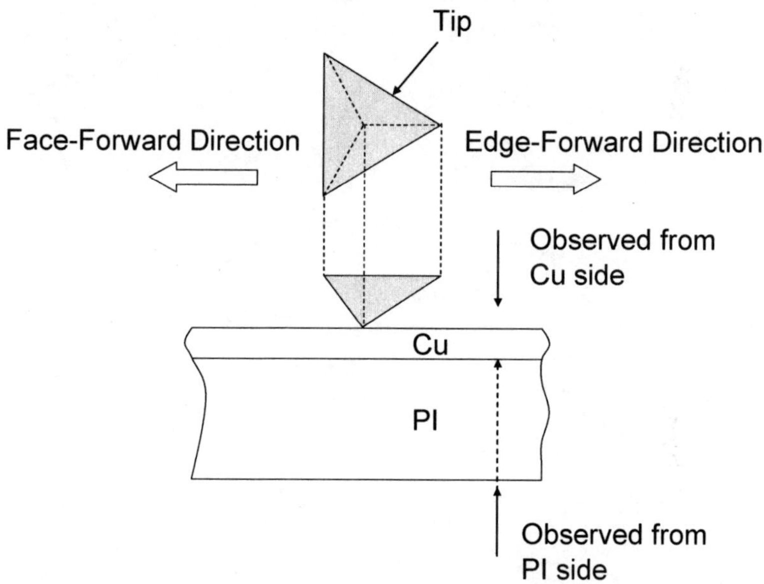

Figure 1: Nanoscratch direction and observe side.

The film surfaces of 170μm in thickness were scratched in the face forward direction (which will be called 'thin-face test') and edge forward direction (thin-edge test) at a scratch speed of 0.1μm/s. The film surfaces of 350μm in thickness were scratched in the edge forward direction at the same speed (thick-edge test). The applied normal load was increased continuously from 0μN to 1000μN for thin-face test and thin-edge test, and 0μN to 700μN for thick-edge test. The scratched surfaces were observed with Scanning Electron Microscope (SEM) and optical microscope from both film and substrate side to observe the interface between Cu films and PI substrates as shown in Figure 1.

3 Results

Figures 2 and 3 show typical scratched surfaces observed by SEM and the relationship between the friction coefficient and normal load for thin-face test and thin-edge test. In the initial stage, the friction coefficient remained constant with some amplitude of vibration (stage 1). After that, the slope of friction coefficient changed and the friction coefficient increased linearly (stage 2). Then the slope of friction coefficient changed to smaller value than that of stage 2 (stage 3). Finally, the friction coefficient rapidly increased (stage 4) in the thin-face test. From the SEM observation, neither cracks nor fracture occurred during stage 1 and stage 2. Then the crack initiated during stage 3. At the beginning of stage 4, the indentation tip reached to the PI substrate and the Cu film peered off the substrate in the thin-face test.

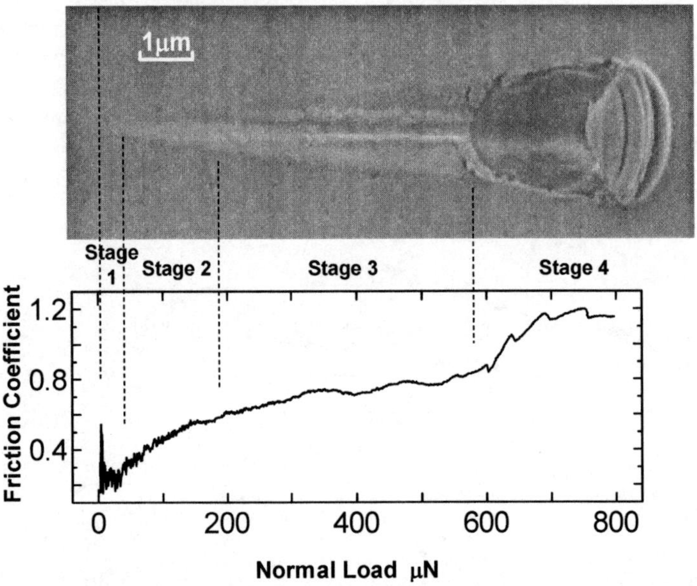

Figure 2: Typical SEM image and relationship between normal load and friction coefficient of thin-face test (as-received).

Figure 4 shows a typical result for thick edge test. The observed photos by optical microscope from Cu and PI side are also shown in Figure 4. In the initial stage, the friction coefficient remained constant with some amplitude of vibration until the normal load reached about 250μN (stage 1) and dropped suddenly. After that, the slope of friction coefficient changed and the friction coefficient increased linearly (stage 2). From the SEM observation, neither cracks nor fracture occurred during stage 1 and stage 2. The scratch damage observed from PI side occurred at the beginning of stage 2. Figure 5 shows another result for thick edge test. For this specimen, sudden drop of the friction coefficient between stage 1 and 2 was not clear, but after stage 1, the slope of friction coefficient changed and the friction coefficient increased linearly (stage 2), showing the same behaviour described for Figure 4.

4 Discussions

Schematic drawings of four stages described above are shown in Figure 6. For all the tests during both stage 1 and stage 2, no cracks or no fractures of films were observed. The scratch damage observed from PI side with optical microscope started at the beginning of stage 2. The observation from PI side allowed us to observe the interface between films and substrates because of a transparency of PI substrates. This indicates that the change of friction coefficient between stage 1 and stage 2 shows the occurrence of interfacial fractures. For thin-face test and thin-edge test, cracks of films were observed during stage 3. It is indicated that this cracks released energy and had the slope of friction coefficient decreased. For thin-face test, fractures of films and interfacial fractures between Cu films and PI substrates were observed in stage 4. For thin-edge test, there was no boundary between stage 3 and stage 4 in the relationship between normal load and friction coefficient. This is because the edge of the tip cleaved the PI substrates for thin edge test. On the contrary, for thin edge test, the film fractured in buckled in stage 4.

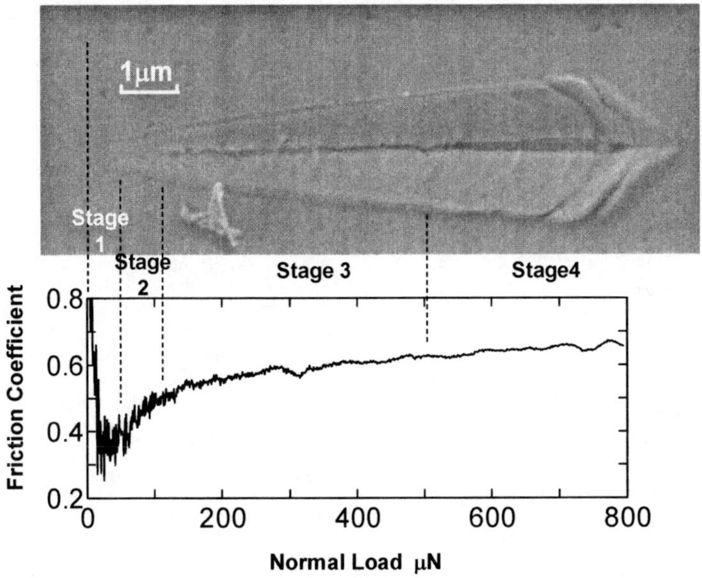

Figure 3: Typical SEM image and relationship between normal load and friction coefficient of thin-edge test (as-received).

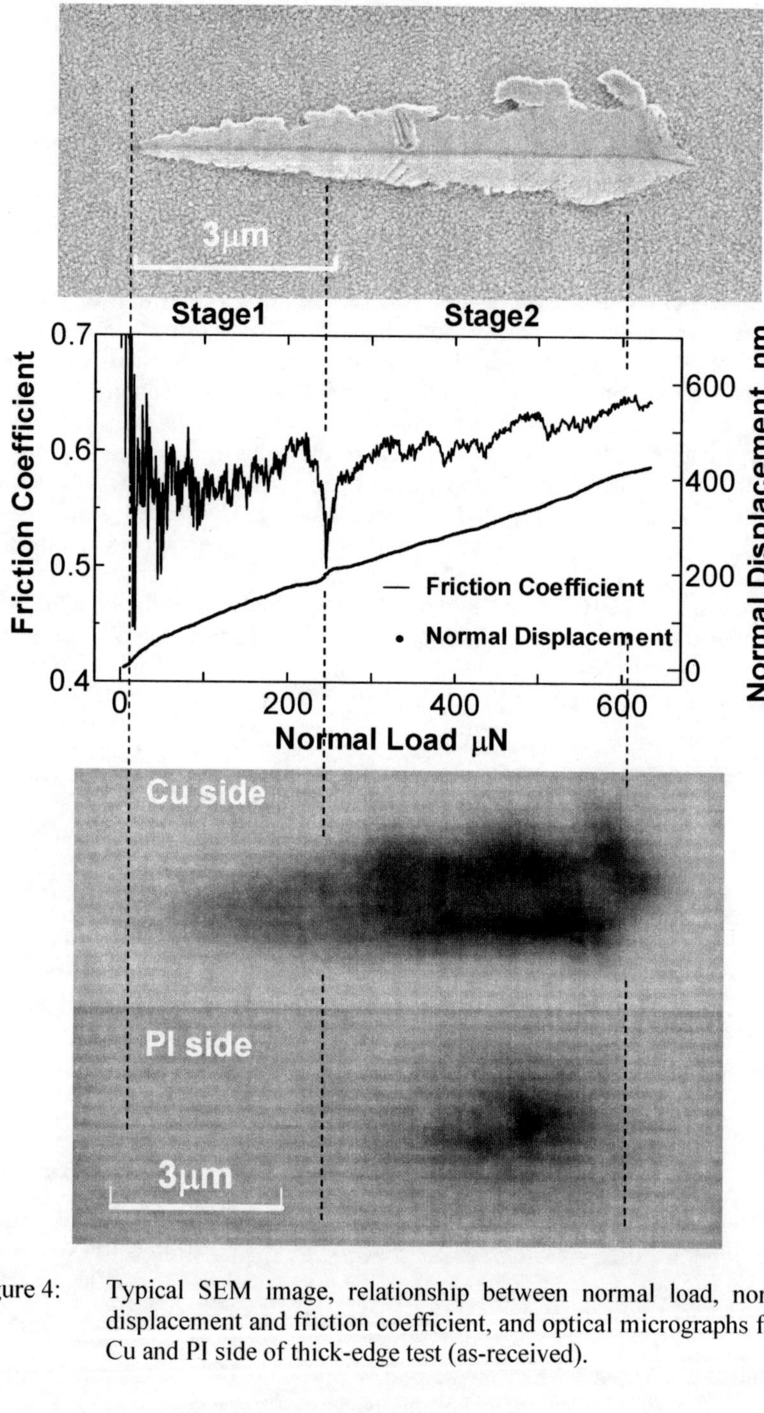

Figure 4: Typical SEM image, relationship between normal load, normal displacement and friction coefficient, and optical micrographs from Cu and PI side of thick-edge test (as-received).

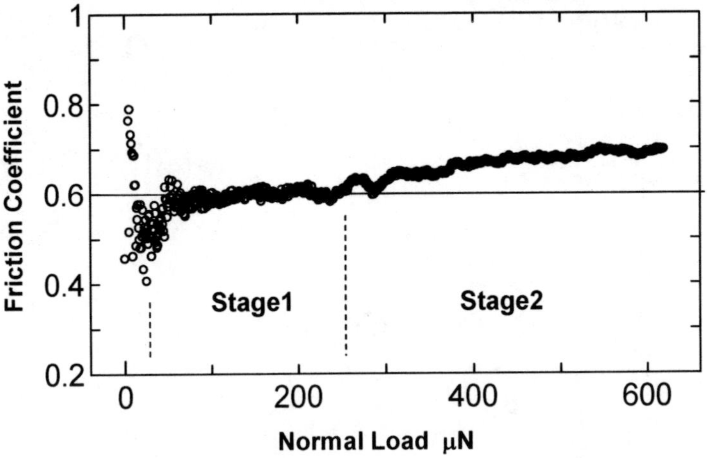

Figure 5: Typical relationship between normal load and friction coefficient.

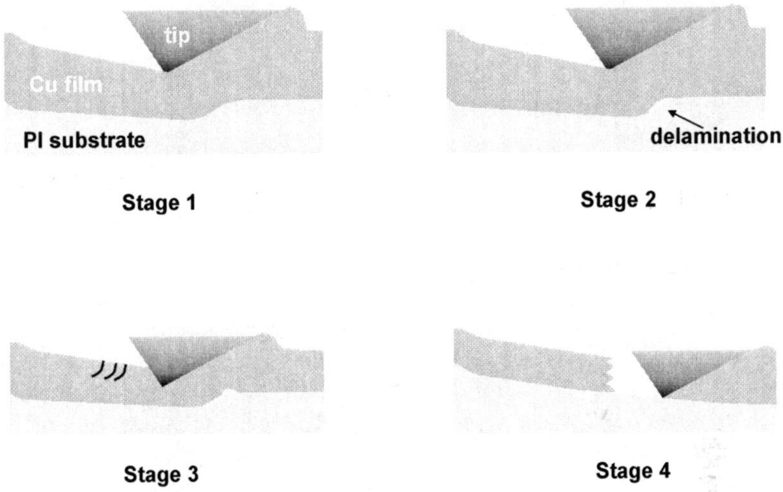

Figure 6: Schematic drawings of four stages.

Considering the discussion above, the normal load at the end of stage 1 can be considered as a critical normal load, a semi-quantitative measure of the film-substrate adhesion. Figure 7 shows the relationship between the critical normal loads and the RMS roughness of un-modified and modified PI substrates measured by thick edge test. The critical normal loads correlate with the RMS roughness of PI substrates. This indicates that the adhesive strength between films and substrates correlate with the roughness of the substrates due to anchor effects. The correlation between the roughness of the surface of the substrates and the adhesive strength was also indicated by Ge et al [9].

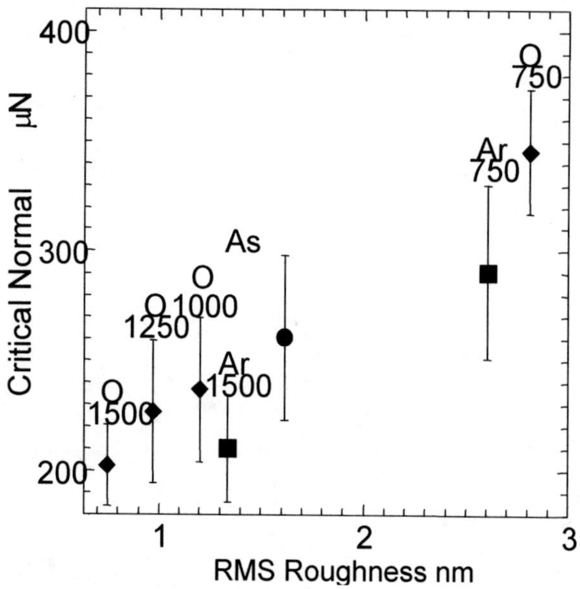

Figure 7: Relationship between roughness of the substrates and critical normal loads.

5 Conclusions

By analyzing the scratch processes of Cu films on PI substrates, it was found that four stages occur as the normal load increases during the scratch test; plastic deformation of Cu films (stage 1), interfacial fractures without fractures of films (stage 2), cracks in films (stage 3), fractures of films and delamination caused by the tip reaching to the PI substrates (stage 4). In the first stage, the friction coefficient remained constant. In the second stage, the slope of friction coefficient increased and the friction coefficient increased linearly. In the third stage, the slope of friction coefficient changed to smaller value than that of the second stage. In the last stage, the friction coefficient rapidly increased because of the scratching in face-forward direction. For the scratch tests in edge forward direction, there was no boundary between the third and the last stage in the relationship between the normal load and the friction coefficient. In our study, the adhesive strength between films and substrates correlated with the roughness of the substrates.

References

[1] Bull, S.J., Rickerby, D.S., Matthews, A., Leyland, A., Pace, A. R. and Valli, J., 'The Use of Scratch Adhesion Testing for the Determination of

Interfacial Adhesion: The Importance of Frictional Drag', Surface and Coatings Technology, 36, pp503-517, 1988.

[2] Holmberg, K., Laukkanen, A., Ronkainen, H. and Wallin, K., 'Tribological analysis of fracture conditions in thin surface coatings by 3D FEM modeling and stress simulations', Tribology International, 38, pp.1035-1049, 2005.

[3] Damayanti, M., Widodo, J., Sritharan, T., Mhaisalkar, S.G., Lu, W., Gan, Z.H., Zeng, K.Y. and Hsia, L.C., 'Adhesion study of low-k/Si system using 4-point bending and nanoscratch test', Materials Science and Engineering, B121, pp.193-198, 2005.

[4] Ye, J., Kojima, N., Ueoka, K., Shimanuki, J., Nasuno, T. and Ogawa, S., 'Nanoscratch evaluation of adhesion and cohesion in SiC/low-k/Si stacked layers', Journal of Applied Physics, vol.95, no.7, pp.3704-3710, 2004.

[5] Mutoh, Y., Xu, J., Miyashita, Y., Kuroishi, T. and Sasaki, Y., 'On Evaluation of Adhesive Strength in Scratch Test of Coating Materials.' Transactions of the Japan Society of Mechanical Engineers A68-670, pp. 909-915, 2002.

[6] Shibutani, T., Yu, Qu., Shiratori, M. and Akai, T., 'Mechanism of Damage Process on Interface between Films in Nanoscratch Test', M&M Shinshu Spring Symposium, No.05-03, pp.43-46 2005.

[7] Li, X. and Bhushan, B., 'Micro/nanomechanical and Tribological Characterization of Ultrathin Amorphous Carbon Coatings', Journal of Materials Research, Vol.14, No.6, pp.2328-2337 1999.

[8] Ektessabi, A. I., Yasui, N. and Okuyama, D., 'Characteristics of an Ion Beam Modification System with a Linear Ion Source', Review of Scientific Instruments, Vol.73, No.2, pp873-876 2002.

[9] Ge, J., Turunen, M.P.K. and Kivilahti, J.K., 'Surface modification and characterization of photodefinable epoxy/copper systems', Thin Solid Films, 440, pp198-207, 2003.

Section 7
New applications

Parametric simulation of SiC Schottky JBS structures

T. Rang & R. Kurel
Department of Electronics, Tallinn University of Technology, Estonia

Abstract

The parametric simulation has been carried out for optimizing the influence of the Schottky contact metal work function (contact surface interface) influence on the distribution of the built-in electrical field and consequently on static and dynamic characteristics of the JBS device. On the basis of simulations and discussions the low-power losses solution for the JBS device has been developed. The results suggest keeping the electric field strength under contact surface as low as possible in order to reduce or even avoid the relatively expensive and complex passivation solutions by the manufacturing of the JBS devices.

The numerical simulator DYNAMIT 2DT-SCHOTTKY developed at the Department of Electronics TUT was used in our research.
Keywords: SiC, JBS structures, contact metal work function, numerical modelling.

1 Introduction

Silicon carbide (SiC) is an outstanding compound semiconductor material with extremely promising physical properties that make it an excellent candidate in high-speed and high-temperature power electronic applications. Metal semiconductor interface is a fundamental aspect in any semiconductor device technology. The research of different characteristics of Schottky contact involved semiconductor structures is important. Furthermore, the 4H- and 6H- polytypes of SiC have some remarkably different parameters, which have direct influence on device characteristics.

The authors of this paper have studied devices with different Schottky barriers for a long time. The changes in barrier heights strongly influence the

WIT Transactions on Engineering Sciences, Vol 55, © 2007 WIT Press
www.witpress.com, ISSN 1743-3533 (on-line)
doi:10.2495/SECM070301

current transport in Schottky interfaces and the investigation results are shown for example in [1, 2]. There are many aspects in simple Schottky devices, which are similar to ones in JBS (e.g. current crowding phenomenon). From another point of view the JBS devices have specific advantages compared to traditional simple *pn-* or Schottky devices. Many experimental results dealing with simulation of JBS devices have been published (e.g. [3-8]). Almost all simulation reports discuss only about the best relation of *pn-* and Schottky areas, and only some of the reports deal with the different Schottky contact properties (barrier height, size), placing of implanted regions, geometrical dimensions and doping concentration of drift region. There have been no papers about JBS devices based on *p*-type SiC. The processes inside of JBS devices have been poorly investigated. The electric field strength distribution and minority carrier distribution in the devices have been studied. Those studies have been done only for certain parts inside the device and in one dimension. In this paper we make an attempt to bridge this gap in some specific aspects.

The simulations have been done with simulation software SIC-DYNAMIT-2DT SCHOTTKY, developed at TUT Department of Electronics [9, 10], aided by bash shell scripts for automation of simulations and gnu-plot for presenting the results.

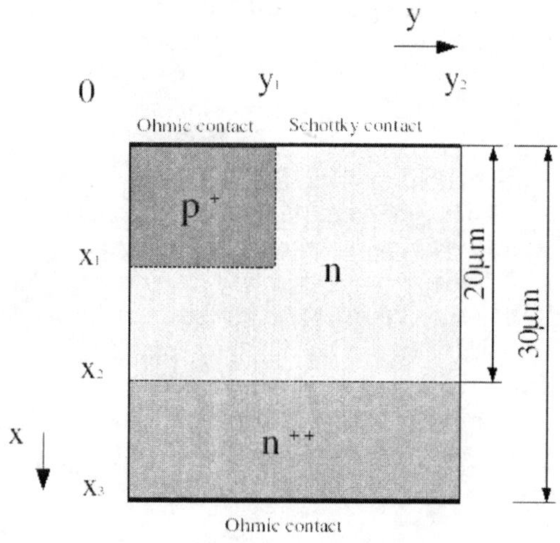

Figure 1: The device under study.

2 Description of the model

The simulations were performed on 6H-SiC and 4H-SiC Junction Barrier Schottky (JBS) structures (Fig. 1). Its dimensions and parameters were changed repeatedly to study the influence of those changes to electrical parameters and characteristics of the device.

The parameters of device are: width of the device (y_2) varies between 1 and 15 μm, width of p^+ area (y_1) between 0.25 and 5 μm, depth (x_1) is chosen to be between 1 and 20 μm and acceptor doping concentration 5×10^{16} cm^{-3}, epitaxial layer (*n* area) depth (x_2) 20 μm with donor doping concentration between 1×10^{15} and 4×10^{15} cm^{-3}, part of the substrate (n^{++} area) depth (x_3–x_2) 10 μm with donor doping concentration 1×10^{19} cm^{-3}, Schottky contact metal work function varies between 5.1 and 5.7 V, and ambient temperature gap lies between 300 and 900 K. The area of the device was chosen to be always 1 cm^2, to equal current and current density. Complementary *p* semiconductor based device data correspond to previously described ones.

Avalanche ionization coefficients of SiC semiconductor were taken from [11].

The Shockley–Roosbroeck system containing Poisson's equation, charge carriers continuity and transport equations with additional Maxwell's total current equation is used. The Electron Hole Scattering (EHS) effect is introduced in *SIC-DYNAMIT-2DT* simulation package. Additional coefficients are used for high electron and hole concentrations. Mobility dependence on doping concentration, temperature, electrical field, and on the concentration of charge carriers is used. Boundary conditions for Ohmic and Schottky interfaces are described as it is shown for example in reference [10].

The detailed description of the model used here is presented in [12], the appropriate model parameters are presented in our latest paper on this topic [13], and therefore these questions will be not discussed here in more detail.

3 Results and discussion

Many practical examples of SiC JBS devices have been introduced. Still it is difficult to find the comparable examples, where the experimental results of manufactured devices match well with the simulation results, particularly with the results of our simulations. The reason for that is rather simple. The so called best device defined on base of our investigations should have very deep emitter area, which is unfortunately very difficult to realize with today's SiC device manufacturing technologies. So, our results presented here have therefore the meaning of the so called best theoretical device. Of the physically manufactured devices, the closest to our results is the one proposed by Rutgers University, USA. The experimental results are presented in the paper [14]. This device is to the best of our knowledge the demonstrator with the best electrical properties presented till today. The geometrical and electrical parameters of this device are presented in [14] as well. To validate our model with best experimental data some critical results from the point of view of device behavior were investigated earlier and the part of comparable results are presented in our earlier work [12]. Here we continue to report our latest simulation results.

Firstly, we comment the full dimensions of the JBS device. The forward characteristics depend not only on relative dimensions (ratio of *pn*-junction area to the Schottky junction area) of the device, but also on the absolute dimensions of the device parts as well. The device with the smallest dimensions can conduct much lower forward current compared with other examined devices. The reason stems directly from the screening behavior of Schottky part (about 0.66 μm) of the device by the *pn*-part. The similar situation is followed with applied reverse voltage too, which means that the *pn* part of the device defines the current at low voltages (Fig. 2). It is important to state that at the smallest ratio of the Schottky and *pn*–junction the Schottky area leakage current is almost completely blocked under 1000 V and JBS device conducts the leakage current of *pn*-junction. The lower breakdown voltage in case of the smallest device (about 15 μm) is caused by higher electric field strength in semiconductor volume near the emitter bottom region (Fig. 3). This is clearly seen on Fig. 4, where the electric field strength dependence on device dimensions is presented. It is seen that in case of the highest simulated dimensions the effect of the lowering electric field strength under Schottky contact is very weak. Similar behavior has been seen also in the case of the smallest devices (about 3 μm). The latter situation concludes from the limited propagation of depletion region of *pn*-junction.

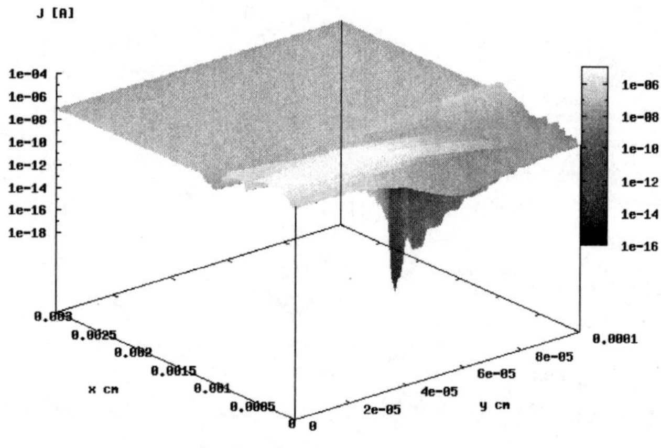

Figure 2: Current crowding to *p* area at U_F=0.2V, device size 1 μm.

Our simulations show clearly that the turn-off time depends weakly on device final dimensions. The small difference in turn-off time values seems to be caused by the ratio of the depleted region to conducting area of Schottky part. Depleted region edge around the emitter area is the same for all dimensions, but the Schottky area dimensions are changing. From this situation we conclude that although the Schottky area ratio for the whole device area is not changing, the conducting Schottky area ratio to the whole device area is higher at larger device dimensions and the turn-off time increases minimally (less than 3% compared to shortest turn-off times of the device).

The next important topic is the variation of the Schottky contact metal work function and its influence on forward and reverse characteristics, and on the distribution of the electrical field strength and the turn off time. The unique *p*-substrate structure is under the investigations. The behaviour of the forward and reverse characteristics is shown in Fig. 5. The work function has remarkable influence on forward characteristics (Fig. 5) through influence on the height of Schottky barrier. The influence is clearly seen at low voltage values as then the majority of the total current is conducted by the Schottky part of the device. On higher voltages most of the current flows through the *pn*-part of the device and the influence of the Schottky barrier loses its importance. Also the influence of the work function on reverse characteristics (Fig. 6) is remarkable. Namely, the lower Schottky contact metal work function causes higher leakage current values in the Schottky part of the device and therefore in the whole device as well.

On the base of the results of the simulations we state that the electric field strength distribution and the turn-off time of the device are almost not influenced by the contact metal work function.

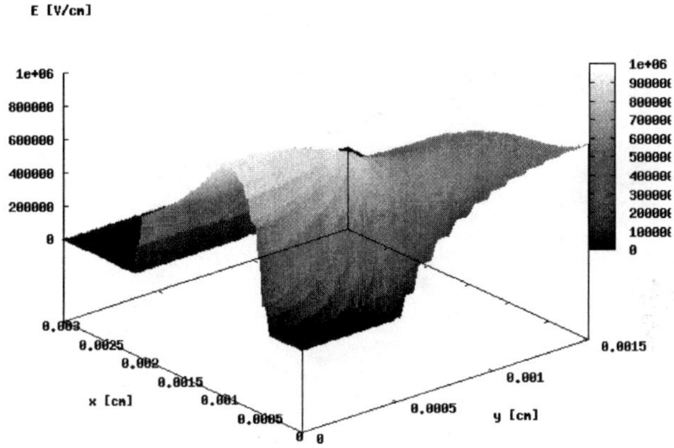

Figure 3: Electric field strength distribution at U_R=1000 V, device size 15 µm.

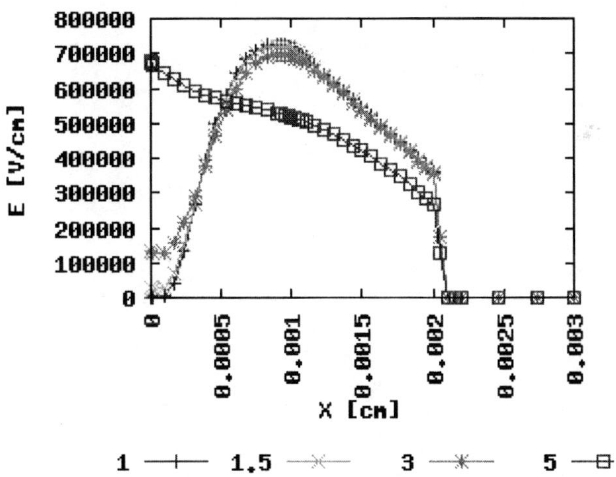

Figure 4: Electric field strength distribution dependence on device dimensions [µm].

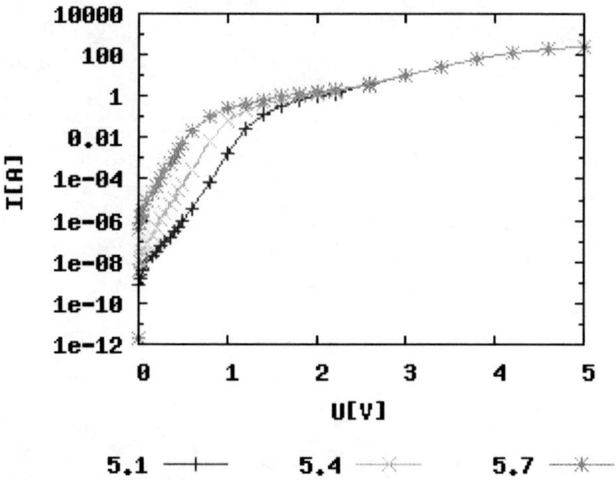

Figure 5: Forward characteristics dependence on contact metal work function [V] for *p*-substrate.

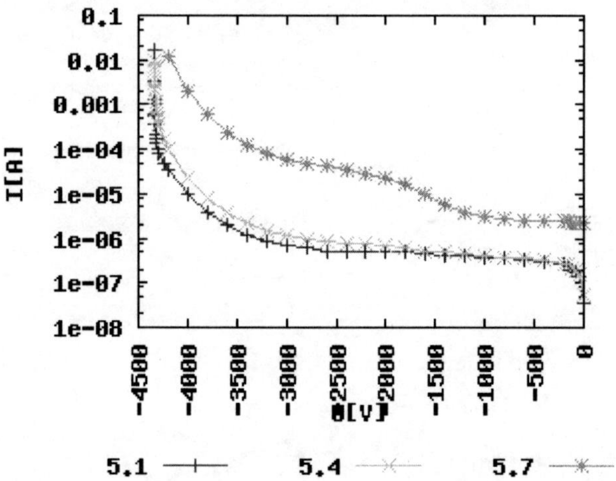

Figure 6: Reverse characteristics dependence on contact metal work function [V] for *p*-substrate.

The next analysis compares the Schottky JBS device with the pure *pn*- and Schottky diodes. Concluding from the working principles of JBS device the forward and reverse characteristics of JBS device are laying between the characteristics of pure Schottky diode and *pn*-diode, but on lower forward voltages values only Schottky part of the device conducts and on higher forward voltages only the *pn*-part of the device conducts. In case of reverse characteristics the breakdown voltage for *pn*-diode is lower compared to JBS device. The reason, why such a phenomenon takes place is that the dimension of the region, where the electric field strength is distributed is very narrow (only about 10 μm) and therefore the break down arrives earlier compared to classically possible situation. This statement is partly supported by the analysis of electric field strength distribution inside the device, where the electric field strength is distributed over a wider area than in case of *pn* diode, but its maximum value is still low and situated under surface. In switching processes the analysis of currents shows clearly that the Schottky part of the device acts against the *pn*-part of the device for a short period of time. Although eventually the turn-off time for Schottky structures is generally shorter than for *pn*-diode, it is important to stress that after as short as 30 ps the current of Schottky diode is already about 8 times smaller than of *pn*-diode, and 90% of reverse voltage is achieved already in 23 ns in case of Schottky diode. For the *pn*-diode this time period is about 600 ns.

Finally, we make some comments about the situation with *p*-substrate versus *n*-substrate. The forward current level for *n*-substrate device is about three orders of magnitude higher almost over the whole forward voltage region. The current values difference at low forward voltages is explained mostly by the lower barrier height of *n*-4H-SiC (about 1.1 eV) compared to *p*-4H-SiC (about 1.4 eV). We have to stress that the metal work function is taken to be the same for both cases. On higher forward voltages the majority of current flows through the *pn*-junction part of the device defining the current level differences for these situations. On reverse voltages the device with *n*-substrate has about two and a half orders of magnitude higher leakage current compared with the device with *p*-substrate. Similarly to the forward bias situation the low reverse voltage situation is also caused by the clear difference in barrier heights for *p* and *n*-type material (Schottky contact metal work functions have equal values for both cases).

The maximum value of the electric field strength is lower in case of *n*-substrate. The reason concludes from the slightly different dimensions of the width of the device's emitter region (in our simulations 1 μm for *p*-substrate JBS device). The turn-off time of *n*-substrate device is about 4 times higher compared to *p*-substrate device (Fig. 7). The reason for this significant difference stems directly from the differences of the values of holes and electron mobility's. For 4H-SiC the mobility of holes is almost 10 times lower than the mobility of electrons.

4 Conclusions

In this paper we have introduced some very new and original simulation results concerning JBS devices. First of all the parametric simulation for determining the best dimensioned device was discussed. After that the original results of *p*-substrate devices taking into account the Schottky contact metal work function and different SiC polytypes have been presented. There is still no *p*-substrate based JBS device examinations published. Finally, the comparison *p*-substrate versus *n*-substrate was presented. Our simulations revealed that *p*-substrate JBS devices have generally no substantial advantages over *n*-substrate devices, but in some particular categories we met strong advantages over *n*-substrate devices.

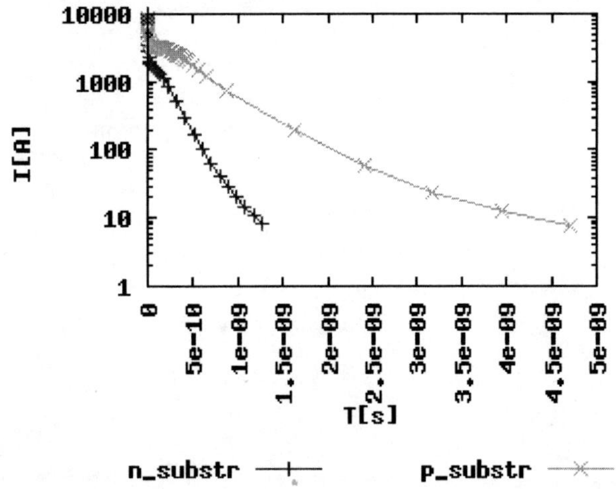

Figure 7: Turn-off time characteristics *n*- versus *p*-substrate.

The new findings we want to stress are the following:

- The deepness of *p*-region is the most important dimension in forming the shape of electrical field.
- Electrical field distribution is important in three aspects: maximum electric field strength determines avalanche breakdown of device, electric field strength under Schottky contact has influence on reverse current through Schottky barrier lowering, and the electrical field near the device surface has great importance on device surface breakdown phenomenon.

The dimensioning of the JBS devices results in the conclusion that there never exists the so called best device for all the conditions. The device has to be chosen keeping in mind mainly low-power losses in high power applications. It is an extremely important from the point of view of manufacturing technologies. The goal is to reduce or to keep the electric field strength under surface very low using the high quality and complex passivation solutions.

The model used in simulations does not include the description of the tunneling mechanisms. Tunneling can influence in particular situations the behavior of the device remarkably. Therefore additional investigation of JBS device behavior with inclusions of tunneling into the model would be an interesting subject for future investigations.

Acknowledgements

The authors wish to thank the Estonian Science Foundation for the support of this research through the research grant G5901. We would also like to thank Dr. Andres Udal for developing *SiC-DYNAMIT-2DT* modelling software used here as a simulating instrument.

References

[1] Kurel, R., Rang, T.: Ambient temperature influence on current suppressing effect caused by self-heating phenomenon at 6H-SiC Schottky interfaces. Proc. of the BEC'98, Oct.7-10, Tallinn, Estonia, 1998, 253-254.

[2] Kurel, R., Rang, T.: Self-heating phenomenon and current suppressing effect at 6H-SiC Schottky interfaces. Proc. of 5th NEXUSPAN Workshop on Thermal Aspects in Microsystem Technology. May 6-8, Budapest, Hungary, 1998, 88-91.

[3] L. Zhu, M. Shanbhag, T.P. Chow, K.A. Jones, M.H. Ervin, P.B. Shah, M.A. Derenge, R.D. Vispute, T. Venkatesan, A. Agarwal, 1kV 4H-SiC JBS Rectifiers Fabricated Using an AlN Capped Anneal, Trans Tech Publications, Materials Science Forum Vols. 433-436 (2003), 2003.

[4] R. Singh, D.C. Capell, A.R. Hefner, J. Lai, J.W. Palmour, High-Power 4H-SiC JBS Rectifiers, IEEE, IEEE Transactions on Electron Devices, Vol. 49, No. 11, November 2002, 0018-9383/02, 2002

[5] R. Pérez, N. Mestres, D. Tournier, X. Jordà, M. Vellvehí, P. Godignon, Temperature Dependence of 4H-SiC JBS and Schottky Diodes after High Temperature Treatment of Contact Metal, Trans Tech Publications, Material Science Forum Vols. 483-485 (2005), 2005

[6] K. Asano, T. Hayashi, R. Saito, Y. Sugawara, High Temperature Static and Dynamic Characteristics of 3.7kV High Voltage 4H-SiC JBS, Power Semiconductors Devices and ICs, 2000. Proceedings. The 12th International Symposium, 2000.

[7] K. Rottner, M. Frischholz, T. Myrtveit, D. Mou, K. Nordgren, E. Henry, C. Hallin, U. Gustafsson, A. Schöner, SiC power devices for high voltage applications, Elsevier Science, Material Science and Engineering, Vol. B61-62, 1999.

[8] T. Ayalew, S. C. Kim, T. Grasser, S. Selberherr, Numerical Analysis of SiC Merged PiN Schottky Diodes, Trans Tech Publications, Materials Science Forum, Vol. 483-485, 2005.

[9] Udal, A.: "Development of Numerical Semiconductor Device Models and their Application in Device Theory and Design", Theses of Tallinn Technical Univ., No.D12, ISBN 9985-59-092-9, 1998, 140 p.

[10] Kurel, R., Udal, A.: Two-dimensional nonisothermal analysis of current crowding effect at nonuniform SiC Schottky contacts using device simulator DYNAMIT-2DT. Proc. of the 8th Baltic Electronics Conference. Oct 6-9 Tallinn, Estonia, 2002, 51-54.

[11] Chow, T.: High voltage SiC power devices, In "High-Temperature Electronics in Europe, International Technology Research Institute", 2000, 99-152.

[12] Kurel, R.: Investigation of Electrical Characteristics of SiC Based Complementary JBS Structures, Doctoral Theses, Tallinn University of Technology, 2005, 98 pgs.

[13] Kurel, R., Rang, T.: Static and dynamic behavior of the SiC complementary JBS structures. Proc. of the 10th Baltic Electronics Conference. Oct 2-4 Tallinn, Estonia, 2006, 59-62.

[14] Wu, J., Fursin, L., Li, Y., Alexandrov, P., Zhao, J. H.: 4,308V, 20.9mO cm² 4H-SiC MPS Diodes Based on a 30μm Drift Layer, Silicon Carbide and Related materials 2003, Eds. Roland Madar, Jean Camassel, Elizabeth Blanquet, Trans Tech Publications, 2004, 1109-1112.

Expansion of capillary force range by probe-tip curvature

K. J. Obata[1], S. Saito[2] & K. Takahashi[1]
[1] *Department of International Development Engineering, Tokyo Institute of Technology, Japan*
[2] *Department of Mechanical and Aerospace Engineering, Tokyo Institute of Technology, Japan*

Abstract

This paper deals with the expansion of capillary force range with a concave probe-tip for micromanipulation. From numerical simulation, we found the following; the concave probe-tip can generate a much larger capillary than a flat one, provided it is designed to fit the convex surface of the object; the more wettable a material is, the greater capillary force it can generate; the magnified capillary force can be reduced/controlled by liquid volume regulation. To prove this, we measured the capillary force for a given gap distance between a spherical object and a concave surface coaxially fabricated in a cylinder. We used three different materials (glass, stainless steel, and polytetrafluoroethylene) to check the influence of contact angles. The liquid volumes were given in the range from one hundredth- to ten-times the radius of the cubed sphere. Comparison between our experimental data and the theoretical prediction expressed in the normalized form shows good agreement, if the liquid volume is larger than a certain value. This suggests that micromanipulation by capillary force should be more practical by using probes with concave tips specifically designed for the object.
Keywords: micromanipulation, hydrostatics, liquid bridge, capillary force.

1 Introduction

Recently, micromanipulation techniques have been in demand to fabricate highly functional micro-devices or micro-electro-mechanical-systems (MEMS). In micromanipulation, the influence of adhesional force is extremely large compared to gravitational force [1, 2]. Furthermore, adhesional force has large dispersion

WIT Transactions on Engineering Sciences, Vol 55, © 2007 WIT Press
www.witpress.com, ISSN 1743-3533 (on-line)
doi:10.2495/SECM070311

because of its dependence on surface condition such as surface roughness, at each contact point. Thus, in order to realize reliable micromanipulation, we need a force that is controllable and greater than the adhesional force, or some mechanism to avoid the dispersion of adhesional force. Saito *et al* [3, 4] have investigated the mechanical force required to slip and roll an object in considering the adhesional effect. The mechanical method, however, might damage the object. Takahashi *et al* [5] have evaluated the force generated by Coulomb interaction, although the electrostatic method might cause a discharge or melt an object [6]. In order to avoid damaging the manipulated object, use of capillary force is considered effective if use of liquid is allowed. Actually, Tanikawa *et al* [7] have picked/placed an object with a micro-hand and a micro-drop, but they have not provided any analysis of the capillary force involved. We have proposed a scheme for micromanipulation based on capillary force by regulation of the liquid volume [8]. In our previous scheme, it is assumed that the object shape is spherical, and the probe and substrate surfaces are flat. The profile of a liquid bridge between two solids was analyzed based on Orr's theory [9], and the force generated for the profile was presented clearly. These analyses have indicated the feasibility of the proposed scheme. The range of the force, however, never seems large enough for practical/reliable micromanipulation. In this study, we propose a probe with a concave-tip as shown in fig. 1. The concaved probe-tip would generate greater capillary force than a flat probe-tip so that it can expand the possibility of picking up manipulation. If a large amount of liquid was supplied, liquid must overflow to a flat surface, and capillary force would be equivalent to the flat probe-tip case for successful placing. Through both numerical estimation and its experimental verification, the magnified range of capillary force is presented as a function of concavity radius, wettability, and liquid volume.

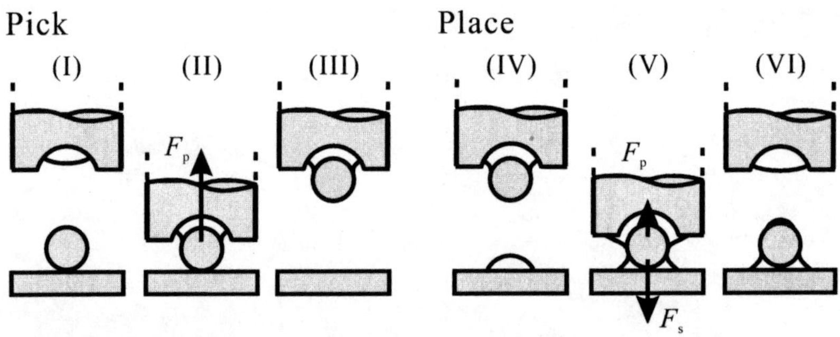

Figure 1: Schematic illustration of manipulation procedure: (I) positioning, (II) lowering, (III) picking up, (IV) positioning, (V) lowering, and (VI) placing.

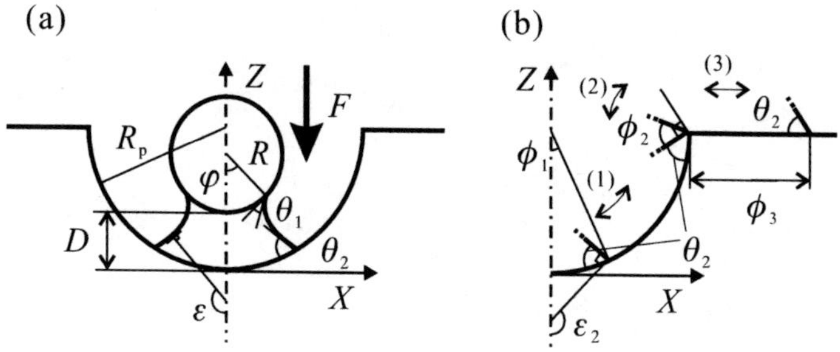

Figure 2: Liquid bridge between a spherical object and a concave shaped probe: (a) whole menisci (b) menisci end at the probe surface.

2 Analysis of the liquid bridge

Fig. 2 shows an axisymmetric model for the analysis of a liquid bridge between a spherical object and a concave probe-tip, where R is the radius of the object, R_p is the radius of the concavity, D is the distance from the probe to the object, φ is the filling angle of the object, F is the attractive force acting on the object, and V is the volume of the liquid bridge between two solids. The meniscus forms contact angles θ_1 at the object and θ_2 at the probe-tip. The profile of meniscus is expressed by the cylindrical coordinates(X,Z). The value of ε expresses the angle between the normal to the meniscus and the vertical axis. We make the following assumptions. (i) The influence of gravity is negligible and the profile of the liquid bridge follows Young–Laplace equation [9]; (ii) the dynamic flow of the liquid is negligible; (iii) the contact angles are determined by Young's equation [2].

Capillary force F can be expressed as the sum of the pressure difference force and the surface tension force:

$$F = -\Delta P \pi {X_1}^2 + 2\pi\sigma X_1 \sin\varepsilon_1, \tag{1}$$

where ΔP is Laplace pressure, i.e., the hydrostatic pressure difference between inside and outside the liquid, σ is the surface tension, X_1 is the X-coordinate at the end-point of the profile on the object, and ε_1 is the ε-angle corresponding to the point X_1.

The Laplace pressure can be expressed by Young–Laplace equation, which relates the pressure difference to the local mean curvature H and the surface tension σ;

$$\Delta P = 2H\sigma. \tag{2}$$

Since ΔP is hydrostatic, and thus, constant at any local point, the surface of the meniscus has the same mean curvature everywhere. As shown by Orr [9], the value of H in equation (2) can be expressed by geometrical parameters as

$$2H = \frac{\mathrm{d}}{\mathrm{d}X}(\sin\varepsilon) + \frac{\sin\varepsilon}{X}. \tag{3}$$

Since the left-hand side of this equation is constant, it can be solved as a two-point-boundary value problem, for which the boundary conditions are the ε-angle and X-coordinates of the menisci end on the solid surfaces. The ε-angles are determined by the slopes of the solid surfaces and the respective contact angles θ_1 and θ_2 (see fig. 2(a) and (b)). Fig. 2(b) shows three boundary states on the probe surface, which appears (1) on the concavity, (2) at the brim, and (3) on the flat surface. If one of the menisci end-point is on the concavity (see fig. 2(b-1)), the boundary conditions and corresponding Z coordinate can be written as

$$\left.\begin{array}{lll}\varepsilon_1 = \theta_1 + \varphi, & X_1 = R\sin\varphi, & Z_1 = D + R(1-\cos\varphi), \\ \varepsilon_2 = \pi - \theta_2 + \phi_1, & X_2 = R_\mathrm{p}\sin\phi_1, & Z_2 = R_\mathrm{p}(1-\cos\phi_1),\end{array}\right\} \tag{4}$$

where ϕ_1 is the filling angle of the concavity ($0 \leq \phi_1 \leq \frac{\pi}{2}$). When the menisci end reaches the brim of the concavity as fig. 2(b-2), i.e., in the case of $\phi_1 = \frac{\pi}{2}$, the boundary condition on the probe surface is

$$\varepsilon_2 = \pi - \theta_2 + \left(\frac{\pi}{2} - \phi_2\right), \quad X_2 = R_\mathrm{p}, \quad Z_2 = R_\mathrm{p}, \tag{5}$$

where ϕ_2 is the angle changing at the brim of the concavity ($0 \leq \phi_2 \leq \frac{\pi}{2}$). In the case of fig. 2(b-3) for $\phi_1 = \phi_2 = \frac{\pi}{2}$, the boundary can be shown as

$$\varepsilon_2 = \pi - \theta_2, \quad X_2 = R_\mathrm{p} + \phi_3, \quad Z_2 = R_\mathrm{p}, \tag{6}$$

where ϕ_3 is the X-displacement overflowed to the flat surface.

The boundary-value problem has the solution [9]. The meniscus profile (X, Z), the distance D, the liquid volume V, and also the capillary force F can be calculated from given four parameters; contact angles θ_1 and θ_2, the filling angle φ, and the parameter $\phi(\equiv \phi_1 + \phi_2 + \phi_3)$. If the volume V is given in advance instead of the parameter ϕ, the value of ϕ must be determined so that V could be equal to the given value. Then, the relation between D and F, which has the conservative liquid volume and given contact angles, can be plotted as a function of the filling angle φ. To generalize the following discussion, all the parameters are normalized as

$$z = \frac{Z}{R}, \quad x = \frac{X}{R}, \quad d = \frac{D}{R}, \quad f = \frac{F}{\pi R\sigma}, \quad v = \frac{V}{R^3}, \quad r_\mathrm{p} = \frac{R_\mathrm{p}}{R}. \tag{7}$$

Fig. 3 shows a relation between the normalized maximum capillary force and the normalized concavity radius for $v = 0.10$. Both horizontal and vertical scale is logarithmic. Note that the variable of horizontal axis is not r_p but $r_\mathrm{p} - 1$. Maximum value of the capillary force is the critical value of the object detachment from the concaved probe-tip. This figure suggests that as the radius of concavity approaches to the sphere, the maximum capillary force increases drastically, and also suggests that if the object and probe have smaller contact angles, capillary force becomes much larger than that for relatively large contact angles.

On the other hand, the capillary force should be reduced for placing manipulation. The solid lines in fig. 4 shows a relationship between the normalized

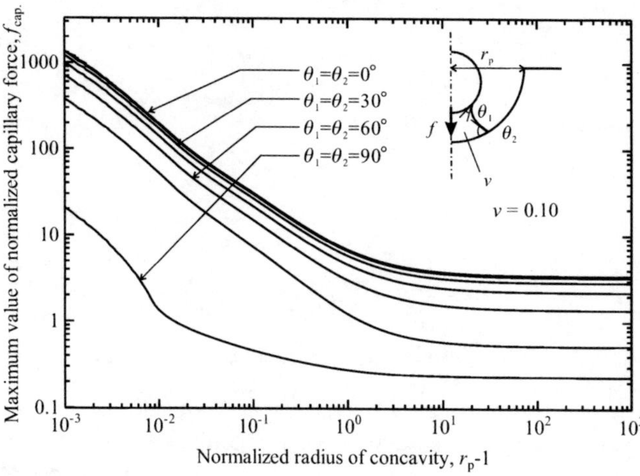

Figure 3: Relation between the normalized maximum capillary force $f_{\text{cap.}}$ and the normalized radius of concave curvature r_{p} for the normalized liquid volume $v = 0.10$.

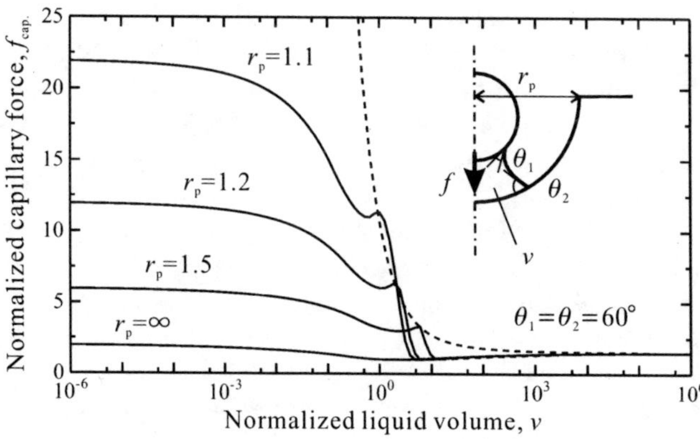

Figure 4: Relation between the normalized maximum capillary force $f_{\text{cap.}}$ and the normalized liquid volume v for $\theta_1 = \theta_2 = 60°$.

maximum capillary force and the normalized liquid volume for $\theta_1 = \theta_2 = 60°$ and $r_{\text{p}} = 1.1, 1.2, 1.5$, and ∞. The infinite value of r_{p} means the sphere-plate model. As approaching r_{p} to 1, the force difference by liquid volume regulation can be expanded. This means that the force control by the liquid volume is valid for reliable micromanipulation.

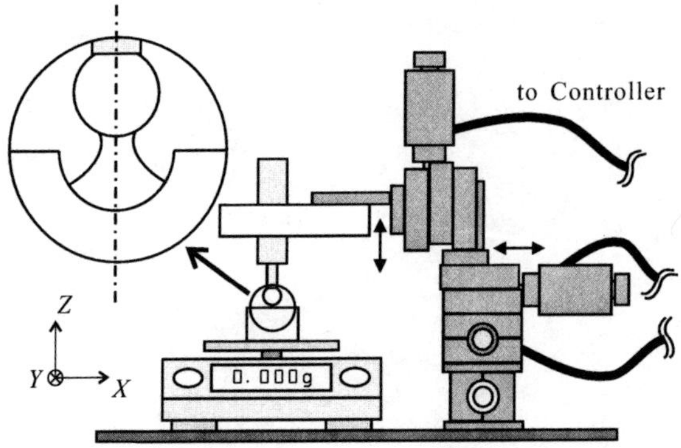

Figure 5: Schematic illustration of experimental system used. An electronic balance and a three-dimensional automated stage are fixed on a baseplate. An object functioning as a concave probe-tip is placed on the balance plate. The bigger circle shows the magnified cross-section of the smaller circle.

3 Measurement of capillary force

Fig. 5 shows the experimental system used for verification of numerical analysis. The experiment was performed in atmosphere. The order of the object size was determined according to the Bond Number ($\rho g R^2/\sigma$, where ρ is the density of the fluid and g is the gravitational acceleration). Since the influence of gravity is negligible as long as the Bond Number is small enough, we adopted $R = 1.984 - 3.175 \times 10^{-3}$ [m] for the object radii corresponding to the bond number $\rho g R^2/\sigma = 0.53 - 1.35$, which might shift the capillary force due to the gravity no more than 10%. The liquid used was purified water with $\sigma = 0.073$ [N/m], which was refined through ion-exchange membrane process. A micro-pipette with a volume resolution of 2×10^{-11} [m^3] was used to determine the volume of the liquid. An electronic balance (Sartorius, TE153S) with a resolution of 10^{-5} [N] was used to measure the force between the object and the probe-tip. Automated precise stages (Suruga Seiki, K701-20LMS) with a resolution of 5×10^{-8} [m] were used to adjust the position of the object and the probe. Probes having concaved tips were fabricated with $R_p = 3.1 \times 10^{-3}$ [m] and 3.3×10^{-3} [m] (of glass anpolytetrafluaroethylene: PTFE); with $R_p = 3.1 \times 10^{-3}$ [m] (of stainless steel). Spherical objects of several sizes were attached to steel rods. With the combination of the radii, the value of r_p can be set to 1.033-1.562. Contact angles for the materials were determined by observing the edge of a water-drop deposited on a plate using a video microscope: 50° for glass, 75° for stainless steal, and 85° for PTFE.

Fig. 6 shows the measurement value of the capillary force f as a function of the distance d for the liquid volume $v = 0.08$. The experimental value of the capillary force is obtained as square marks in an approaching process, and as triangle marks

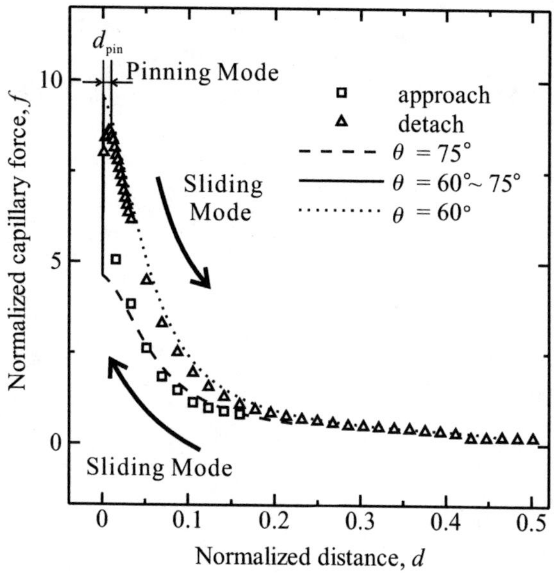

Figure 6: Relation between the normalized capillary force f and the normalized distance d for the normalized radius of concave curvature $r_{\text{p}} = 1.19$ and the normalized liquid volume $v = 0.08$. Squares, and triangles denote data of force measurement for approaching and detaching process, respectively.

in a detaching process. Such hysteresis occurs due to the inequality of the contact angle between these processes. The broken line and the dotted line denote the theoretical value of the capillary force for the fixed contact angle $\theta_1 = \theta_2 = 60°$ and 75° (Sliding Mode according to Pitois [10]). The solid line denotes the value of the force for the contact angle transition from 60° to 75° (Pinning Mode according to Pitois [10]). In this case, because of contact angle hysteresis, the observed capillary force never achieves the maximum value of the theoretical prediction.

The relations between the normalized maximum capillary force $f_cap.$ and the normalized radius of concaved curvature r_{p} for a given normalized liquid volume $v = 0.08$ are shown in fig. 7. Lines denote the values estimated from the numerical analysis for contact angles $\theta_1 = \theta_2 = 50°, 75°$, and $85°$. Circle, square, and triangle marks denote the values actually measured in the experiment. Both horizontal and vertical scales are logarithmic. Note that the variable of the horizontal axis is $r_{\text{p}} - 1$ instead of r_{p}. The experimental results are in good agreement with the theoretical predictions for all three materials. The force $f_{\text{cap.}}$ drastically increases as r_{p} approaches 1,which suggests that a probe with the concave dimension closer to the convex dimension of the object can generate much larger capillary force. Probes made of the material with small contact angle generate much larger capillary force. For r_{p} closer to 1, the larger differences are found between the measured values and the theoretical predictions. We presume the reason for this is that the error of the positional adjustment has relatively larger influence on the generated force as the concave radius approaches the convex radius of the object.

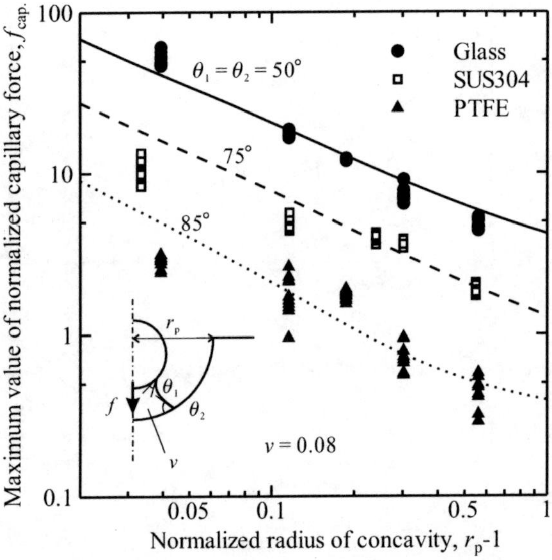

Figure 7: Relation between the normalized maximum capillary force $f_{cap.}$ and the normalized radius of concave curvature r_p for the normalized liquid volume $v = 0.08$. Circles, squares, and triangles denote data of force measurement for glass, stainless steel, and polytetrafluoroethylene (PTFE), respectively.

As shown in fig. 8, magnified capillary force can be controlled by the regulation of liquid volume. Circle, triangle, and square marks are expressing the experimental values. These are in good agreement with theoretical predictions for $v > 0.1$. In the case of $v < 0.1$, the experimental value of capillary force and the calculation considering contact angle hysteresis (broken lines) are almost the same. The broken lines are calculated by neglecting the capillary force during the pinning mode ($0 \leq d \leq d_{\rm pin}$). Assuming the pinning mode distance $d_{\rm pin}$ is constant, the capillary force is reduced with decreasing liquid volume. This means that too small supply of liquid causes less capillary force generated. In order to realize efficient and reliable manipulation, the normalized liquid volume should be controlled from 0.1 to 10.

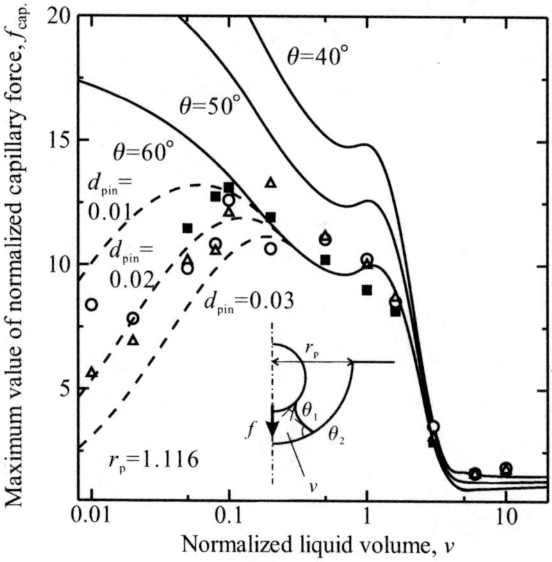

Figure 8: Relation between the normalized maximum capillary force $f_{\text{cap.}}$ and the normalized liquid volume v for glass specimen and $r_{\text{p}} = 1.116$.

4 Conclusion

Through both analysis and measurement of capillary force, this study clarifies the most important factors for reliable capillary micromanipulation by concave probe, i.e. the material wettability, the concave shape and dimension, and the amount of liquid supply. The shape of the probe-tip can be designed as shown in fig. 1 so that the range of the capillary force can be extended due to the change of the apparent contact angle for a given liquid volume. The more wettable a material is, the greater capillary force it can generate. The magnified capillary force can be reduced/controlled by liquid volume regulation. This suggests that micromanipulation by capillary force has a great potential for a wide range of applications. In the capillary force measurement, the contact angle hysteresis can be observed, and we presume that it reduces the maximum value of the capillary force in case of relatively small liquid volume. For actual manipulation, a mechanism that is able to supply proper amounts of liquid needs to be developed.

References

[1] K.L. Johnson, K. Kendall, and A.D. Roberts, Surface energy and the contact of elastic solids. *Proc. R. Soc. Lond. A.*, **324**, pp. 301-313, 1971

[2] Jacob N. Israelachvili, Intermolecular and Surface Forces. *Academic Press*, New York, pp. 301-322, 1985

[3] Shigeki Saito, Hideki T. Miyazaki, and Tomomasa Sato, Micro-object Pick

and Place Operation under SEM based on Micro-physics. *J. Robotics and Mechatronics*, **14**, pp. 227-237, 2002

[4] Shigeki Saito, Hideki T.Miyazaki, Tomomasa Sato, and Kunio Takahashi, Kinematics of mechanical and adhesional micromanipulation under a scanning electron microscope. *J. Applied Physics*, **92(9)**, pp. 5140-5149, 2002

[5] Kunio Takahashi, Hideaki Kajihara, Masataka Urago, and Shigeki Saito, Voltage required to detach an adhered particle by Coulomb interaction for micromanipulation. *J. Applied Physics*, **90(1)**, pp. 432-437, 2001

[6] Shigeki Saito, Hideo Himeno, and Kunio Takahashi, Electrostatic detachment of an adhering particle from a micromanipulated probe. *J. Applied Physics*, **93(4)**, pp. 2219-2224, 2003

[7] Tamio Tanikawa, Yoshiyuki Hashimoto, and Tatsuo Arai, Micro Drops for Adhesive Bonding of Micro Assemblies and Making a 3-D Structure 'Micro Scarecrow'. *Proceedings of the IEEE/RSJ Intl. Conference on Intelligent Robotics and Systems*, pp. 776-781, 1998

[8] Kenichi J. Obata, Tomoyuki Motokado, Shigeki Saito, and Kunio Takahashi *J. Fluid Mech.*, **498** pp. 113-121, 2004

[9] F.M. Orr, L.E. Scriven, and A.P. Rivas, Pendular rings between solids: meniscus properties and capillary force. *J. Fluid Mech.*, **67(4)**, pp. 723-742, 1975

[10] Oliver Pitois and Xavier Chateau, Small Particle at a Fluid Interface: Effect of Contact Angle Hysteresis on Force and Work of Detachment, *Langmuir*, **18**, pp. 9751-9756, 2002

Author Index